环境工程概论

——专业英语教程

Overview to Environmental Engineering

官 涤 主 编

哈爾濱工業大學出版社

内 容 提 要

本书以介绍环境工程专业知识、培养良好外语能力为出发点，从环境与可持续发展的角度，介绍了给水与废水处理、大气污染与控制、固体废物管理、危险废物管理、噪声污染与控制、辐射污染与控制、环境世界观与可持续性等内容，重点介绍环境工程的基本概念和基本原理。全书以英文编写，每一章节均由课文、工程实例与复习题组成，内容参考多种国外原版教材，为便于使用，书后附有常用环境术语表。

本书图文并茂，内容清晰，适用于我国高等院校给水排水、环境工程专业本科双语和专业英语教学，也可供环境保护爱好者、相关环境研究人员学习参考。

图书在版编目(CIP)数据

环境工程概论——专业英语教程/官涤主编.—哈尔滨:哈尔滨工业大学出版社,2011.6(2017.7 重印)

ISBN 978－7－5603－3294－9

Ⅰ.环…　Ⅱ.①官…　Ⅲ.①环境工程－概论
Ⅳ.①XT

中国版本图书馆 CIP 数据核字(2011)第 106097 号

策划编辑　王桂芝　贾学斌
责任编辑　翟新烨
封面设计　卞秉利
出版发行　哈尔滨工业大学出版社
社　　址　哈尔滨市南岗区复华四道街 10 号　邮编 150006
传　　真　0451－86414749
网　　址　http://hitpress.hit.edu.cn
印　　刷　哈尔滨圣铂印刷有限公司
开　　本　787mm×1092mm　1/16　印张 16.75　字数 386 千字
版　　次　2011 年 7 月第 1 版　2017 年 7 月第 2 次印刷
书　　号　ISBN 978－7－5603－3294－9
定　　价　33.00 元

前　言

近 20 年来，全球的环境工程技术发展迅速，随着环境意识的深入，人们对环境问题越来越重视，环境工程与技术的高等教育达到一个新的平台。在这一背景下编写适应环境工程专业本科生的英语教材很有必要。本书重点介绍环境工程的基本概念和基本原理，共分 7 章。第 1 章主要概念环境工程的背景知识、技术和相关政策发展；第 2 章主要介绍给水处理系统、废水处理方法以及污泥的处理与处置；第 3 章主要介绍空气污染物及其控制处理技术；第 4 章主要概述固体废物和危险废物的综合管理、处理与处置方法；第 5 章主要概述噪声的特性与影响、评估系统和控制方法；第 6 章主要介绍辐射的特性与暴露、防护方法；第 7 章阐述环境世界观的类型与相应的社会可持续发展方向。本书注重材料的新颖与实用性统一，尽可能考虑实际教学中的课时安排与教学要求、学生需求等诸方面情况，力图在宏观上介绍当环境工程领域的发展趋势以及该课程的基本内容。为兼顾教学和学生自学需要，每章后面留有习题。课文内容包括环境工程专业英语词汇和相当数量的常用科技词汇，词汇复现率较高，有助于提高学生正确、快速地阅读英语科技文献能力。

本书由哈尔滨工程大学官涤、丁学忠、米海蓉，吉林建筑工程学院陆海，哈尔滨供水工程有限责任公司王志军等合作编写。书中第 1 章由米海蓉编写，第 2、4、7 章由官涤编写，第 3 章由陆海编写，第 5 章由丁学忠编写，第 6 章由王志军编写。全书由官涤担任主编并统稿，丁学忠、陆海、王志军担任副主编。参加本书编写的还有哈尔滨供排水集团有限责任公司王志滨、黑龙江省环境保护科学研究院任伊滨、牡丹江师范学院李继光、牡丹江大学孙慧等，在此一并致谢。

本书选材于原版英文教科书及相关专业著作，书中的引文，编者尽力与原文作者取得联系，但仍有部分联系不上，在此深表歉意。由于本教材涉及面广泛，限于编者水平，书中难免有疏漏不妥之处，敬请有关专家和读者批评批正。

编　者

2011 年 5 月

Contents

Chapter 1 Introduction

1.1 Introduction to Environmental Engineering

Engineering is a profession that applies mathematics and science to utilize the properties of matter and sources of energy to create useful structures, machines, products, systems, and processes. Environmental engineering is a relatively new profession with a long and honorable history. The roots of this profession reach into several major disciplines including civil engineering, public health, ecology, chemistry, and meteorology. From each foundation, the environmental engineering profession draws knowledge, skill, and professionalism.

The Environmental engineering division of the American Society of Civil Engineers (ASCE) has published the following statement of purpose:

Environmental engineering is manifest by sound engineering thought and practice in the solution of problems of environmental sanitation, notably in the provision of safe, palatable, and ample public water supplies; the proper disposal of or recycle of wastewater and solid wastes; the adequate drainage of urban and rural areas for proper sanitation; and the control of water, soil, and atmospheric pollution and the social and environmental impact of these solutions. Furthermore it is concerned with engineering problems in the field of public health, such as control of arthropod-borne diseases, the elimination of industrial health hazards, and the provision of adequate sanitation in urban, rural, and recreational areas, and the effect of technological advances on the environment (ASCE, 1977).

Thus, we may consider what environmental engineering is not. It is not concerned primarily with heating, ventilating, or air conditioning(HVAC), nor is it concerned primarily with landscape architecture. Neither should it be confused with the architectural and structural engineering functions associated with built environments, such as homes, offices, and other workplaces.

The general mission of colleges and universities is to allow students to mature intellectually and socially and to prepare for careers that are rewarding. The chosen vocation is ideally an avocation as well. It should be a job that is enjoyable and one approached with enthusiasm even after experiencing many of the ever-present bumps in the road. Designing a water treatment facility to provide clean drinking water to a community can serve society and become a personally satisfying undertaking to the environmental engineer. Environmental engineers now are employed in virtually all heavy industries and utility companies in the world, in any aspect of public works construction and management, by the Environmental Protection Agency(EPA) and other national agencies, and by the consulting firms used by these agencies. In addition, every state and most local governments have agencies dealing with air quality, water quality and water resource management, soil quality, forest and natural resource management, and agricultural management that employ environmental engineers. Pollution control engineering has also become an exceedingly profitable venture.

Environmental engineering has a proud history and a bright future. It is a career that may be challenging, enjoyable, personally satisfying, and monetarily rewarding. Environmental engineers are committed to high standards of interpersonal and environmental ethics. They try to be part of the solution while recognizing that all people including themselves are part of the problem.

1.2 Overview of Environmental Technology

Before beginning a study of the many different topics that make up environmental terminology and technology, it would be helpful to have an understanding of the overall goals, problems, and alternative solutions available to practitioners in this field.

To present an overview of such a broad subject, we can consider an engineering project involving the subdivision and development of a tract of land into a new community, which will include residential, commercial, and industrial centers, whether the project owner is a governmental agency or a private developer, a wide spectrum of environmental problems will have to be considered and solved before construction of the new community can begin. Usually, the project owner retains the services of an independent environmental consulting firm to address these problems.

1.2.1 Water Supply

One of the first problems project developers and consultants must consider is the provision of a potable water supply, one that is clean wholesome, safe to drink, and available in adequate quantities to meet the anticipated demand in the new community. Some of the questions that must be answered are as follows:

(1) Is there an existing public water system nearby with the capacity to connect with and serve the new development? If not,

(2) Is it best to build a new centralized treatment and distribution system for the whole community, or would it be better to use individual well supplies? If a centralized treatment facility is selected,

(3) What types of water treatment processes will be required to meet federal and state drinking water standards? (Water from a river or a lake usually requires more extensive treatment than groundwater does, to remove suspended particles and bacteria.) Once the source and treatment processes are selected,

(4) What would be the optimum hydraulic design of the storage, pumping, and distribution network to ensure that sufficient quantities of water can be delivered to consumers at adequate pressures?

1.2.2 Sewage Disposal and Water Pollution Control

When running water is delivered into individual homes and businesses, there is an obvious need to provide for the disposal of the used water, or sewage. Sewage contains human wastes, wash water, and dishwater, as well as a variety of chemicals if it comes from an industrial or commercial area. It also carries microorganisms that may cause disease and organic material that can damage lakes and streams as it decomposes.

It will be necessary to provide the new community with a mean for safety disposing of the sewage, to prevent water pollution and to protect public and environmental health. Some of the technical questions that will have to be addressed include the following:

(1) Is there a nearby municipal sewage system with the capacity to handle the additional flow from the new community? If not,

(2) Are the local geological conditions suitable for on-site subsurface disposal of the wastewater (usually septic systems), or it is necessary to provide a centralized sewage treatment plant for new community and to discharge the treated sewage to nearby stream? If treatment and surface discharge are required,

(3) What is the require degree or level of wastewater treatment to prevent water pollution? With a secondary treatment level, which removes at least 85 percent of biodegradable pollutants, be adequate? Or will some forms of advanced treatment be required to meet federal and state discharge standards and stream quality criteria? (Some advanced treatment facilities can remove more than 99 percent of the pollutants.)

(4) Is the flow of industrial wastewater an important factor?

(5) Is it possible to use some type of land disposal of the treated sewage, such as spray irrigation, instead of discharging the flow into a stream?

(6) What methods will be used to treat and dispose of the sludge, or biosolids, that is removed from the wastewater?

(7) What is the optimum layout and hydraulic design of a sewage collection system that will convey the wastewater to the central treatment facility with a minimum need for pumping?

1.2.3 Air Pollution Control

Major sources of air pollution include fuel combustion for power generation, certain industrial and manufacturing processes, and automotive traffic. Project developers can exercise the most control over traffic. Private industry will have to apply appropriate air pollution control technology at individual facilities to meet federal and state standards.

The volume of traffic in the area will obviously increase leading to an increase in exhaust fumes from cars and other vehicles. Proper layout of roads and traffic-flow patterns, however, can minimize the amount of stop-and-go traffic, thus reducing the amount of air pollution in the development.

Usually, the developer's consultant will have to prepare an environmental impact statement (EIS), which will describe the traffic plan and estimate the expected levels of air pollutants. It will have to be shown that air quality standards will not be violated, for the project to gain approval from regulatory agencies. (In addition to air pollution, the completed EIS will address all other environmental effect related to the proposed project.)

1.2.4 Solid and Hazardous Waste Management

The development of a new community (or growth of an exciting community) will certainly lead to the generation of more municipal refuse and industrial waste materials. Ordinarily, the collection and disposal of solid waste is a responsibility of the local municipality. However, some of the waste from industrial sources may be particularly dangerous, requiring special handling and disposal methods.

There is a definite relationship between public and environmental health and the proper handling and disposal of solid wastes. Improper garbage disposal practices can lead to the spread of diseases such as typhus and plague due to the breeding of rats and flies.

If municipal refuse is improperly disposed of on land in a "garbage dump," it is also very likely that surface and groundwater resources will be polluted with leachate (leachate is a contaminated liquid that seeps through the pile of refuse into nearby streams as well as into the ground). On the other hand, incineration of the refuse may cause significant air pollution problems if proper controls are not applied or are ineffective.

Hazardous wastes such as poisonous or ignitable chemicals from industrial processes, must receive special attention with respect to storage, collection, transport, treatment, and final disposal. This is particularly necessary to protect the quality of groundwater, which is the source of water supply for about half the population in the United States. In recent years, an increasing number of water supply wells have been found to be contaminated with synthetic organic chemicals, many of which are thought to cause cancer and other illnesses in humans. Improper disposal of these hazardous materials, usually by illegal burial in the ground, is the cause of the contamination.

Some of the general questions related to the disposal of solid and hazardous wastes from the new community include the following:

(1) Is there a materials recycling facility (MRF, or "murf") serving the area? What will be the waste storage, collection, and recycling requirements (for example, will source separation of household refuse be necessary)?

(2) Will a waste processing facility (such as one that provides for shredding, pulverizing, baling, composting, or incineration) be needed to reduce the waste volume and improve its handling characteristics?

(3) Is there a suitable sanitary landfill serving the area, and will it have sufficient capacity to handle the increased amounts of solid waste for a reasonable period of time? (Despite the best efforts to recycle solid waste or reduce its volume, some material will require final disposal in the ground in

an environmentally sound manner) If not,

(4) Is there a suitable site for construction and operation of a new landfill to serve the area? (A modern sanitary landfill site must meet strict requirements with respect to topography, geology, hydrology, and other environmental conditions.)

(5) Will commercial or industrial establishments be generating hazardous waste, and, if so, what provisions must be made to collect, transport, and process that material? Is there a secure landfill for final disposal available, or must a new one be constructed to serve the area?

1.2.5 Noise and Radiation Pollution Control

Noise can be considered to be a type of air pollution in the form of waste energy—sound vibrations. Noise pollution will result from the construction activity, causing a temporary or short-term impact. The builders may have to observe limitations on the types of construction equipment and the hours of operation to minimize this negative effect on the environment. A long-term impact with respect to the generation of noise will be caused by the increased amount of vehicular traffic. This is another environmental factor that the consulting will have to address in the EIS.

Radiation is commonly defined as energy that flows through matter or through a vacuum. Ionizing radiation, in particular, plays invaluable roles in medical diagnosis and therapy, industrial process control, research, and numerous other areas. This ionizing energy flux may have adverse impacts on biological matter. Radiation sickness, cancer, shortened life, or immediate death may result from varying exposures. Radiation doses to living tissue are measured as grays or rads, units of absorbed energy, and sieverts or rems, units of relative biological damage. The general concern of the environmental engineer is radiation from anthropogenic sources, particularly the radioactive matters from nuclear power plants, radiation from natural sources like mine tailings, and radon, because it is ubiquitous. Wastes are handled in much the same manner as the hazardous wastes.

1.2.6 Other Environmental Factors

Not to be overlooked as all environmental factors in any land development project is the potential impact on local vegetation and wildlife. The destruction of woodlands and meadows to make room for new buildings and roads can lead to significant ecological problems, particularly if there are any rare or endangered species in the area. Cutting down trees and paving over meadows can cause short-term impacts related to soil erosion and stream sedimentation. On a long-term basis, it will cause the displacement of wildlife to other suitable habitats, presuming, of course, that such habitats are available nearby, otherwise, several species may disappear from the area entirely.

Human activity in wetland areas, including marshes and swamps, can be very damaging to the environment. Coastal wetlands are habitats for many different species of organisms, and the tremendous biological productivity of these wetland environments is a very important factor in the food chain for many animals. When wetlands are drained, filled in, or dredged for building and land development projects, the life cycle of many organisms is disrupted. Many species may be destroyed as a result of habitat loss or loss of a staple food source. Wetlands also play important roles in filtering and cleaning water and in serving as reservoir for floodwaters. There is a definite need to control or restrict construction activities in wetland environments and to implement a nationwide wetlands protection program.

Environmental concerns related to general sanitation in a new community include food and beverage protection, insect and rodent control, industrial hygiene and occupational safety, and the cleanliness of recreation areas such as public swimming pools. These concerns are generally the responsibility of local departments.

1.3 The Development of Environmental Policy

Public policy is the general principal by which government branches—the legislative, executive, and judicial—are guided in their management of public affairs. The legislature (Congress) is directed to declare and shape national policy by passing legislation, which is the same as enacting law. The executive (president) is directed to enforce the law while the judiciary (the court system interprets the law when a dispute arises (see Figure 1.1)).

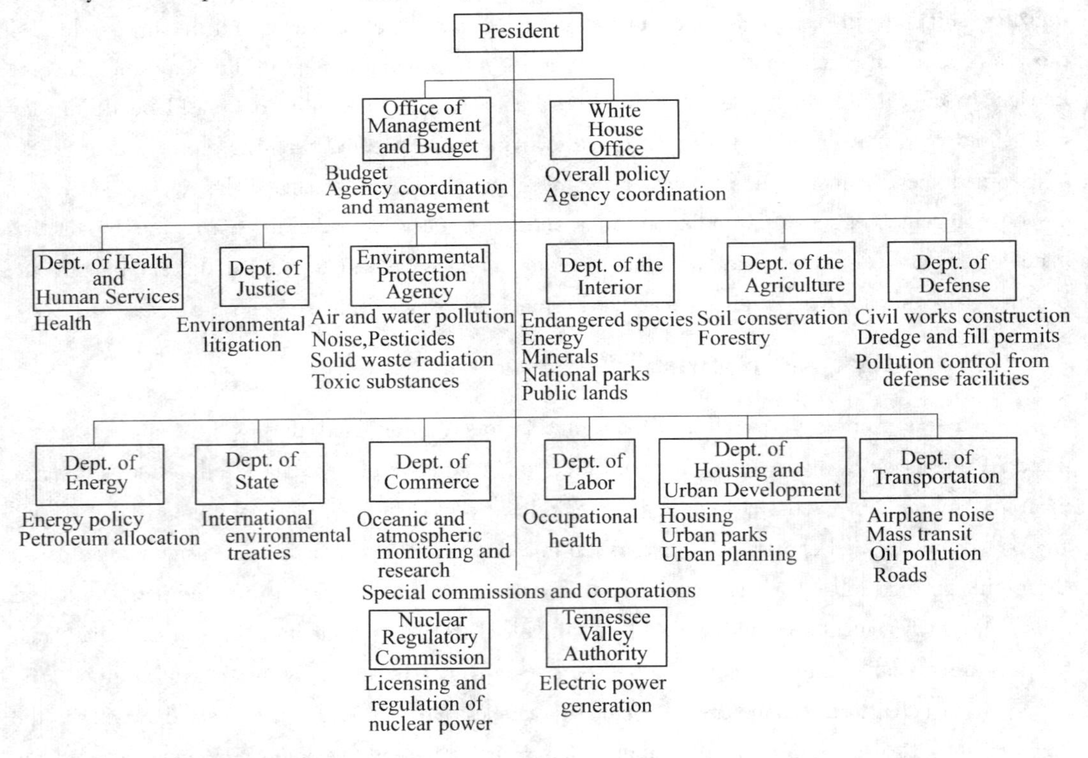

Figure 1.1 Major Agencies of the Executive Branch in the U.S.

(Major agencies of the executive branch are shown with their environmental responsibility)

When Congress considers certain conduct to be against public policy and against the public good, it passes legislation in the form of acts or statutes. Congress specifically regulates, controls, or prohibits activity in conflict with public policy and attempts to encourage desirable behavior. Through legislation, Congress regulates behavior, selects agencies to implement new programs, and sets general procedural guidelines. When Congress passes environmental legislation, it also declares and shapes the national environmental policy, thus fulfilling its policy-making function (see Figure 1.2).

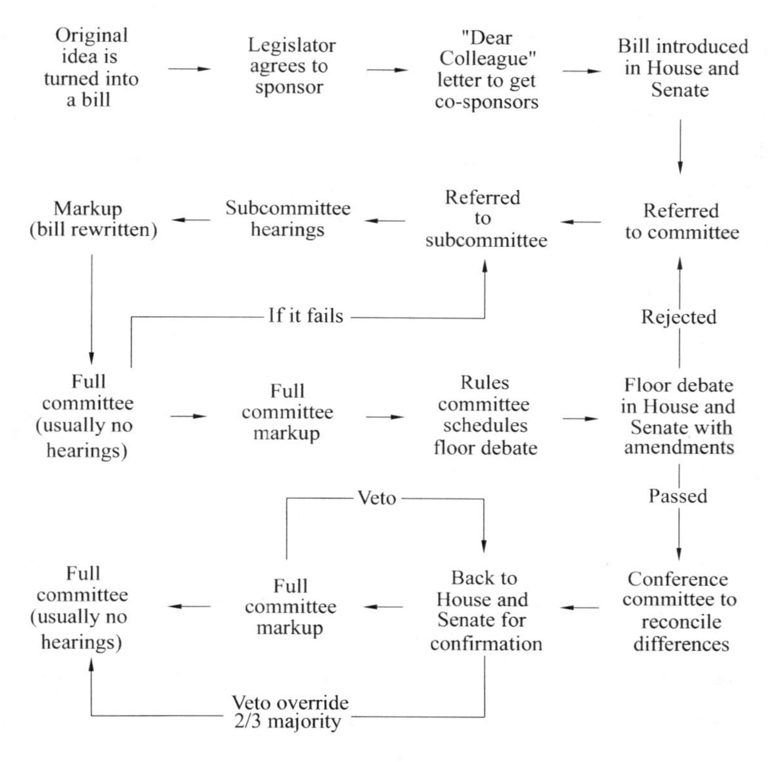

Figure 1.2 Passage of a Law

(This figure illustrates the path of a bill in the U.S. Congress from organization to becoming a law)

Over 90 years ago, President Teddy Roosevelt declared that nothing short of defending your country in wartime "compares in leaving the land even better land for our descendants than it is for us." The environmental issues that Roosevelt strongly believed in, however, did not become major political issues until the early 1970s.

While the publication of Rachel Carson's *Silent Spring* in 1962 is considered to be the beginning of the modern environmental movement, the first Earth Day on April 22, 1970, was perhaps the single event that put the movement into high gear. In 1970, as a result of mounting public concern over environmental deterioration—cities clouded by smog, rivers on fire, waterways choked by raw sewage—many nations, including the United States, began to address the most obvious, most acute environmental problems.

Public opinion polls indicate that a permanent change in national priorities followed Earth Day

1970. When polled in May 1971, 25 percent of the U. S. public declared protecting the environment to be an important goal—a 2 500 percent increase over the proportion in 1969.

During the 1970s, many important pieces of environmental legislation were enacted in the United States (see Figure 1.2). Many of the identified environmental problems were so immediate, so obvious, that it was relatively easy to see what had to be done and to summon the political will to do it (see Table 1.1).

Just as it was beginning to gain momentum, however, the environmental movement began to decline. When the energy crisis threatened to stall the North American economy in the early 1970s, environmental concerns quickly faded. By 1974, President Gerald Ford had proposed accelerating his administration's leasing program for offshore gas and oil drilling. A turn around in environmental policy was even more pronounced in the 1980s during the Reagan administration. Former Vice President Walter Mondale was fond of noting that President Ronald Reagan "would rather take a polluter to lunch than to court." During the mid-1980s, the environment was not a priority in the Reagan administration.

The period from 1970 to 1990, however, did bring forth some very tangible accomplishments in environmental policy. Among the most visible and quantifiable is the expansion of protected areas. During this period, federal parklands in the United States—excluding Alaska—increased 800 000 hectares (2 million acres), to 10.5 million (25.9 million acres). In Alaska, 18.3 million additional hectares (45 million acres) were protected, bringing the state's total to over 232 million hectares (573 million acres). Also, the extent of the water-ways included in the National Wild and Scenic Rivers System increased by more than 12 times, to some 15 000 kilometers (9 300 miles).

By the late 1980s, however, a new environmental awareness and concern began to surface as a major political issue. This was in part due to a number of highly visible environmental problems that appeared nightly on the evening news. Images of toxic waste (including hospital waste, such as used syringes) washing up on the nation's beaches and of the pristine waters of Alaska covered in oil from the Exxon Valdez spill made an impact on the public. Once again, the public reacted by organizing and putting pressure on the political system, and, as in 1970, the politicians began to respond. For the first time in the history of the United States, the environment became a key issue in a presidential campaign. In 1988, the environmental records of the two major candidates were hotly debated. Environmentalism was evolving as a major public issue. By the 1992 U. S. presidential election, the environment was established as a major campaign issue, a trend that continues today.

In many respects, the environmental movement of the 1970s and 1980s came of age in the 1990s. The linking of politics and science, and emotionalism and logic in a new environmental movement represented a significant integration of human thinking. It has been said that politics have always forged science, Prioritization of issues and political will determine where money will be spent. By the mid-1990s, it appeared that political will in the United States to address environmental concerns was on the rise.

Table 1.1 Major U.S. Environmental and Resource Conservation Legislation

Wildlife conservation

Anadromous Fish Conservation Act of 1965
Fur Seal Act of 1966
National Wildlife Refuge System Act of 1966, 1976, 1978
Species Conservation Act of 1966, 1969
Marine Mammal Protection Act of 1972
Marine Protection, Research, and Sanctuaries Act of 1972
Endangered Species Act of 1973, 1982, 1985, 1988, 1995
Fishery Conservation and Management Act of 1976, 1978, 1982, 1996
Whale Conservation and Protection Study Act of 1976
Fish and Wildlife Improvement Act of 1978
Fish and Wildlife Conservation Act of 1980 (Nongame Act)
Fur Seal Act Amendments of 1983

Land use and conservation

Taylor Grazing Act of 1934
Wilderness Act of 1964
Multiple Use Sustained Yield Act of 1968
Wild and Scenic Rivers Act of 1968
National Trails System Act of 1968
National Coastal Zone Management Act of 1972, 1980
Forest Reserves Management Act of 1974, 1976
Forest and Rangeland Renewable Resources Act of 1974, 1978
Federal Land Policy and Management Act of 1976
National Forest Management Act of 1976
Soil and Water Conservation Act of 1977
Surface Mining Control and Reclamation Act of 1977
Antarctic Conservation Act of 1978
Endangered American Wilderness Act of 1978
Alaskan National Interests Lands Conservation Act of 1980
Coastal Barrier Resources Act of 1982
Food Security Act of 1985
Emergency Wetlands Resources Act of 1986
North American Wetlands Conservation Act of 1989
Coastal Development Act of 1990
California Desert Protection Act of 1994
Federal Agriculture Improvement and Reform Act of 1996

General

National Environmental Policy Act of 1969 (NEPA)
International Environmental Protection Act of 1983

Energy

Energy Policy and Conservation Act of 1975
National Energy Act of 1978, 1980
Northwest Power Act of 1980
National Appliance Energy Conservation Act of 1987
Energy Policy Act of 1992

Water quality

Refuse Act of 1899
Water Quality Act of 1965
Water Resources Planning Act of 1965
Federal Water Pollution Control Acts of 1965, 1972
Ocean Dumping Act of 1972
Safe Drinking Water Act of 1974, 1984, 1996
Clean Water Act of 1977, 1987
Great Lakes Toxic Substance Control Agreement of 1986
Great Lakes Critical Programs Act of 1990
Oil Spill Prevention and Liability Act of 1990

Air quality

Clean Air Act of 1963, 1965, 1970, 1977, 1990

Noise control

Noise Control Act of 1965
Quiet Communities Act of 1978

Resources and solid waste management

Solid Waste Disposal Act of 1965
Resources Recovery Act of 1970
Waste Reduction Act of 1990

Toxic substances

Toxic Substances Control Act of 1976
Resource Conservation and Recovery Act of 1976
Comprehensive Environmental Response, Compensation, and Liability (Superfund) Act of 1980, 1986, 1990
Nuclear Waste Policy Act of 1982

Pesticides

Food, Drug, and Cosmetics Act of 1938
Federal Insecticide, Fungicide, and Rodenticide Control Act of 1972, 1988
Food Quality Protection Act of 1996

1.4 Organization of This Text

The second chapter, Water and Wastewater Treatment, examines some characteristics of water that affect its quality, explains how to treat water for public consumption. And wastewater treatment is another subject of this chapter, which introduces the removal methods of pollutants that reduce the quality of the lake or stream.

The third chapter, Air pollution and control, introduces the health effects and other environmental impacts of air pollution, as well as some methods of control the pollutions from automobile emissions, chemical odors, factory smoke, and similar materials are directly related to the number of people living in an area and the kinds of activities.

The forth chapter, Solid Waste Management, presents functional elements, collection, disposal, and recycling of solid waste, which fundamental needs of urban society. Methods of dealing with hazardous waste sites and managing the wastes continually generating are discussed.

The fifth chapter, Noise Pollution and Control, introduces the waste product generated in conjunction with various anthropogenic activities, explains the knowledge concerning noise, its effects, and its abatement and control, also examines transport processes that carry pollutants from their source of people.

The sixth chapter, Radiation Pollution and Protection, conducts the examination of ionizing radiation, introduces the health effects of radiation and managment techniques for both radioactive wastes and X-rays.

The final chapter, Environmental Worldviews and Sustainability, addresses the different environmental worldviews as human-centered, life-centered and earth-centered one, as well as the sustainable development in societies with such worldviews.

All sections include problems to be addressed individually by the reader or collectively in a classroom setting.

Review Questions

1. Define Environmental Engineering.

2. Give a brief definition of environmental technology, including mention of basic activities and objectives.

3. Why do industries pollute?

4. What are some of the enforcement options in U.S. environmental policy?

Chapter 2 Water and Wastewater Treatment

2.1 Drinking Water Quality

Precipitation in the form of rain, hail, or sleet contains very few impurities. It may contain trace amounts of mineral matter, gases, and other substances as it forms and falls through the earth's atmosphere. Once precipitation reaches the earth's surface, many opportunities are presented for the introduction of mineral and organic substances, microorganisms, and other forms of pollution (contamination). When water runs over or through the ground surface, it may pick up particles of soil. This is noticeable in the water as cloudiness or turbidity. It also picks up particles of organic matter and bacteria. As surface water seeps downward into the soil and through the underlying material to the water table, most of the suspended particles are filtered out. This natural filtration may be partially effective in removing bacteria and other particulate materials. However, the chemical characteristics of the water may change and vary widely when it comes in contact with mineral deposits. As surface water seeps down to the water table, it dissolves some of the minerals contained in the soil and rocks. Groundwater, therefore, often contains more dissolved minerals than surface water.

2.1.1 Drinking Water Contaminants

Roughly 1 000 contaminants have been detected in the public water supply in the United States, and virtually every major water source is vulnerable to pollution. The common contaminants of drinking water are shown in Table 2.1. About 60 percent of the U.S. population relies on surface water from rivers, lakes, and reservoirs that may contain industrial and agricultural wastes and pesticides washed off fields by rain. The other 40 percent uses groundwater that may be tainted by chemicals slowly seeping in from toxic-waste dumps, agricultural activities, and leaking sewage and septic systems. In some areas where groundwater supplies are being gradually depleted, the chemical pollutants are becoming more concentrated.

Most pollutants are probably not concentrated enough to pose significant health hazards; however, there are exceptions. The most widespread danger in water is lead, which can cause high blood pressure and an array of other health problems. Lead is especially hazardous to children, since it impairs the development of brain cells. The contamination comes from old lead pipes and solder that have been used in plumbing for years. These materials are gradually being replaced in homes and water systems. Individuals may want to have their water tested for lead by an official lab. If the level is too high, they can investigate ways to deal with the problem or switch to bottled water for drinking and cooking. Even then, caution is called for—some bottled waters contain many of the same contaminants that tap water does.

The other four types of contamination in the U.S. water supply, along with their source and risk, are shown in the table. Regardless of the problems, however, the water supply in the United States is among the cleanest in the world. Consequently, in the development of a water supply system, it is necessary to examine carefully all the factors that might adversely affect the intended use of a water supply source.

Table 2.1 Common Contaminants of Drinking Water

Substance	Source	Health Effects
Persistent chlorinated organic compounds	Used as solvents in industry; past use as pesticides	Various, including reproductive problems and cancer
Trihalomethanes	Produced by chemical reactions when water is disinfected by chlorination	Liver and kidney damage and possible cancer
Nitrates	Primarily from fertilizer and effluent from concentrated livestock raising	Can react to reduce oxygen uptake by blood particularly a problem for children
Lead	Old piping and solder in public water distribution systems, homes, and other buildings	Nerve damage, learning difficulties in children, birth defects, possible cancer
Pathogenic bacteria, protozoa, and viruses	Leaking septic tanks and sewers; contamination of water supply from birds and animals; inadequate disinfection	Acute gastrointestinal illness and other serious health problems

2.1.2 Drinking Water Physical Characteristics

1. Turbidity

The presence of suspended material such as clay, silt, finely divided organic material, plankton, and other particulate material in water is known as turbidity. The unit of measure is a Turbidity Unit (TU) or Nephlometric Turbidity Unit (NTU). It is determined by reference to a chemical mixture that produces a refraction of light. Turbidities in excess of 5 TU are easily detectable in a glass of water and are usually objectionable for aesthetic reasons. Clay or other inert suspended particles in drinking water may not adversely affect health, but water containing such particles may require treatment to make it suitable for its intended use. Following a rainfall, variations in the groundwater turbidity may be considered an indication of surface or other introduced pollution.

2. Color

The color is imparted by dissolved organic matters from decaying vegetation or some inorganic materials, such as colored soils (red soil) etc.. The algae or other aquatic plants may also impart color. Again it is more objectionable from aesthetics point of view than the health. The standard unit of color is that which is produced by one milligram of platinum cobalt dissolved in one liter of distilled water.

3. Taste and Odour

The dissolved inorganic salts or organic matter or the dissolved gases may impart taste and odour to the water. The water must not contain any undesirable or objectionable taste or odour. The extent of taste or odour is measured by a term called odour intensity which is related with threshold odour, which represents the dilution ratio at which the odour is hardly detectible. The water to be tested is gradually diluted with odour free water and the mixture at which the detection of taste and odour is just lost is determined. The number of times the sample is diluted is known as the threshold number. Thus if 20 mL of water is made 100 mL(until it just looses its taste or odour) then the threshold number is 5. For domestic water supplies the water should be free from any taste and odour so the threshold number should be 1 and should not exceed 3.

4. Temperature

The most desirable drinking waters are consistently cool and do not have temperature fluctuations of more than a few degrees. Groundwater and surface water from mountainous areas generally meet these criteria. Most individuals find that water having a temperature between 10 ~ 15 ℃ is most palatable.

5. Specific conductivity of water

The specific conductivity of water is determined by means of a portable dionic water tester and is expressed as micromhos per cm at 25 ℃. Mho is the unit of conductivity and is equal to 1 amper/1 volt. The specific conductivity is multiplied by a co-efficient (generally 0.65) so as to directly obtain the dissolved salt content in ppm.

2.1.3 Drinking Water Chemical Characteristics

1. pH

pH is the negative logarithm of hydrogen ion concentration present in water. The higher values of pH mean lower hydrogen ion concentrations and thus represent alkaline water and vice versa. The neutral water has same number of H^+ and OH^- ions. The concentration of both ions in neutral water is 10^{-7} moles per liter. If the pH of water is more than 7, it is alkaline; and if it is less than 7, it is acidic. Generally, the alkalinity in water is caused by the presence of bicarbonates of calcium and magnesium, or by the carbonates or hydroxides of sodium potassium calcium and magnesium. Some of the compounds which cause alkalinity also cause hardness. Acidity is caused by the presence of mineral acids, free carbon dioxide, sulphates of iron and aluminium etc..

2. Hardness

Hardness of water is caused by certain dissolved salts of calcium and magnesium which form scum with soap and reduce the formation of foam which helps in removing the dirt from clothes. These salts keep on depositing on the surface of boilers and thus form a layer known as scale which reduces the efficiency of the boilers. The hardness is known as temporary hardness if it is due to the bicarbonates of calcium and magnesium as this can be easily removed by boiling water or adding lime to it. By boiling the carbon dioxide gas escapes and the insoluble carbonates are deposited (which cause scaling). If sulphates, chlorides and nitrates are present they cannot be easily removed by boiling and so such water requires water softening methods and this type of hardness is known as permanent hardness.

3. Chloride and Fluoride

Most waters contain some chloride. The amount present can be caused by the leaching of marine sedimentary deposits or by pollution from sea water, brine, or industrial or domestic wastes. Chloride concentrations in excess of about 250 mg/L usually produce a noticeable taste in drinking water. Domestic water should contain less than 100 mg/L of chloride. In some areas, it may be necessary to use water with a chloride content in excess of 100 mg/L. In these cases, all of the other criteria for water purity must be met.

In some areas, water sources contain natural fluoride. Where the concentrations approach optimum levels, beneficial health effects have been observed. In such areas, the incidence of dental caries has been found to be below the levels observed in areas without natural fluoride. The optimum fluoride level for a given area depends upon air temperature, since temperature greatly influences the amount of water people drink. Excessive fluoride in drinking water supplies may produce fluorosis (mottling) of teeth, which increase as the optimum fluoride level is exceeded.

4. Metals

Small amounts of iron frequently are present in water because of the large amount of iron in the geologic materials. The presence of iron in water is considered objectionable because it imparts a brownish color to laundered goods and affects the taste of beverage such as tea and coffee. Exposure of the body to lead, however brief, can be seriously damaging to health. Prolonged exposure to relatively small quantities may result in serious illness or death. Lead taken into the body in quantities in excess of certain relatively low "normal" limits is a cumulative poison. Manganese imparts a brownish color to water and to cloth that is washed in it. It flavors coffee and tea with a medicinal taste.

The presence of sodium in water can affect persons suffering from heart, kidney, or circulatory ailments. When a strict sodium-free diet is recommended, any water should be regarded with suspicion. Home water softeners may be of particular concern because they add large quantities of sodium to the water. Zinc is found in some natural waters, particularly in areas where zinc ore deposits have been mined. Zinc is not considered detrimental to health, but it will impart an undesirable taste to drinking water. Arsenic occurs naturally in the environment, and it is also widely used in timber treatment, agricultural chemicals (pesticides), and manufacturing of gallium arsenide waters, glass, and alloys. Arsenic in drinking water is associated with lung and urinary bladder cancer.

5. Toxic Substances

Toxic organic substances include pesticides, insecticides, and solvents. Like the inorganic substances, there effects may be acute to chronic. Nitrates (NO_3), cyanides (CN), and heavy metals constitute the major classes of inorganic substances of health concern. Methemoglobinemia (infant cyanosis or "blue baby syndrome") has occurred in infants who have been water or fed formula prepared water having high concentrations of nitrate. CN ties up the hemoglobin sites that bind oxygen to red blood cells. This results in oxygen deprivation. A characteristic symptom is that the patient has a blue skin color. This condition is called cyanosis. CN causes chronic effects on the thyroid and central nervous system. The toxic heavy metals include arsenic (As), barium (Ba), cadmium (Cd), chromium (Cr), lead (Pb), mercury (Hg), selenium (Se), and silver (Ag). The heavy metals have a wide range of effects. They may be acute poisons (As and Cr^{6+} for example), or they may produce chronic disease (Pb, Cd, and Hg for example).

2.1.4 Drinking Water Microbiological Characteristics

From the public health standpoint, the bacteriological quality of water is as important as the chemical quality. A large number of infectious diseases may be transmitted by water, among them typhoid and cholera. Water for drinking and cooking purposes must be made free from disease-producing organisms (pathogens). These organisms include viruses, bacteria, protozoa, and helminthes (worms). Some organisms which cause disease in people originate with the fecal discharges of infected individuals. Others are from the fecal discharge of animals. Unfortunately, the specific disease-producing organisms present in water are not easily identified.

The techniques for comprehensive bacteriological examination are complex and time-consuming. It has been necessary to develop tests that indicate the relative degree of contamination in terms of an easily defined quantity. The most widely used test estimates the number of microorganisms of the coliform group. This grouping includes two genera: *Escherichia coli* and *Aerobacter aerogenes*. While *E. coli* are common inhabitants of the intestinal tract, *Aerobacter* are common in the soil, on leaves and on grain; on occasion they cause urinary tract infections. The test for these microorganisms is called the Total Coliform Test. Current research indicates that testing for *Escherichia coli* specifically may be warranted. Some agencies prefer the examination for *E. coli* as a better indicator of biological contamination than total coliforms.

2.1.5 Drinking Water Radiological Characteristics

The development and use of atomic energy as a power source and the mining of radioactive materials, as well as naturally occurring radioactive materials, have made it necessary to establish limiting concentrations for the intake into the body of radioactive substances, including drinking water. The effects of human exposure to radiation or radioactive materials are harmful, and any unnecessary exposure should be avoided. Humans have always been exposed to natural radiation from water, food and air. The amount of radiation to which the individuals is normally exposed varies with the amount of background radioactivity. Water with high radioactivity is not normal and is confined in great degree to areas where nuclear industries are situated.

2.1.6 Water Classification

Potable water is most conveniently classified as to its source, that is, groundwater or surface water. Generally, groundwater is uncontaminated but may contain aesthetically or economically undesirable impurities. Surface water must be considered to be contaminated with bacteria, viruses, or inorganic substances which could present a health hazard. Surface water may also have aesthetically unpleasing characteristics for potable water. Table 2.2 shows a comparison between groundwater and surface water.

Table 2.2 General characteristics of groundwater and surface water

Ground	Surface
Constant composition	Varying composition
High mineralization	Low mineralization
Little turbidity	High turbidity
Low or no color	Color
Bacteriological safe	Microorganisms present
No dissolved oxygen	Dissolved oxygen
High hardness	Low hardness
H_2S, Fe, Mn	Tastes and odors Possible chemical toxicity

Groundwater is further classified as to its source-deep or shallow wells. Municipal water quality factors of safety, temperature, appearance, taste and odor, and chemical balance are most easily satisfied by a deep well source. High concentrations of calcium, iron, manganese, and magnesium typify well waters. Some supplies contain hydrogen sulfide, while others may have excessive concentrations at chloride, sulfate or carbonate. Shallow wells are recharged by a nearby surface watercourse. They may have qualities similar to the deep wells, or they may take on the characteristics of the surface recharge water. A sand aquifer between the shallow well supply and the surface water-course may act as an effective filter for removal of organic matter and as a heat well for buffering temperature changes. To predict water quality from shallow wells, careful studies of the aquifer and nature of recharge water are necessary.

Surface water supplies are classified as to whether they come from a lake, reservoir, or river. Generally, a river has the lowest water quality and a reservoir the highest. Water quality in rivers depends upon the character of the watershed. River quality is largely influenced by pollution (or lack thereof) from municipalities, industries, and agricultural practices. The characteristics of a river can be highly variable. During rains or periods of runoff, turbidity may increase substantially. Many rivers will show an increase in color and taste and in odor-producing compounds. In warm months, algal blooms frequently cause taste and odor problems. Reservoir and lake sources have much less day-to-day variation than rivers. Additionally, quiescent conditions will reduce both the turbidity and, on occasion, the color. As in rivers, summer algal blooms can create taste and odor problems in lakes and reservoirs.

2.2 Water Treatment Systems

Many aquifers and isolated surface waters are high in water quality and may be pumped from the supply and transmission network directly to any number of end uses, including human consumption, irrigation, industrial processes, or fire control. However, clean water sources are the exception in many parts of the world, particularly regions where the population is dense or where there is heavy agricultural use. In these places, the water supply must receive varying degrees of treatment before distribution.

Impurities enter water as it moves through the atmosphere, across the earth's surface, and between soil particles in the ground. These background levels of impurities are often supplemented by human activities. Chemicals from industrial discharges and pathogenic organisms of human origin, if allowed to enter the water distribution system, may cause health problems. Excessive silt and other solids may make water aesthetically unpleasant and unsightly. Heavy metal pollution, including lead, zinc, and copper, may be caused by corrosion of the very pipes that carry water from its source to the consumer.

The method and degree of water treatment are important considerations for environmental engineers. Generally speaking, the characteristics of raw water determine the treatment method. Most public water systems are relied on for drinking water as well as for industrial consumption and fire fighting, so that human consumption, the highest use of the water, defines the degree of treatment. Thus, we focus on treatment techniques that produce potable water.

A typical water treatment plant is diagrammed in Figure 2.1. Coagulation, rapid mixing, flocculation, sedimentation, filtration, and disinfection are employed to remove color, turbidity, taste and odors, organic matter, and bacteria. The raw (untreated) surface water enters the plant via low-lift pumps or gravity. Usually screening has taken place prior to pumping. During mixing, chemicals called coagulants are added and rapidly dispersed through the water. The chemical reacts with the desired impurities and forms precipitates (flocs) that are slowly brought into contact with one another during flocculation. The objective of flocculation is to allow the flocs to collide and "grow" to a settleable size. The particles are removed by gravity (sedimentation). This is done to minimize the amount of solids that are applied to the filters. Filtration is the final polishing (removal) of particles. During filtration the water is passed through sand or similar media to screen out the fine particles that will not settle. Disinfection is the addition of chemicals (usually chlorine) to kill or reduce the number of pathogenic organisms. Storage may be provided at the plant or located within the community to meet peak demands and to allow the plant to operate on a uniform schedule. The high-lift pumps provide sufficient pressure to convey the water to its ultimate destination. The precipitated chemicals, original turbidity, and suspended material are removed from the sedimentation basins and from the filters. These residuals must be disposed of properly.

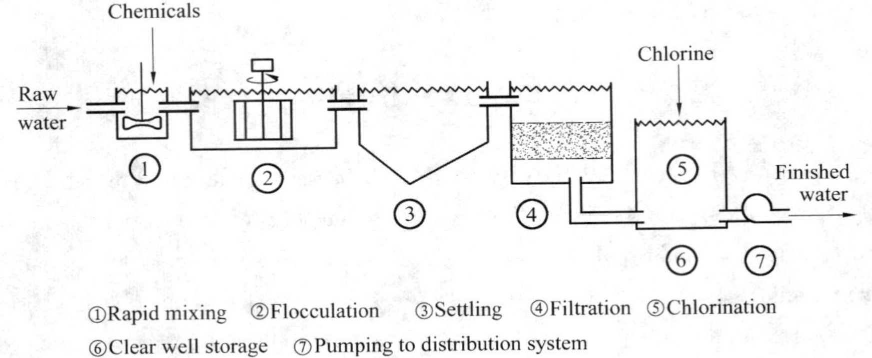

Figure 2.1 Flow diagram of a conventional surface water treatment plant

2.2.1 Coagulation

The object of coagulation (and subsequently flocculation) is to turn the small particles. These flocs are then conditioned so that they will be readily removed in subsequent processes. Technically, coagulation applies to the removal of colloidal particles. However the term has been applied more loosely to removal of dissolved ions, which is actually precipitation. Coagulation in this chapter will refer to colloid removal only. We define coagulation as a method to alter the colloids so that they will be able to approach and adhere to each other to form larger floc particles.

During coagulation a positive ion is added to water to reduce the surface charge to the point where the colloids are not repelled from each other. A coagulant is the substance (chemical) that is added to the water to accomplish coagulation. There are three key properties of a coagulant:

(1) Trivalent cation. The colloids most commonly found in natural waters are negatively charged, hence a cation is required neutralize the charge. A trivalent cation is the most efficient cation.

(2) Nontoxic. This requirement is obvious for the production of a safe water.

(3) Insoluble in the neutral pH range. The coagulant that is added must precipitate out of solution so that high concentrations of the ion are not left in the water. Such precipitation greatly assists the colloid removal process.

The two most commonly used coagulant are aluminum (Al^{3+}) and ferric iron (Fe^{3+}). Both meet the above three requirements, and their reactions are outlined here.

1. Aluminum

Aluminum can be purchased as either dry or liquid alum[$Al_2(SO_4)_3 \cdot 14H_2O$]. Commercial alum has an average molecular weight of 594. Liquid alum is sold as approximately 48.8 percent alum(8.3% Al_2O_3) and 51.2 percent water. Dry alum costs about 50 percent more than an equivalent amount of liquid alum so that only users of very amounts of alum purchase dry alum. When alum is added to a water containing alkalinity, the following reaction occurs:

$$Al_2(SO_4)_3 \cdot 14H_2O + 6HCO_3^- \rightleftharpoons 2Al(OH)_3 \cdot 3H_2O(s) + 6CO_2 + 8H_2O + 3SO_4^- \quad (2.1)$$

Such that each mole of alum added uses six moles of alkalinity and produces six moles of carbon dioxide. The above reaction shifts the carbonate equilibrium and decreases the pH. However, as long as sufficient alkalinity is present and $CO_2(g)$ is allowed to evolve, the pH is not drastically reduced and is generally not an operational problem. When sufficient alkalinity is not present to neutralize the sulfuric acid production, the pH may be greatly reduced:

$$Al_2(SO_4)_3 \cdot 14H_2 \rightleftharpoons 2AI(OH)_3 - 3H_2O(s) + 3H_2SO_4 + 2H_2O \quad (2.2)$$

If the second reaction occurs, lime or sodium carbonate may be added to neutralize the acid.

Two important factors in coagulant addition are pH and dose. The optimum dose and pH must be determined from laboratory tests. The optimal pH range for alum is approximately 5.5 to 6.5 with adequate coagulation possible between pH 5 to pH 8 under some condition.

An important aspect of coagulation is that the aluminum ion does not really exist as Al_3^+ and the final product is more complex than $Al(OH)_3$. When the alum is added to the water, it immediately dissociates, resulting in the release of all aluminum ion surrounded by six water molecules. The aluminum ion immediately starts reacting with the water, forming large $Al \cdot OH \cdot H_2O$ complexes. Some have suggested that it forms$[Al_8(OH)_{20} \cdot 28H_2O]^{4+}$ as the product that actually coagulates. Regardless of the actual species produced, the complex is a very large precipitate that removes many of the colloids by enmeshment as it falls through the water. This precipitate is referred to as a floc. Floc formation is one of the important properties of a coagulant for efficient colloid removal. The final product after coagulation has three water molecules associated with it in the solid form as indicated in the equations.

2. Iron

Iron can be purchased as either the sulfate salt ($Fe_2(SO_4)_3 \cdot xH_2O$) or the chloride salt ($FeCl_3 \cdot xH_2O$). It is available in various forms, and the individual supplier should be consulted for the specifics of the product. Dry and liquid forms are available. The properties of iron with respect to forming large complexes, dose, and pH curves are similar to those of alum. An example of the reaction of $FeCl_3$ in the presence of alkalinity is

$$FeCl_3 + 3HCO_3^- + 3H_2 \rightleftharpoons Fe(OH)_3 \cdot 3H_2O(s) + 3CO_2 + 3Cl^- \quad (2.3)$$

And without alkalinity

$$FeCl_3 + 6H_2O \rightleftharpoons Fe(OH)_3 \cdot 3H_2O(s) + 3HCl \quad (2.4)$$

forming hydrochloric acid which in turn lowers the pH. Ferric salts generally have a wider pH range for effective coagulation than aluminum, that is, pH ranges from 4 to 9.

3. Coagulant Aids

The four basic types of coagulant aids are pH adjusters, activated silica, clay, and polymers. Acids and alkalies are both used to adjust the pH of the water into the optimal range for coagulation. The acid most commonly used for lowering the pH is sulfuric acid. Either lime $[Ca(OH)_2]$or soda ash (Na_2CO_3) are used to raise the pH.

When activated silica is added to water, it produces a stable solution that has a negative surface charge. The activated silica can unite with the positively charged aluminum or with iron flocs, resulting in a larger, denser floc that settles faster and enhance enmeshment. The addition of activated silica is especially useful for treating highly colored, low-turbidity waters because it adds weight to the floc. However, activation of silica does require proper equipment and close operational control, and many plants are hesitant to use it.

Clays can act much like activated silica in that they have a slight negative charge and can add weight to the flocs. Clays are also most useful for colored, low-turbidity water, but are rarely used.

Polymers can have a negative charge (anionic), positive charge (cationic), positive and negative charge (polyamphotype), or no charge (nonionic). Polymers are long-chained carbon compounds of high molecular weight that have many active sites. The active sites adhere to flocs, joining them together and producing a larger, tougher floc that settle better. This process is called interparticle bridging. The type of polymer, dose, and point of addition must be determined for each water, and requirements may change within a plant on a seasonal, or even daily, basis.

2.2.2 Mixing and Flocculation

Clearly, if the chemical reactions in coagulating water are going to take place, the chemical must be mixed with the water. Mixing and flocculation are the physical methods necessary to accomplish the chemical processes of coagulation.

Mixing, or rapid mixing as it is called, is the process whereby the chemicals are quickly and uniformly dispersed in the water. Ideally, the chemicals would be instantaneously dispersed throughout the water. During coagulation, chemical reactions that take place in rapid mixing form precipitates, may be aluminum hydroxide or iron hydroxide. The precipitates formed in these processes must be brought into contact with one another so that they can agglomerate and form larger particles, called flocs. This contacting process is called flocculation and is accomplished by slow, gentle mixing.

In the treatment of water and wastewater the degree of mixing is measured by the velocity gradient, G. The velocity gradient is best thought of as the amount of shear taking place; that is, the higher the G value, the more violent the mixing. The velocity gradient is a function of the power input into a unit volume of water. It is given by

$$G = \sqrt{\frac{P}{\mu V}} \tag{2.5}$$

where G = velocity gradient, s^{-1}

P = power input, W

V = volume of water in mixing tank, m^3

μ = dynamic viscosity, Pa·s

While rapid mix is the most important physical factor affecting coagulant efficiency, flocculation is the most important factor affecting particle-removal efficiency. The objective of flocculation is to bring the particles into contact so that they will collide, stick together, and grow to a size that will readily settle. Enough mixing must be provided to bring the floc into contact and to keep the floc from settling in the flocculation basin, the velocity gradient must be controlled within a relatively narrow range. Flexibility should also be built into the flocculator so that the plant operator can vary the G value by a factor of two to three. The heavier the floc and the higher the suspended solids concentration, the more mixing is required to keep the floc in suspension. This is reflected in Table 2.3. An increase in the floc concentration (as measured by the suspended solids concentration) also increases the required G. With water temperatures of approximately 20℃, modern plants provide about 20 minutes of flocculation time (θ) at plant capacity. With lower temperatures, the detention time is increased. At 15℃ the detention time is increased by 7 percent, at 10℃ it is increased 15 percent, and at 5℃ it is increased 25 percent. Flocculation is normally accomplished with an axial-flow impeller, a baffled chamber (see Figure 2.2), or a paddle flocculator (see Figure 2.3).

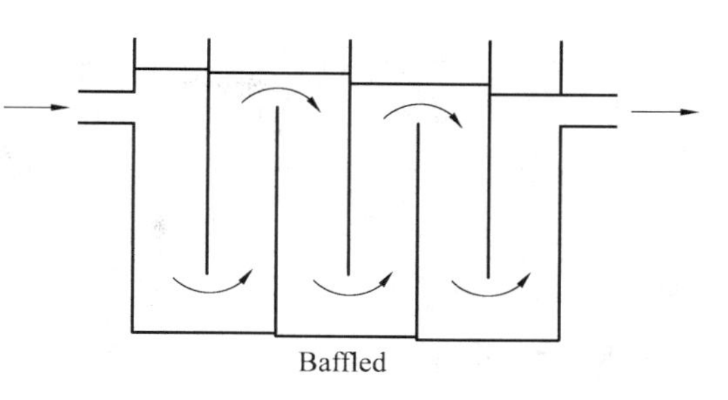

Figure 2.2 Baffled chamber flocculator

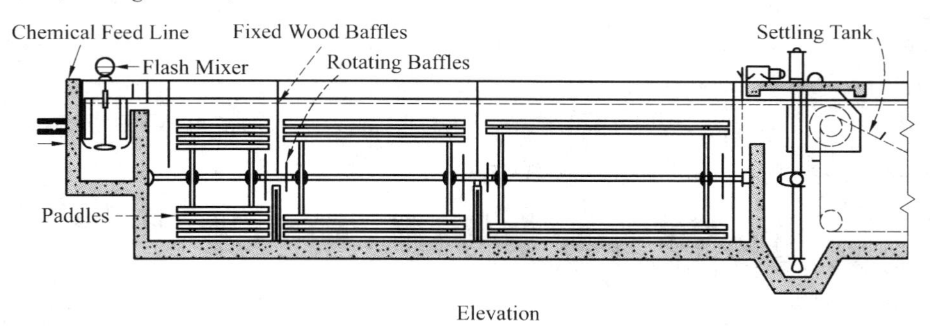

Figure 2.3 Paddle flocculator

Table 2.3 G values for flocculation

Type	G/s	Gθ (unitless)
Low-turbidity, color removal coagulation	20 ~ 70	60 000 to 200 000
High-turbidity, solids removal coagulation	30 ~ 80	36 000 to 96 000

2.2.3 Sedimentation

When the flocs have been formed, they must be separated from the water. This is invariably done in gravity settling basin (also called a clarifier) that allows the heavier-than-water particles to settle to the bottom. Sedimentation basins, designed to approximate uniform flow and to minimize turbulence, are usually rectangular or circular with either a radial or upward water flow pattern. Regardless of the type of basin, the design can be divided into four zones: inlet, settling, outlet, and sludge storage. While our intent is to present the concepts of sedimentation and to design a sedimentation tank, a brief discussion of all four zones is helpful in understanding the sizing of the settling zone. A schematic showing the four zones is shown in Figure 2.4.

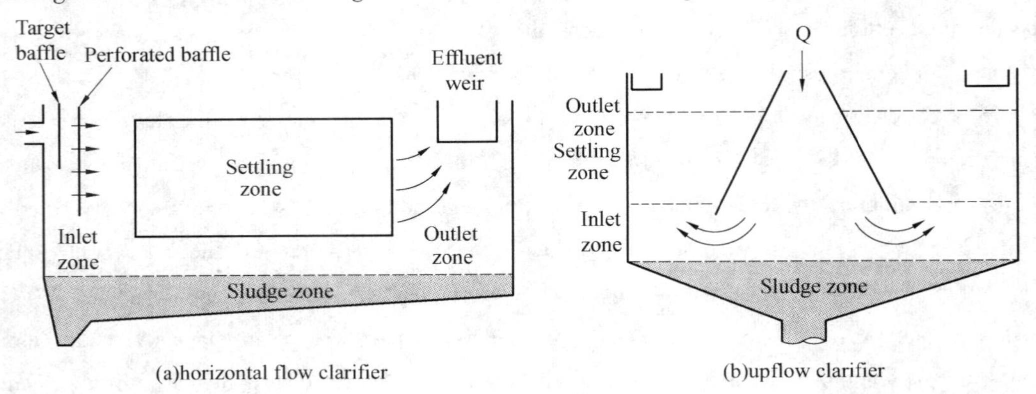

Figure 2.4 Zones of sedimentation

The purpose of the inlet zone is to evenly distribute the flow and suspended particles across the cross section of the settling zone. The inlet zone consists of a series of inlet and baffles placed about 1 m into the tank and extending the full depth of the tank. Following the baffle system, the flow takes on a flow pattern determined by the inlet structure. At some point the flow pattern is evenly distributed, and the water velocity slowed to the design velocity of the sedimentation-zone. At that point the inlet zone ends and the settling zone begins. With a well-designed inlet baffle system, the inlet zone extends approximately 1.5 m down the length of the tank. Proper inlet zone design may well be the most important aspect of removal efficiency.

With improper design, the inlet velocities may never subside to the settling-zone design velocity. Typical design numbers are usually conservative enough that an inlet zone length does not have to be added to the length calculated for the settling zone. In an accurate design, the inlet and settling zones are each designed separately and their lengths added together.

The configuration and depth of the sludge storage zone depends upon the method of cleaning, the frequency of cleaning, and the quantity of sludge estimated to be produced. All these variables can be evaluated and a sludge storage zone designed. In lieu of these design details, some genera guidelines can be presented. With a well-flocculated solid and good inlet design, over 75 percent of the solids may settle in the first fifth of the tank.

If the tank is long enough, storage depth can be provided by bottom slope; if not, a sludge hopper is necessary at the inlet end. Mechanically-cleaned basins may be equipped with a bottom scraper, such as shown in Figure 2.5. The sludge is continuously scraped to a hopper where it is pumped out. For mechanically cleaned basins, a one percent slope toward the sludge withdrawal point is used. A sludge hopper is designed with sides sloping with a vertical to horizontal ratio of 1.2:1 to 2:1.

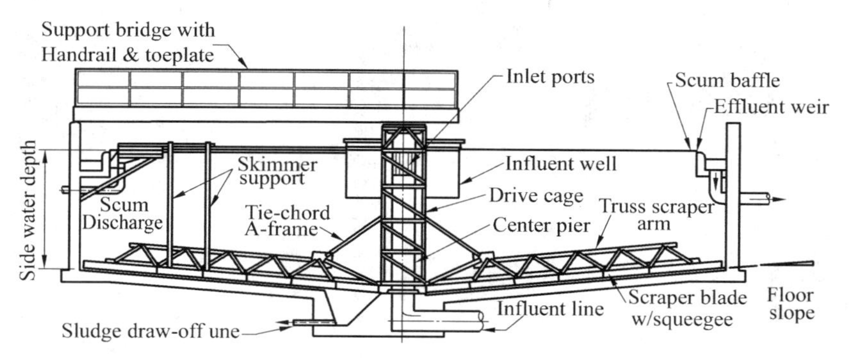

Figure 2.5 Schematic diagram of circular sludge scraper

The outlet zone is designed so as to remove the settled water from the basin without carrying away any of the floc particles. A fundamental property of water is that the velocity of flowing water is proportional to the flow rate divided by the area through which the water flows, that is,

$$v = Q/A_c \tag{2.6}$$

where v = water velocity, m/s

Q = water flow, m^3/s

A_c = cross-sectional area, m^2

Within the sedimentation tank, the flow is going through a very large area (basin depth times width); consequently, the velocity is slow. To remove the water from the basin quickly, it is desirable to direct the water into a pipe or small channel for easy transport, which will produce a significantly higher velocity. If a pipe was to be placed at the end of the sedimentation basin, all the water would "rush" to the pipe. This rushing would create high velocity profiles in the basin, which would tend to raise the settled floc from the basin and into the effluent water. This phenomenon of washing out the floc is called scouring, and one way to create scouring is with an improper outlet design. Rather than put a pipe at the end of the sedimentation basin, it is desirable to first put a series of troughs, called weirs, which provide a large area for the water to flow through and minimize the velocity in the sedimentation tank near the outlet zone. The weirs then feed into a central channel or pipe for transport of the settled water. Figure 2.6 shows various weir arrangements. The length of weir required is a function of the type of solids. The heavier the solids, the harder it is to scour them and the higher the allowable outlet velocity. Therefore, heavier particles require a shorter length of weir than do light particles. Each state generally has a set of standards which must be followed, but Table 2.4 shows typical values for weir loadings. The units for weir overflow rates are $m^3/d \cdot m$, which is water flow (m^3/d) per unit length of weir(m).

(a)rectangular

(b)circular

Figure 2.6 Weir arrangements

Table 2.4 Typical weir overflow rates

Type of floc	Weir overflow rate($m^3/d \cdot m$)
Light alum floc (low-turbidity water)	143 ~ 179
Heavier alum floc (higher-turbidity water)	179 ~ 268
Heavy floc from lime softening	268 ~ 322

In a rectangular basin, the weirs should cover at least one-third, and preferably up to one-half of the basin length. Spacing may be as large as 5 to 6 m on-centers but is preferably on the order of one-half this distance.

2.2.4 Filtration

The water leaving the sedimentation tank still contains floc particles. The settled water turbidity is generally in the range from 1 to 10 TU with a typical value being 2 TU. In order to reduce this turbidity to less than 0.3 TU, a filtration process is normally used. Water filtration is a process for separating suspended or colloidal impurities from water by passage through a porous medium, usually a bed of sand or other medium. Water fills the pores (open spaces) between the sand particles, and the impurities are left behind, either clogged in the open spaces or attached to the sand itself.

There are several methods of classifying filters. One way is to classify them according to the type of medium used, such as sand, coal (called anthracite), dual media (coal plus sand), or mixed media (coal, sand, and garnet). Another common way to classify the filters is by allowable loading rate. Loading rate is the flow rate of water applied per unit area of the filter. It is the velocity of the water approaching the face of the filter:

$$v_a = \frac{Q}{A_s} \tag{2.7}$$

where v_a = face velocity, m/d

= loading rate, $m^3/d \cdot m^2$

Q = flow rate onto filter surface, m^3/d

A_s = surface area of filter, m^2

Based on loading rate, the filters are described as being slow sand filters, rapid sand filters, or high-rate sand filters.

Slow sand filters were first introduced in the 1800s. The water is applied to the sand at a loading rate of 2.9 to 7.6 $m^3/d \cdot m^2$. As the suspended or colloidal material is applied to the sand, the particles begin to collect in the top 75 mm and to clog the pore spaces. As the pores become clogged, water will no longer pass through the sand. At this point the top layer of sand is scraped off, cleaned, and replaced. Slow sand filters require large areas of land and are operator intensive.

Rapid sand filters have graded (layered) sand within the bed. The sand grain size distribution is selected to optimize the passage of water while minimizing the passage of particulate matter. Rapid sand filters are cleaned in place by forcing water backwards through the sand. This operation is called backwashing. The washwater flow rate is such that the sand is expanded and the filtered particles are removed from the bed. After backwashing, the sand settles back into place. The largest particles settle first, resulting in a fine sand layer on top and a coarse sand layer on the bottom. Rapid sand filters are the most common type of filter used in water treatment today. Figure 2.7 shows a cutaway drawing of a rapid sand filter. The bottom of the filter consists of a support media and collection system. The support media is designed to keep the sand in the filter and prevent it from leaving with the filtered water. Layers of graded gravel (large on bottom, small on top) traditionally have been used for the support. The under drain blocks collect the filtered water.

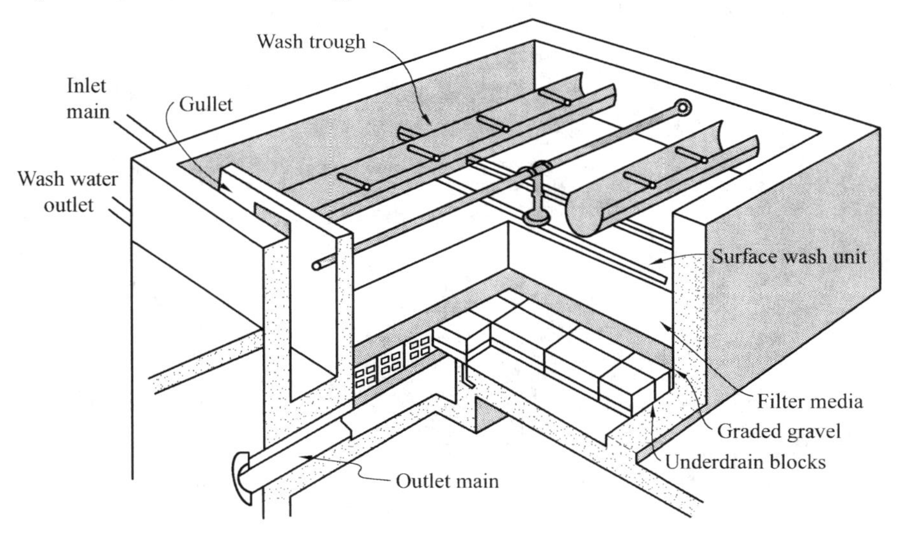

Figure 2.7 Typical gravity filter box

On top of the support media is a layer of graded sand. The sand depth varies between 0.5 m and 0.75 m. If a dual media filter is used, the sand is about 0.3 m thick and the coal about 0.45 m thick. Approximately 0.7 m to 1 m above the top of the sand are the washwater troughs. The washwater troughs collect the backwash water used to clean the filter. The troughs are placed high enough above the sand layer so that sand will not be carried out with the back washwater. Generally, a total depth of 1.8 m to 3 m is allowed above the sand layer for water to build up above the filter. This depth of water provides sufficient pressure to force the water through the sand during filtration.

Figure 2.8 shows a slightly simplified version of a rapid sand filter. Water from the settling basins enters the filter and seeps through the sand and gravel bed, through a false floor, and out into a clear well that acts as a storage tank for finished water. During filtration, valves A and C are open. During filtration, the filter bed will become more and more clogged. As the filter clogs, the water level will rise above the sand as it becomes harder to force water through the bed. Eventually, the water level will rise to the point that the filter bed must be cleaned. This point is called terminal head loss. When this occurs, the operator turns off valves A and C. This stops the supply of water from the sedimentation tank and prevents any more water from entering the clear well. The operator then opens valves E and B. This allows a large flow of washwater (clean water stored in an elevated tank or pumped from a clear well) to enter below the filter bed. This rush of water forces the sand bed to expand and sets individual sand particles in motion. By rubbing against each other, the light colloidal particles that were trapped in the pore spaces are released and escape with the washwater. The washwater is a waste stream that must be treated. After a few minutes the washwater is shut off and filtration resumed.

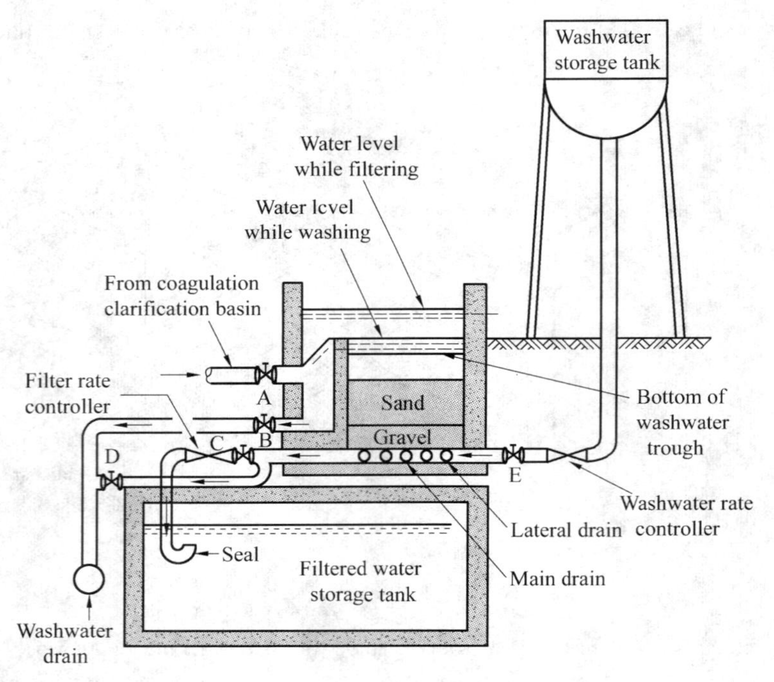

Figure 2.8 Operation of a rapid sand filter

The size distribution or variation of a sample of granular material is determined by sieving the sample through a series of standard sieves (screens). One such standard series is the U.S.

Standard Sieve Series. The U.S. Standard Sieve Series (see Table 2.5) is based on a sieve opening of 1 mm. Sieves in the "fine series" stand successively in the ratio of $(2)^{1/4}$ to one another, the largest opening in this series being 5.66 mm and the smallest 0.037 mm. All material that passes through the smallest sieve opening the series is caught in a pan that acts as the terminus of the series.

Table 2.5 U.S. Standard Sieve Series

Sieve designation number	Size of opening /mm	Sieve designation number	Size of opening /mm
200	0.074	20	0.84
140	0.105	(18)	(1.00)
100	0.149	16	1.19
70	0.210	2	1.68
50	0.297	8	2.38
40	0.42	6	3.36
30	0.59	4	4.76

Source: Fair and Geyer, 1954.

Experience has suggested that, for silica sand, the effective size should be in the range of 0.35 mm to 0.55 mm with a maximum of about 1.0 mm. The uniformity coefficient should range between 1.3 and 1.7. Smaller effective sizes result in product water that is lower in turbidity, but they also result in higher pressure losses in the filter and shorter operating cycles between cleanings.

2.2.5 Disinfection

Disinfection is used in water treatment to reduce pathogens (disease-producing microorganisms) to an acceptable level. Disinfection is not the same as sterilization. Sterilization implies the destruction of all living organisms. Drinking water need not be sterile. Three categories of human enteric pathogens are normally of consequence: bacteria, viruses, and amebic cysts. Purposeful disinfection must be capable of destroying all three.

1. Chlorination

Chlorination, addition of chlorine or chlorine compounds to water, is considered to be the single most important process for preventing the spread of waterborne disease. Molecular chlorine, Cl_2, is a greenish-yellow gas at ordinary room temperature and pressure. In gaseous form, it is very toxic, and even in low concentrations it is a severe irritant. But when the chlorine is dissolved in low concentrations in clean water, it is not harmful, and if it is properly applied, objectionable tastes and odors due to the chlorine and its by-products are not noticeable to the average person.

Chlorine is commercially available in gaseous form or in the form of solid and liquid compounds called hypochlorites. For the disinfection of relatively large volumes of water, the gaseous form of chlorine is generally the most economical, but for smaller volumes, the use of hypochlorite compounds is more common.

Gaseous chlorine is stored and shipped in pressurized steel cylinders. Under pressure, the chlorine is actually in liquid form in the cylinder; when it is released from the cylinder, it vaporizes into a gas. The cylinders may range in capacity from 45 kg (100 lb) to about 1 000 kg (1 ton). Very large water (or wastewater) treatment plants may use special railroad tank cars filled with chlorine. A device called an all-vacuum chlorinator is considered to provide the safest type of chlorine feed installation. It is mounted directly on the chlorine cylinder. The gaseous chlorine is always under a partial vacuum in the line that carries it to the point of application; chlorine leaks cannot occur in that line. A typical vacuum chlorine feed system is shown in Figure 2.9. The vacuum is formed by water flowing through the ejector unit at high velocity. There are other types of chlorinators, some of which have the chlorine or concentrated chlorine solutions conveyed relatively long distances under pressure. These present somewhat greater risks of chlorine leaks. In any chlorine feed installation, safety factors are very important because of the toxicity of the gas.

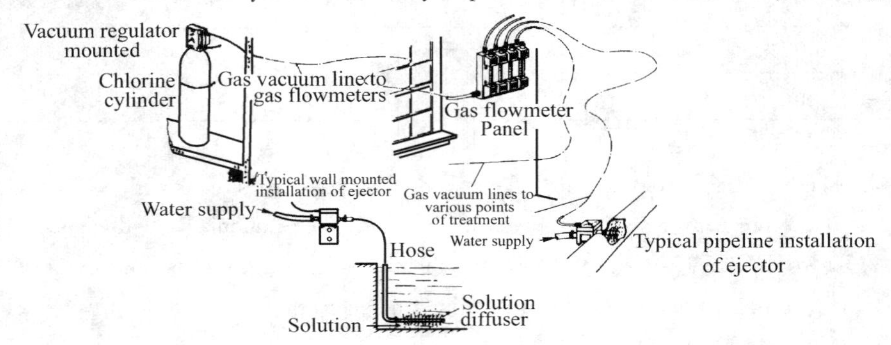

Figure 2.9 Typical vacuum-feed chlorination system

Hypochlorites are usually applied to water in liquid form by means of small pumps, such as the one illustrated in Figure 2.10. These are positive-displacement-type pumps, which deliver a specific amount of liquid on each stroke of a piston or flexible diaphragm. Two types of hypochlorite compounds are available for disinfection: sodium hypochlorite and calcium hypochlorite. Sodium hypochlorite is available only in liquid form and contains up to 15 percent available chlorine. It is usually diluted with water before being applied as a disinfectant. Calcium hypochlorite is a dry compound, available in granular or tablet form; it is readily soluble in water. Calcium hypochlorite solutions are more stable than solutions of sodium hypochlorite, which deteriorate over time.

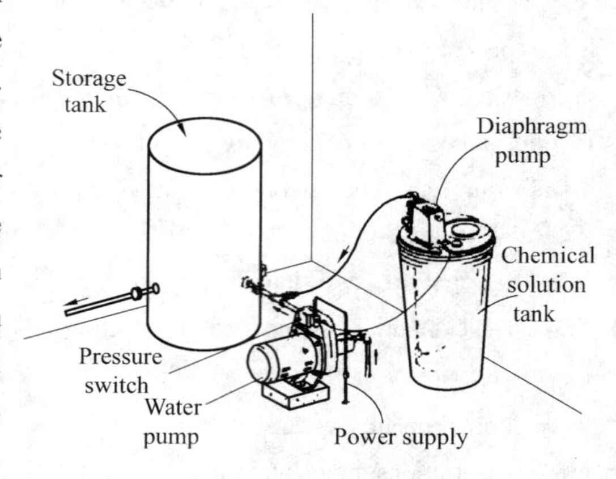

Figure 2.10 A typical hypochlorinator installation

In addition to pH, the effectiveness of chlorine and chlorine compounds in destroying bacteria depends on the chlorine concentration and the contact time. Contact time is the time period during which the free or combined chlorine is acting on the microorganisms. At pH values close to 7 (neutral conditions), a free chlorine residual of 0.2 mg/L with a 10-min contact time has about the same disinfecting power as 1.5 mg/L of combined chlorine residual with a 1-h contact time.

Natural waters often contain trace amounts of organic compounds, primarily from natural sources such as decaying vegetation. These substances can react with the chlorine to form compounds called trihalomethanes (THMs), which may cause cancer in humans. Chloroform is an example of a THM compound. The EPA has set standards that limit the maximum amount of THM compounds in drinking waters. One way to prevent THM formation is to make sure that the chlorine is added to the water clarification and the removal of most of the organics. Also, alternative methods of disinfection are available that do not use chlorine.

2. Other Methods of Disinfection

Ozone (O_3) is a highly reactive gas at ordinary temperature and pressures, and acts as a very potent disinfectant when mixed with water. It has been used for over 90 years in European countries as an alternative to chlorine, which sometimes leaves a noticeable taste and odor in drinking water. Ozone can be produced by passing a very high voltage electric current through air or oxygen. However, because it is very unstable and cannot be stored, it must be manufactured on site, where it is used. And because it does not leave a measurable residual in water after the initial contact time, some chlorine (although in relatively smaller amounts) must be used to ensure continued disinfection as the water flows throughout the network of water distribution pipes.

In addition to the ability of ozone to act as a disinfectant without causing taste and odor problems, it does not react to form THM compounds. Ozone is also a stronger disinfectant than chlorine and is able to inactivate most viruses in addition to bacteria. (It is approved by the EPA for disinfection.) It can assist as a coagulant when used with alum, thus reducing the amount of chemicals needed to adjust the final pH of the water to make it noncorrosive. Because it greatly aids the coagulation process, ozone can also facilitate the application of a direct filtration process and eliminate the need for large sedimentation basins. Despite these advantages, the high cost of its production and application compared to that for chlorine has discouraged widespread use of ozone for disinfection. A flow diagram showing major treatment steps at the Haworth plant is shown in Figure 2.11.

Ultraviolet(UV) light can be also used for disinfection. UV light is electromagnetic radiation just beyond the blue end of the light spectrum, outside the range of visible light. It has a much higher energy level than visible light, and in large doses it destroys bacteria and viruses. The UV energy is absorbed by genetic material in the microorganisms, interfering with their ability to reproduce and survive. UV light can be generated by a variety of lamps; submerged, low-pressure mercury lamps are best suited for use in disinfection systems because they generate a large fraction of UV energy that gets absorbed. Ultraviolet disinfection systems do not involve chemical handing, as

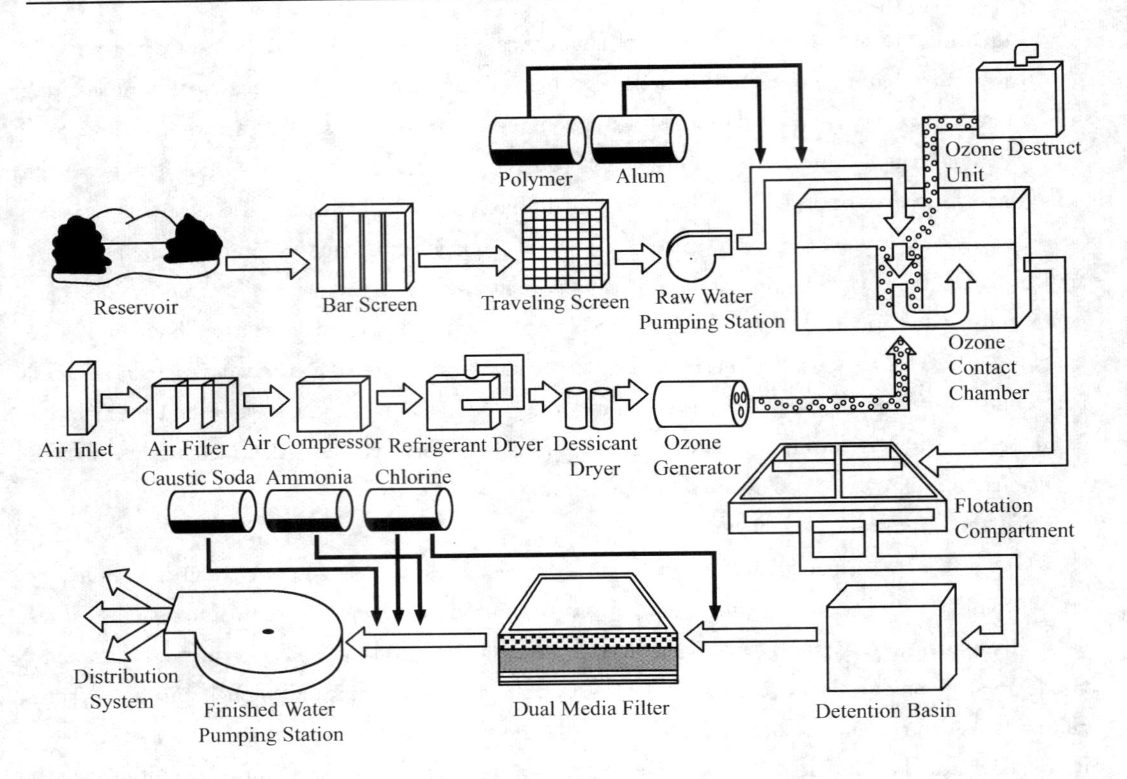

Figure 2.11 A flow diagram of the ozone water purification process at United Water Resources Haworth water treatment plant

do chlorine or ozone systems, thereby minimizing chemical safety concerns. Like ozone, though, UV radiation leaves no measurable residual in the water. Advances in UV germicidal lamp technology are making UV disinfection a more reliable and economical option for disinfection.

2.2.6 Adsorption

Adsorption is a mass transfer process wherein a substance is transferred from the liquid phase to the surface of a solid where it is bound by chemical or physical forces. Generally, in water treatment, the adsorbent (solid) is activated carbon, either granular (GAC) or powdered (PAC). PAC is fed to the raw water in a slurry and is generally used to remove taste and odor-causing substances or to provide some removal of synthetic organic chemicals (SOCs). GAC is added to the existing filter system by replacing the anthracite with GAC, or an additional contactor is built and is placed in the flow scheme after primary filtration. The design of the GAC contactor is very similar to a filter box, although deeper. At present, the applications of adsorption in water treatment are predominately for taste and odor removal. However, adsorption is increasingly being considered for removal of SOCs, Volatile organic chemicals(VOCs), and naturally occurring organic matter, such as THM precursors and disinfection by produts(DBPs).

Biologically derived earthy-musty odors in water supplies are a widespread problem. Their occurrence interval and concentration vary greatly from season to season and is often unpredictable. As mentioned, one of the most popular methods for removing these compounds is the addition of PAC to the raw water. The dose is generally less than 10 mg/L. The advantage of PAC is that the capital equipment is relatively inexpensive and it can be used on an as-needed basis. The disadvantage is that the adsorption is often incomplete. Sometimes even doses of 50 mg/L are not sufficient.

As an alternative for taste and odor control, many plants have replaced the anthracite in the filters with GAC. The GAC will last from one to three years and then must be replaced. It is very effective in removing many taste and odor compounds. Concern about SOCs in drinking water has motivated interest in adsorption as a treatment process for removal of toxic and potentially carcinogenic compounds present in minute, but significant, quantities. Few other processes can remove SOCs to the required low levels. Generally, GAC is used for SOC removal either as a filter media replacement or as a separate contactor. The data for how long the GAC will last for any given SOC are somewhat limited and must be evaluated on a case-by-case basis. However, if the GAC is to be used continuously for SOC removal, then a separate contactor may be warranted.

GAC has been proposed to be used to remove naturally occurring organic matter that would, in tarn, reduce the formation of DBPs. Testing has shown that GAC will remove these organics. It must operate in a separate contactor since the depth of conventional filter is inadequate. The GAC will typically last 90 to 120 days until it loses its adsorptive capacity. Because of its short life, the GAC needs to be regenerated by burning in a high-temperature furnace. This can be done on-site or can be done by returning the GAC to the manufacture. GAC has also been considered for removal of THMs. However, the capacity is very low and the carbon may only last up to 30 days. GAC is not considered practical for THM removal.

2.2.7 Membranes

A membrane is a thin layer of material that is capable of separating materials as a function of their physical and chemical properties when a driving force is applied across the membrane. In the case of water treatment, the driving force is supplied by using a high pressure pump and the membrane type is selected based on the constituents to be removed.

In the membrane process, the feed stream is divided into two streams, the concentrate or reject stream, and the permeate or product stream as shown in Figure 2.12, The membrane is at the heart of every membrane process and can be considered a barrier between the feed and product water that does not allow certain contaminants to pass, as represented by Figure 2.13. The performance or efficiency of a given membrane is determined by its selectivity and the applied flow. The efficiency is called the flux and is defined as the volume flowing through the membrane per unit area and time, $m^3/(m^2 \cdot s)$. The selectivity of a membrane toward a mixture is express by its retention, R, and is given by

$$R = \frac{c_p - c_f}{c_p} \times 100\% \tag{2.8}$$

where c_p = contaminant concentration in feed

c_f = contaminant concentration in permeate

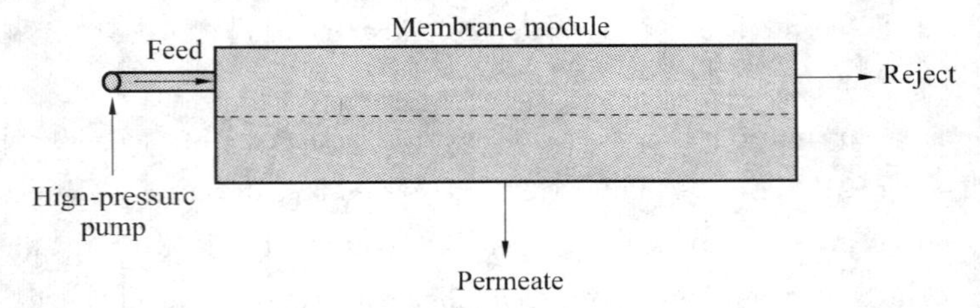

Figure 2.12 Schematic representation of a membrane process

The value of R varies between 100 percent (complete retention) and 0 percent (no retention).

Membrane technology is becoming increasingly popular as an alternative treatment technology for drinking water. The anticipation of more stringent water quality regulations, a decrease in availability of adequate water resources, and an emphasis on water for reuse has made membrane processes more viable as a water treatment process. As advances are being made in membrane technology, capital and operation and maintenance costs continue to decline, further endorsing the use of membrane treatment techniques.

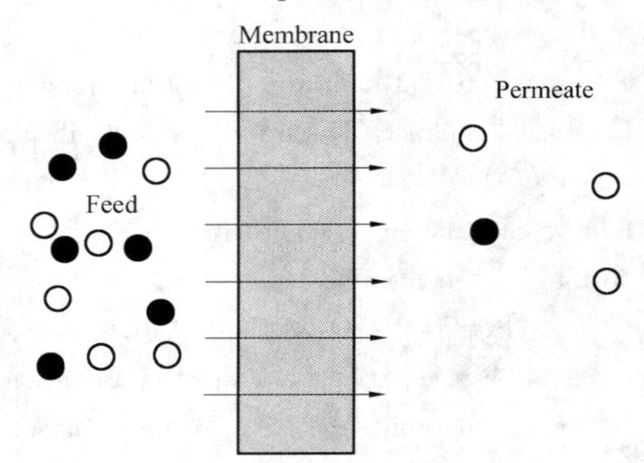

Figure 2.13 Schematic representation of contaminants separated by a membrane

2.3 Wastewater Constituents

Every community produces both liquid and solid wastes and air emissions. The liquid waste—wastewater—is essentially the water supply of the community after it has been used in a variety of applications (see Figure 2.14). From the standpoint of sources of generation, wastewater may be defined as a combination of the liquid or water-carried wastes removed from residences, institutions, and commercial and industrial establishments, together with such groundwater, surface water, and stormwater as may be present.

When wastewater accumulates and is allowed to go septic, the decomposition of the organic matter it contains will lead to nuisance conditions including the production of malodorous gases. In addition, untreated wastewater contains numerous pathogenic microorganisms that dwell in the human intestinal tract. Wastewater also contains nutrients, which can stimulate the growth of aquatic

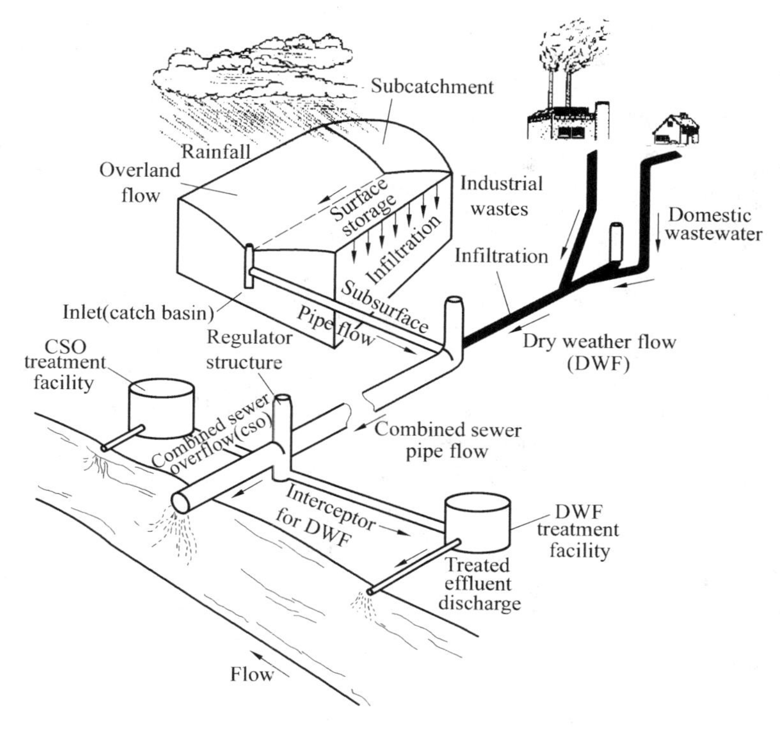

Figure 2.14 Schematic diagram of a wastewater management infrastructure

plants, and may contain toxic compounds or compounds that potentially may be mutagenic or carcinogenic. For these reasons, the immediate and nuisance-free removal of wastewater from its sources of generation, followed by treatment, reuse, or dispersal into the environment is necessary to protect public health and the environment.

The important constituents of concern in wastewater treatment are listed in Table 2.6. Secondary treatment standards for wastewater are concerned with the removal of biodegradable organics, total suspended solids, and pathogens. Many of the more stringent standards that have been developed recently deal with the removal of nutrients, heavy metals, and priority pollutants. When wastewater is to be reused, standards normally include additional requirements for the removal of refractory organics, heavy metals, and in some cases, dissolved inorganic solids.

Table 2.6 Principal constituents of concern in wastewater treatment

Constituents	Reason for importance
Suspended solids	Suspended solids can lead to the development of sludge deposition and anaerobic conditions when untreated wastewater is discharged in the aquatic environment.
Biodegradable organics	Composed principally of proteins, carbohydrates, and fats. Biodegradable organics are measured most commonly in terms of BOD and COD. If discharged untreated to the environment their biological stabilization can lead to the depletion of natural oxygen resources and to the development of septic conditions.
Pathogens	Communicable diseases can be transmitted by the pathogenic organisms that may be present in wastewater.

续表 2.6

Constituents	Reason for importance
Nutrients	Both nitrogen and phosphorus, along with carbon, are essential nutrients for growth. When discharged to the aquatic environment, the nutrients can lead to the growth of undesirable aquatic life. When discharged in excessive amount on land, they can also lead to the pollution of groundwater.
Priority pollutants	Organic and inorganic compounds selected on the basis of their known or suspected carcinogenicity, metogenicity, teratogenicity or high acute toxicity. Many of these compounds are found in wastewater.
Refractory organics	These organics tend to resist conditional methods of wastewater treatment. Typical examples include surfactants, phenols, and agricultural pesticides.
Heavy metals	Heavy metals are usually added to wastewater from commercial and industrial activities and may have to be removed if the wastewater is to be reused.
Dissolved inorganics	Inorganic constituents such as calcium, sodium, and sulfate are added to the original domestic water supply as a result of water use and may have to be removed if the wastewater is to be reused.

2.3.1 Physical Characteristics

The most important physical characteristic of wastewater is its total solids content, which is composed of floating matter, settleable matter, colloidal matter, and matter in solution. Other important physical characteristics include turbidity, color, transmittance, temperature, conductivity, and density, specific gravity and specific weight.

1. Solids

Wastewater contains a variety of solid materials varying from rags to colloidal material. In the characterization of wastewater, coarse materials are usually removed before the sample is analyzed for solids. The various solids classifications are identified in Table 2.7. The interrelationship between the various solids fractions found in wastewater is illustrated graphically on Figure 2.15.

Table 2.7 Definitions of solids found in wastewater

Test	Description
Total solids(TS)	The residue remaining after a wastewater sample has been evaporated and dried at a specified temperature(103 to 105 ℃)
Total volatile solids(TVS)	Those solids that can be volatiled and burned off when the TS are ignited (500 ± 50 ℃)
Total fixed solids(TFS)	The residue that remains after TS are ignited(500 ± 50 ℃)
Total suspended solids(TSS)	Portion of the TS retained on a filter with a specified pore size, measured after being dried at a specified temperature(105 ℃). The filter used most commonly for the determination of TSS is the Whatman glass fiber filter, which has a nominal pore size of about 1.58 μm.

Continued Table 2.7

Test	Description
Volatile suspended solids(VSS)	Those solids that can be volatilized and burned off when the TSS are ignited (500 ± 50 ℃)
Fixed suspended solids(FSS)	The residue that remains after TSS are ignited (500 ± 50 ℃)
Total dissolved solids(TDS)	Those solids that pass through the filter, and are then evaporated and dried at specified temperature. It comprised of colloidal and dissolved solids. Colloids are typically in the size range from 0.001 to 1 μm
Total volatile dissolved solids (VDS)	Those solids that can be volatilized and burned off when the TDS are ignited (500 ± 50 ℃)
Fixed dissolved solids(FDS)	The residue that remains after TDS are ignited (500 ± 50 ℃)
Settleable solids	Suspended solids, expressed as milliliters per liter, that will settle out of suspension within a specified period of time

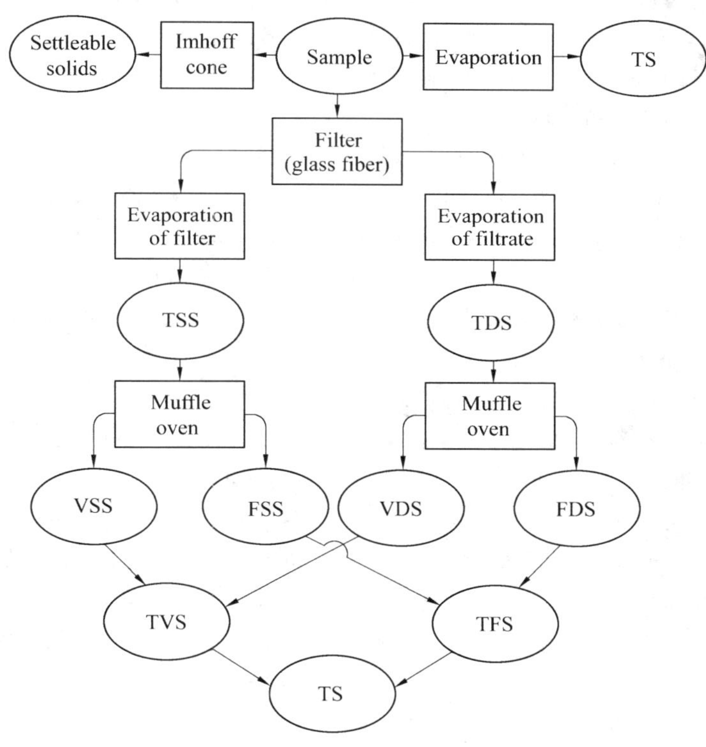

Figure 2.15 Interrelationships of solids found in water and wastewater. In much of the water quality literature, the solids passing through the filter are called dissolved solids

2. Turbidity

Turbidity, a measure of the light-transmitting properties of water, is another test used to indicate the quality of waste discharges and natural waters with respect to colloidal and residual suspended matter. The results of turbidity measurements are reported as nephelometric turbidity units (NTU). Colloidal matter will scatter or absorb light and thus prevent its transmission. It should be noted that the presence of air bubbles in the fluid will cause erroneous turbidity readings. In general, there is no relationship between turbidity and the concentration of total suspended solids in untreated wastewater. There is, however, a reasonable relationship between turbidity and total suspended solids for the settled and filtered secondary effluent from the activated sludge process.

3. Color and Odors

Fresh wastewater is usually a brownish-gray color. However, as the travel time in the collection system increases, and more anaerobic conditions develop, the color of the wastewater changes sequentially from gray to dark gray, and ultimately to black. When the color of the wastewater is black, the wastewater is often described as septic. Some industrial wastewaters may also add color to domestic wastewater. Odors in domestic wastewater usually are caused by gases produced by the decomposition of organic matter or by substances added to the wastewater. The most characteristic odor of stale or septic wastewater is that of hydrogen sulfide, which is produced by anaerobic microorganisms that reduce sulfate to sulfide. Industrial wastewater may contain either odorous compounds or compounds that produce odors during the process of wastewater treatment.

4. Absorption/Transmittance

Percent transmittance is affected by all substances in wastewater that can absorb or scatter light. Unfiltered and filtered transmittance is mearsured in wastewater in connection with the evalution and design of UV disinfection systems. The principal wastewater characteristics that affect the percent transmission include selected inorganic compounds (e.g., copper, iron, etc.), organic compounds (e.g., organic dyes, humic substances, and conjugated ring compounds such as benzene and toluene), and TSS.

5. Temperature

The temperature of wastewater is commonly higher than that of the local water supply, because of the addition of warm water from households and industrial activities. As the specific heat of water is much greater than that of air, the observed wastewater temperatures are higher than the local air temperatures during most of the year and are lower only during the hottest summer months. Depending on the location and time of year, the effluent temperatures can be either higher or lower than the corresponding influent values. The temperature of water is a very important parameter because of its effect on chemical reactions and reaction rates, aquatic life, and the suitability of the water for beneficial uses.

6. Conductivity

The electrical conductivity (EC) of a water is a measure of the ability of a solution to conduct an electrical current. Because the electrical current is transported by the ions in solution, the conductivity increases as the concentration of ions increases. In effect, the measured EC value is used as a surrogate measure of total dissolved solid (TDS) concentration. At present, the EC of a water is one of the important parameter used to determine the suitability of a water for irrigation. The salinity of treated wastewater to be used for irrigation is estimated by measuring its electrical conductivity.

2.3.2 Chemical Constituents

The chemical constituents of wastewater are typically classified as inorganic and organic. Inorganic chemical constituents of concern include nutrients, nonmetallic constituents, metals, and gases. Organic constituents of interest in wastewater are classified as aggregate and individual. Aggregate organic constituents are comprised of a number of individual compounds that cannot be distinguished separately. Both aggregate and individual organic constituents are of great significance in the treatment, disposal, and reuse of wastewater.

1. Chlorides

Chloride is a constituent of concern in wastewater as it can impact the final reuse applications of treated wastewater. Chlorides in natural water result from the leaching of chloride-containing rocks and soils with which the water comes in contact, and in coastal areas from saltwater intrusion. In addition, agricultural, industrial, and domestic wastewaters discharged to surface waters are a source of chlorides. Because conventional methods of waste treatment do not remove chloride to any significant extent, higher than usual chloride concentrations can be taken as an indication that a body of water is being used for waste disposal.

2. Alkalinity

Alkalinity in wastewater results from the presence of the hydroxides [OH^-], carbonates [CO_3^{2-}], and bicarbonates [HCO_3^-] of elements such as calcium, magnesium, sodium, potassium, and ammonia. Of these, calcium and magnesium bicarbonates are most common. Borates, silicates, phosphates, and similar compounds can also contribute to the alkalinity. The alkalinity in wastewater helps to resist changes in pH caused by the addition of acids. The concentration of alkalinity in wastewater is important where chemical and biological treatment is to be used, in biological nutrient removal, and where ammonia is to be removed by air stripping.

3. Nitrogen and Phosphorus

The elements nitrogen and phosphorus, essential to the growth of microorganisms, plants, and animals, are known as nutrients or biostimulants. Trace quantities of other elements, such as iron, are also needed for biological growth, but nitrogen and phosphorus are, in most cases, the major nutrients of importance. Because nitrogen is an essential building block in the synthesis of protein, nitrogen data will be required to evaluate the treatability of wastewater by biological processes. Insufficient nitrogen can necessitate the addition of nitrogen to make the waste treatable. Phosphorus is also essential to the growth of algae and other biological organisms. Because of noxious algal blooms that occur in surface waters, there is presently much interest in controlling the amount of phosphorus compounds that enter surface waters in domestic and industrial waste discharges and natural runoff.

4. Metallic Constituents

Trace quantities of many metals, such as cadmium (Cd), chromium (Cr), copper (Cu), iron

(Fe), lead (Pb), manganese (Mn), mercury (Hg), nickel (Ni), and zinc (Zn) are important constituents of most waters. Although macro and micro amounts of metals are required for proper growth, the same metals can be toxic when present in elevated concentrations. Therefore, it is frequently desirable to measure and control the concentrations of these substances. Calcium, magnesium, and sodium are of importance in determining the sodium adsorption ratio (SAR), which is used to assess the suitability of treated effluent for agricultural use. Where composted sludge is applied in agricultural applications, arsenic, cadmium, copper, lead, mercury, molybdenum, nickel, selenium, and zinc must be determined.

5. Biochemical Oxygen Demand (BOD)

The most widely used parameter of organic pollution applied to both wastewater and surface water is the 5-day BOD (BOD_5). This determination involves the measurement of the dissolved oxygen used by microorganisms in the biochemical oxidation of organic matter. BOD test results are now used (1) to determine the approximate quantity of oxygen that will be required to biologically stabilize the organic matter present, (2) to determine the size of waste treatment facilities, (3) to measure the efficiency of some treatment processes, and (4) to determine compliance with wastewater discharge permits. Because it is likely that the BOD test will continue to be used for some time, it is important to know the details of the test and its limitations.

6. Chemical Oxygen Demand (COD)

The COD test is used to measure the oxygen equivalent of the organic material in wastewater that can be oxidized chemically using dichromate in an acid solution. Some of the reasons for the observed difference between ultimate carbonaceous BOD and COD are as follows: (1) many organic substances which are difficult to oxidize biologically, such as lignin, can be oxidized chemically, (2) inorganic substances that are oxidized by the dichromate increase the apparent organic content of the sample, (3) certain organic substances may be toxic to the microorganisms used in the BOD test, and (4) high COD values may occur because of the presence of inorganic substances with which the dichromate can react.

7. Disinfection Byproducts (DBPs)

It has been found that when chlorine is added to water containing organic matter a variety of organic compounds containing chlorine are formed DBPs. Although general present in low concentrations, they are of concern because many of them are known as suspected potential human carcinogens. Typical classes of compounds include trihalomethanes (THMs), haloacetic acids (HAAs), trichlorophenol, and aldehydes.

2.3.3 Biological Characteristics

Organisms found in surface water and wastewater include bacteria, fungi, algae, protozoa, plants and animals, and viruses. The biological characteristics of wastewater are of fundamental importance in the control of diseases caused by pathogenic organisms of human origin, and because of the extensive and fundamental role played by bacteria and other microorganisms in the decomposition and stabilization of organic matter, both in nature and in wastewater treatment plants. Living single-cell microorganisms that can only be seen with a microscope are responsible for the activity in biological wastewater treatment. The basic functional and structural unit of all living matter is the cell. Living organisms are divided into either prokaryote or eukaryote cells as a function of their genetic information and cell complexity.

1. Prokaryotes

The prokaryotes have the simplest cell structure and include bacteria, blue green algae (cyanobacter), and archaea. The archaea are separated from bacteria due to their DNA composition and unique cellular chemistry, such as differences in the cell wall and ribosome structure. Many archaea are bacteria that can grow under extreme conditions of temperature and salinity, and also include methanogenic methane-producing bacteria, important in anaerobic treatment processes. The prokaryote organisms are generally much smaller compared to eukaryote organisms. The absence of a nuclear membrane to contain the cell DNA is also a distinguishing feature of the prokaryota organisms.

2. Eukaryotes

In contrast to the prokaryotes, the eukaryotes are much more complex and contain plants and animals and single-celled organisms of importance in wastewater treatment including protozoa, fungi, and green algae. The eukaryotic organisms have much more complex internal structures. These include the endoplasmic reticulum, which is a distinct organelle that contains the sites of ribosomes with internal membranes. The golgi bodies are also distinct membrane structures and contain sites for the secretion of enzymes and other macromolecules. The mitochondrion is a complex internal membrane structure where respiration occurs for eukaryotic cells, and is lacking in prokaryotic cells. While the prokaryotes can have photosynthetic pigments, they do not contain chloroplasts, which are used in photosynthesis by green algae.

3. Pathogenic Organisms

Pathogenic organisms found in wastewater may be excreted by human beings and animals who are infected with disease or who are carriers of a particular infectious disease. The pathogenic organisms found in wastewater can be classified into four broad categories, bacteria, protozoa, helminths, and viruses. Bacteria pathogenic organisms of human origin typically cause diseases of the gastrointestinal tract, such as typhoid and paratyphoid fever, dysentery, diarrhea, and cholera. Viruses, composed of a nucleic acid core (RNA or DNA) surrounded by an outer coat of protein and glycoprotein, are obligate intracellular parasites that require the machinery of a host cell to support their growth. Viruses are classified separately according to the host infected. Bacteriophage, as the name implies, are viruses that infect bacteria.

2.4 Physical Unit Operation of Wastewater Treatment

Operations used for the treatment of wastewater in which change is brought about by means of or through the application of physical forces are known as physical unit operations. Because physical unit operations were derived originally from observations of the physical world, they were the first treatment methods to be used. Today, physical unit operations, as shown on Figure 2.16, are a major part of most wastewater treatment systems. The unit operations most commonly used in wastewater treatment include screening, coarse solids reduction (comminution, maceration, and screenings grinding), flow equalization, mixing and flocculation, grit removal, sedimentation, high-rate clarification, accelerated gravity separation (vortex separators), flotation, oxygen transfer, packed-bed filtration, membrane separation, aeration, biosolid dewatering, and volatilization and stripping of volatile organic compounds (VOCs).

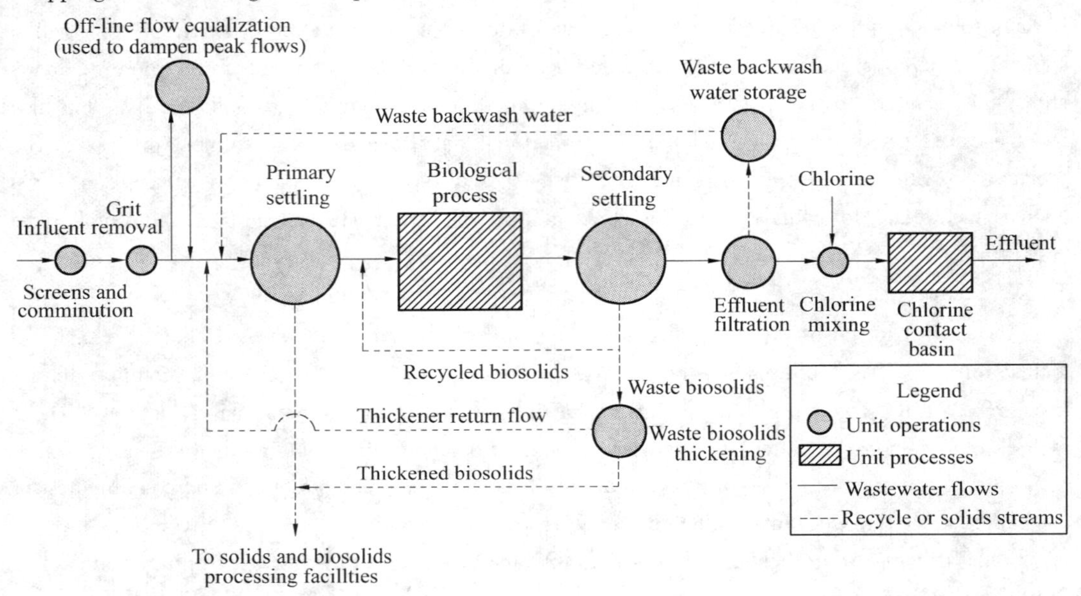

Figure 2.16 Location of physical unit operations in wastewater treatment plant flow diagram

2.4.1 Screening

The first unit operation generally encountered in wastewater treatment plants is screening. A screen is a device with openings, generally of uniform size, that is used to retain solids found in the influent wastewater to the treatment plant or in combined wastewater collection systems subject to overflows, especially from stormwater. As shown in Figure 2.17, two general types of screens, coarse screens and fine screens, are used in preliminary treatment of

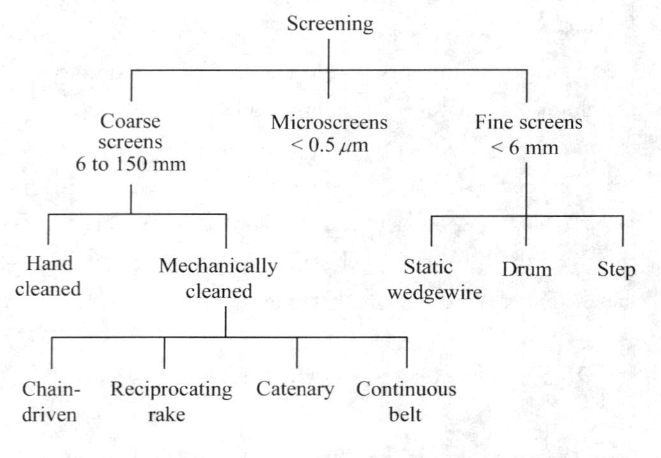

Figure 2.17 Definition sketch for types of screens used in wastewater treatment

wastewater. Coarse screens have clear openings ranging from 6 to 150 mm; fine screens have clear openings less than 6 mm. Microscreens, which generally have screen openings less than 50 μm, are used principally in removing fine solids from treated effluents. The screening element may consist of parallel bars, rods or wires, grating, wire mesh, or perforated plate, and the openings may be of any shape but generally are circular or rectangular slots. A screen composed of parallel bars or rods is often called a "bar rack" or a coarse screen and is used for the removal of coarse solids. Fine screens are devices consisting of perforated plates, wedgewire elements, and wire cloth that have smaller openings. The materials removed by these devices are known as screenings.

1. Coarse Screens (Bar Racks)

In wastewater treatment, coarse screens are used to protect pumps, valves, pipelines and other appurtenances from damage or clogging by rags and large objects. Industrial waste-treatment plants may or may not need them, depending on the character of the wastes. According to the method used to clean them, coarse screens are designated as either hand-cleaned or mechanically cleaned.

Hand-cleaned coarse screens are used frequently ahead of pumps in small wastewater pumping stations and sometimes used at the headworks of small- to medium-sized wastewater-treatment plants. Often they are used for standby screening in bypass channels for service during high-flow periods, when mechanically cleaned screens are being repaired, or in the event of a power failure. Normally, mechanically cleaned screens are provided in place of hand-cleaned screens to minimize manual labor required to clean the screens and to reduce flooding due to clogging. The design of mechanically cleaned bar screens has evolved over the years to reduce the operating and maintenance problems and to improve the screenings removal capabilities. Many of the newer designs include extensive use of corrosion-resistant materials including stainless steel and plastics. Mechanically cleaned bar screens are divided into four principal types: (1) chain-driven, (2) reciprocating rake, (3) catenary, and (4) continuous belt.

Chain-driven mechanically cleaned bar screens can be divided into categories based on whether the screen is raked to clean from the front (upstream) side or the back (downstream) side and whether the rakes return to the bottom of the bar screen from the front or back. In general, front cleaned, front return screens (see Figure 2.18) are more efficient in terms of retaining captured solids, but they are less rugged and are susceptible to jamming by solids that collect at the base of the rake. Most of the chain-operated screens share the disadvantage of submerged sprockets that require frequent operator attention and are difficult to maintain.

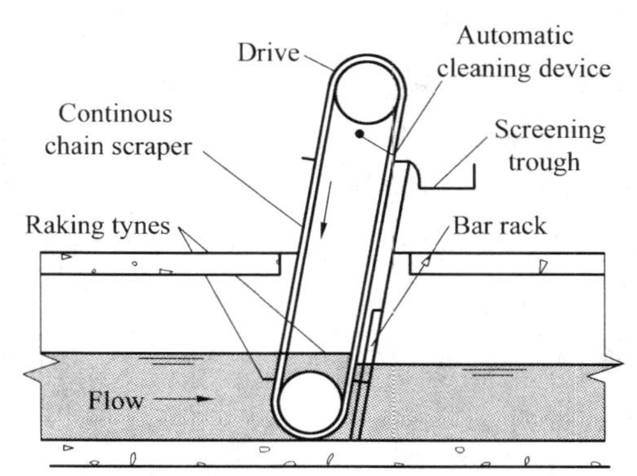

Figure 2.18 Chain-driven screens

The reciprocating-rake-type bar screen (see Figure 2.19) imitates the movements of a person raking the screen. The rake moves to the base of the screen, engages the bars, and pulls the screenings to the top of the screen where they are removed. Most screen designs utilize a cogwheel drive mechanism for the rake. A major advantage is that all parts requiring maintenance are above the waterline and can be easily inspected and maintained without dewatering the channel.

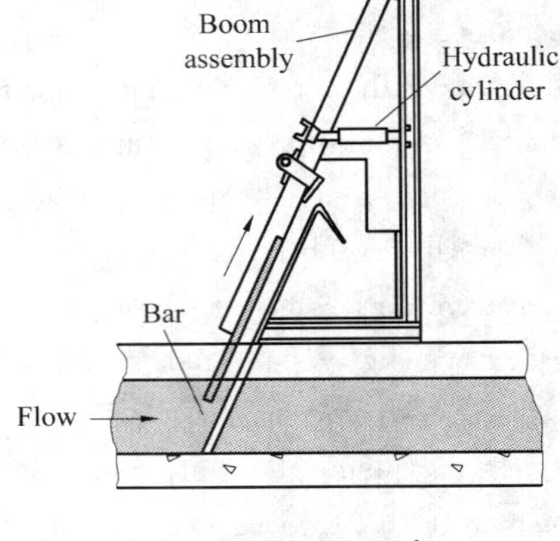

Figure 2.19 Reciprocating rake screens

A catenary screen is a type of front cleaned, front return chain-driven screen, but it has no submerged sprockets. In the catenary screen (see Figure 2.20), the rake is held against the rack by the weight of the chain. If heavy objects become jammed in the bars, the rakes pass over them instead of jamming. The screen, however, has a relatively large "footprint" and thus requires greater space for installation.

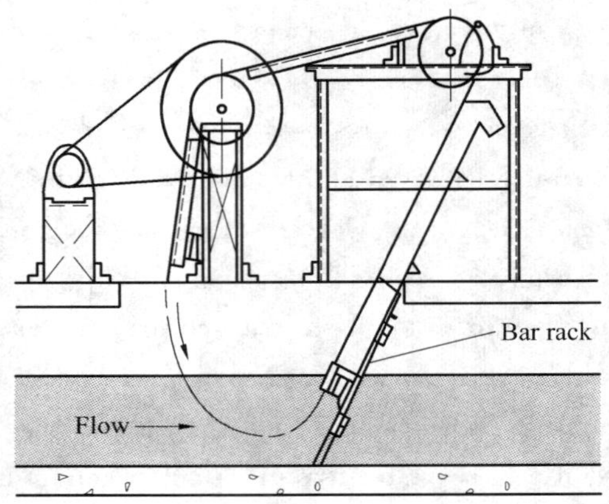

Figure 2.20 Catenary screens

The continuous belt screen is a continuous, self-cleaning screening belt that removes fine and coarse solids (see Figure 2.21). A large number of screening elements (rakes) are attached to the drive chains; the number of screening elements depends on the depth of the screen channel. Because the screen openings can range from 0.5 to 30 mm, it can be used as either a coarse or a fine screen. Hooks protruding from the belt elements are provided to capture large solids such as cans, sticks, and rags.

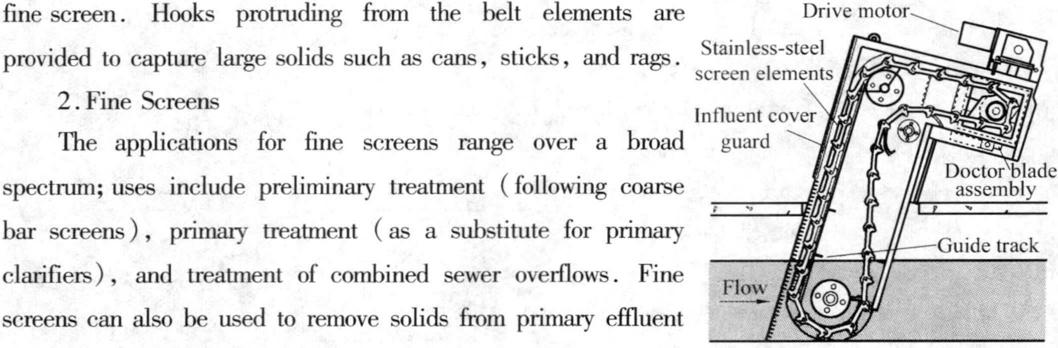

Figure 2.21 Continuous belt screens

2. Fine Screens

The applications for fine screens range over a broad spectrum; uses include preliminary treatment (following coarse bar screens), primary treatment (as a substitute for primary clarifiers), and treatment of combined sewer overflows. Fine screens can also be used to remove solids from primary effluent that could cause clogging problems in trickling filters.

Fine screens used for preliminary treatment are of the (1) static (fixed), (2) rotary drum, or (3) step type. Typically, the openings vary from 0.2 to 6 mm. Examples of fine screens are illustrated on Figure 2.22. Fine screens may be used to replace primary treatment at small wastewater-treatment plants, up to 0.13 m^3/s in design capacity.

Stainless-steel mesh or special wedge-shaped bars are used as the screening medium. Provision is made for the continuous removal of the collected solids, supplemented by water sprays to keep the screening medium clean.

(a) static wedgewire (b) drum (c) step

Figure 2.22 Typical fine screens

In step screens, screenings are moved up the screen by means of movable and fixed vertical plates

3. Microscreens

Microscreening involves the use of variable low-speed (up to 4 r/min), continuously backwashed, rotating-drum screens operating under gravity-flow conditions. The filtering fabrics have openings of 10 to 35 μm and are fitted on the drum periphery. The wastewater enters the open end of the drum and flows outward through the rotating-drum screening cloth. The principal applications for microscreens are to remove suspended solids from secondary effluent and from stabilization-pond effluent.

Typical suspended solids removal achieved with microscreens ranges from 10 to 80 percent, with an average of 55 percent. Problems encountered with microscreens include incomplete solids removal and inability to handle solids fluctuations. Reducing the rotating speed of the drum and less frequent flushing of the screen have resulted in increased removal efficiencies but reduced capacity.

4. Screenings Handling, Processing and Disposal

In mechanically cleaned screen installations, screenings are discharged from the screening unit directly into a screenings grinder, a pneumatic ejector, or a container for disposal; or onto a conveyor for transport to a screenings compactor or collection hopper. Belt conveyors and pneumatic ejectors are generally the primary means of mechanically transporting screenings. Belt conveyors offer the advantages of simplicity of operation, low maintenance, freedom from clogging, and low cost. Belt conveyors give off odors and may have to be provided with covers. Pneumatic ejectors are less odorous and typically require less space; however, they are subject to clogging if large objects are present in the screenings.

Screenings compactors can be used to dewater and reduce the volume of screenings (see Figure 2.23). Such devices, including hydraulic ram and screw compactors, receive screenings directly from the bar screens and are capable of transporting the compacted screenings to a receiving hopper. Compactors can reduce the water content of the screenings by up to 50 percent and the volume by up to 75 percent.

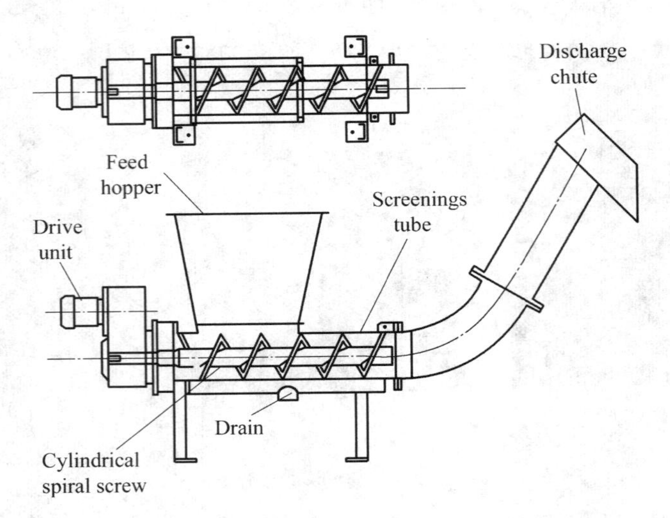

Figure 2.23 Typical device used for compacting screenings

Means of disposal of screenings include (1) removal by hauling to disposal areas (landfill) including codisposal with municipal solid wastes, (2) disposal by burial on the plant site (small installations only), (3) incineration either alone or in combination with sludge and grit (large installations only), and (4) discharge to grinders or macerators where they are ground and returned to the wastewater. The first method of disposal is most commonly used. In some cases, screenings are required to be lime stabilized for the control of pathogenic organisms before disposal in landfills.

2.4.2 Coarse Solids Reduction

As an alternative to coarse bar screens or fine screens, comminutors and macerators can be used to intercept coarse solids and grind or shred them in the screen channel. High-speed grinders are used in conjunction with mechanically cleaned screens to grind and shred screenings that are removed from the wastewater. The solids are cut up into a smaller, more uniform size for return to the flow stream for subsequent removal by downstream treatment operations and processes. Comminutors, macerators, and grinders can theoretically eliminate the messy and offensive task of screenings handling and disposal. The use of comminutors and macerators is particularly advantageous in a pumping station to protect the pumps against clogging by rags and large objects and to eliminate the need to handle and dispose of screenings. They are particularly useful in cold climates where collected screenings are subject to freezing.

1. Comminutors

Comminutors are used most commonly in small wastewater-treatment plants, less than $0.2\ m^3/s$. Comminutors are installed in a wastewater flow channel to screen and shred material to sizes from 6 to 20 mm without removing the shredded solids from the flow stream. A typical comminutor uses a stationary horizontal screen to intercept the flow (see Figure 2.24) and a rotating or oscillating arm that contains cutting teeth to mesh with the screen. The cutting teeth and the shear bars cut coarse material. The small sheared particles pass through the screen and into the down-stream channel. Comminutors may create a string of material, namely, rags, which can collect on downstream treatment equipment.

Figure 2.24 Typical comminutors used for particle size reduction of solids

2. Macerators

Macerators are slow-speed grinders that typically consist of two sets of counter-rotating assemblies with blades (see Figure 2.25). The assemblies are mounted vertically in the flow channel. The blades or teeth on the rotating assemblies have a close tolerance that effectively chops material as it passes through the unit. Macerators can be used in pipeline installations to shred solids, particularly ahead of wastewater and sludge pumps, or in channels at smaller wastewater-treatment plants. Sizes for pipeline applications typically range from 100 to 400 mm in diameter.

3. Grinders

High-speed grinders, typically referred to as hammermills, receive screened materials from bar screens. The materials are pulverized by a high-speed rotating assembly that cuts the materials passing through the unit. The cutting or knife blades force screenings through a stationary grid or louver that encloses the rotating assembly. Washwater is typically used to keep the unit clean and to help transport materials back to the wastewater stream. Discharge from the grinder can be located either upstream or downstream of the bar screen.

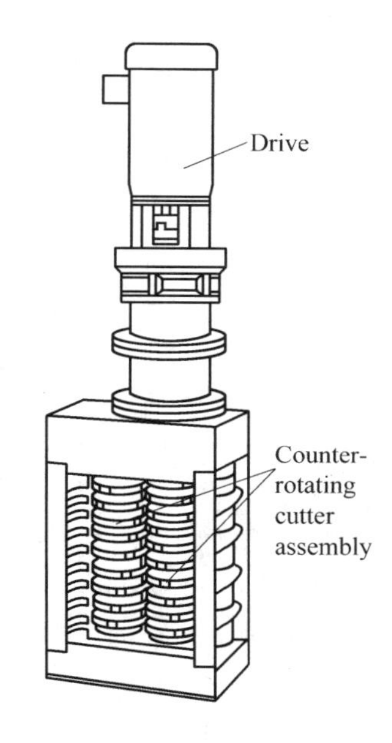

Figure 2.25 Schematic of in-channel type slow-speed grinder/macerator

2.4.3 Flow Equalization

Flow equalization is a method used to overcome the operational problems caused by flowrate variations, to improve the performance of the downstream processes, and to reduce the size and cost of downstream treatment facilities. Flow equalization simply is the damping of flowrate variations to achieve a constant or nearly constant flowrate and can be applied in a number of different situations, depending on the characteristics of the collection system. The principal applications are for the equalization of (1) dry-weather flows to reduce peak flows and loads, (2) wet-weather flows in sanitary collection systems experiencing inflow and infiltration, or (3) combined stormwater and sanitary system flows.

The application of flow equalization in wastewater treatment is illustrated in the two flow diagrams given on Figure 2.26. In the in-line arrangement (Figure 2.26(a)), all of the flow passes through the equalization basin. This arrangement can be used to achieve a considerable amount of constituent concentration and flowrate damping. In the off-line arrangement (Figure 2.26(b)), only the flow above some predetermined flow limit is diverted into the equalization basin.

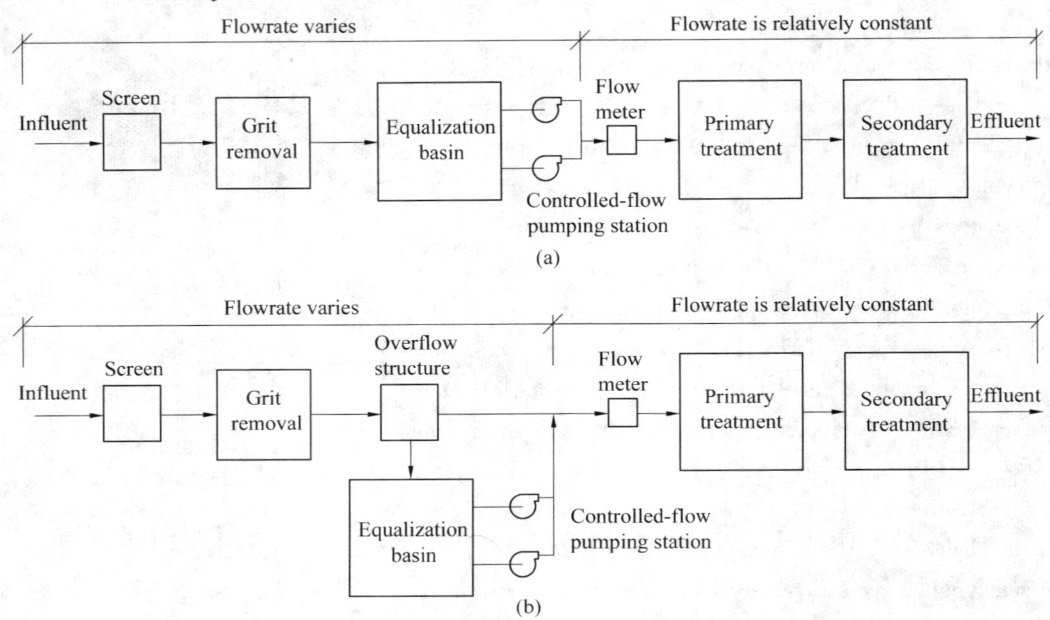

Figure 2.26　Typical wastewater treatment plant flow diagram incorporating flow equalization (a) in-line equalization; (b) off-line equalization

The principal benefits that are cited as deriving from application of flow equalization are: (1) biological treatment is enhanced, because shock loadings are eliminated or can be minimized, inhibiting substances can be diluted and pH can be stabilized (2) the effluent quality and thickening performance of secondary sedimentation tanks following biological treatment is improved through improved consistency in solids loading; (3) effluent filtration surface area requirements are reduced, filter performance is improved, and more uniform filter-backwash cycles are possible by lower hydraulic loading; and (4) in chemical treatment, damping of mass loading improves chemical feed control and process reliability.

2.4.4　Grit Removal

Removal of grit from wastewater may be accomplished in grit chambers or by the centrifugal separation of solids. Grit chambers are designed to remove grit, consisting of sand, gravel, cinders, or other heavy solid materials that have subsiding velocities or specific gravities substantially greater than those of the organic putrescible solids in wastewater. Grit chambers are most commonly located after the bar screens and before the primary sedimentation tanks. Primary sedimentation tanks function for the removal of the heavy organic solids. In some installations, grit chambers precede the screening facilities. Generally, the installation of screening facilities ahead of the grit chambers makes the operation and maintenance of the grit removal facilities easier. Grit chambers are provided to (1) protect moving mechanical equipment from abrasion and accompanying abnormal wear; (2) reduce formation of heavy deposits in pipelines, channels, and conduits; and (3) reduce the frequency of digester cleaning caused by excessive accumulations of grit. The removal of grit is essential ahead of centrifuges, heat exchangers, and high-pressure diaphragm pumps.

There are three general types of grit chambers: horizontal flow, of either a rectangular or a square configuration; aerated; or vortex type. In the horizontal-flow type, the flow passes through the chamber in a horizontal direction and the straight-line velocity of flow is controlled by the dimensions of the unit, an influent distribution gate, and a weir at the effluent end. The aerated type consists of a spiral-flow aeration tank where the spiral velocity is induced and controlled by the tank dimensions and quantity of air supplied to the unit. The vortex type consists of a cylindrical tank in which the flow enters tangentially creating a vortex flow pattern; centrifugal and gravitational forces cause the grit to separate. Design of grit chambers is commonly based on the removal of grit particles having a specific gravity of 2.65 and a wastewater temperature of 15.5 ℃.

1. Square Horizontal-Flow Grit Chambers

Square horizontal-flow grit chambers, such as those shown on Figure 2.27, have been in use for over 60 years. Influent to the units is distributed over the cross section of the tank by a series of vanes or gates, and the distributed wastewater flows in straight lines across the tank and overflows a weir in a free discharge. Where square grit chambers are used, it is generally advisable to use at least two units. In square grit chambers, the solids are removed by a rotating raking mechanism to a sump at the side of the tank. Settled grit may be moved up an incline by a reciprocating rake mechanism or grit may be pumped from the tank through a cyclone degritter to separate the remaining organic material and concentrate grit. The concentrated grit then may be washed again in a classifier using a submerged reciprocating rake or an inclined-screw conveyor. By either method, organic solids are separated from the grit and flow back into the basin, resulting in a cleaner, dryer grit.

2. Aerated Grit Chambers

In aerated grit chambers, air is introduced along one side of a rectangular tank to create a spiral flow pattern perpendicular to the flow through the tank (see Figure 2.28). The heavier grit particles that have higher settling velocities settle to the bottom of the tank. Lighter, principally organic, particles remain in suspension and pass through the tank. The velocity of roll or agitation governs the size of particles of a given specific gravity that will be removed. If the velocity is too great, grit will be carried out of the chamber; if it is too small, organic material will be removed with the grit. Fortunately, the quantity of air is easily adjusted. With proper adjustment, almost 100 percent removal will be obtained, and the grit will be well washed. Grit that is not well washed and contains organic matter is an odor nuisance and attracts insects.

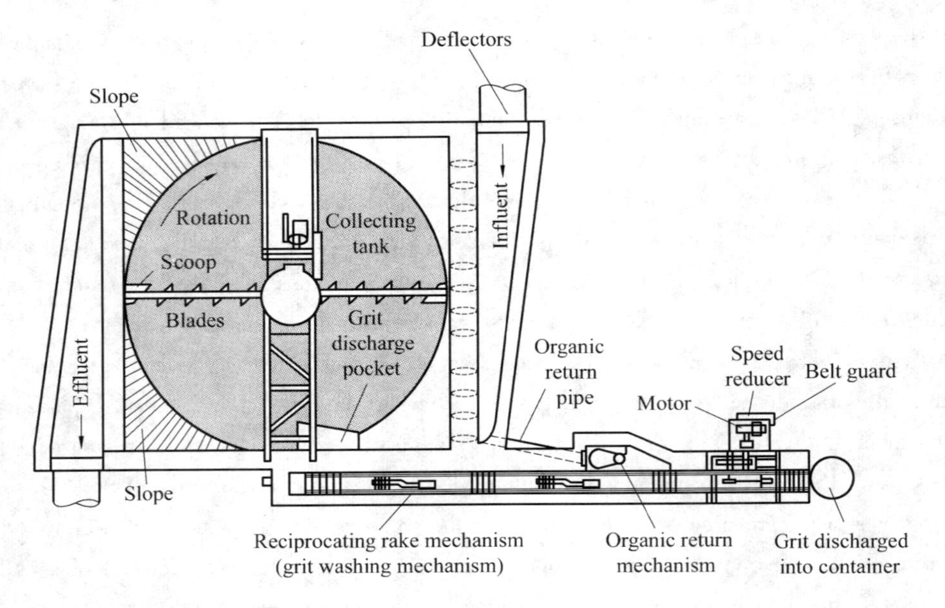

Figure 2.27 Schematic of the typical square horizontal-flow grit chambers

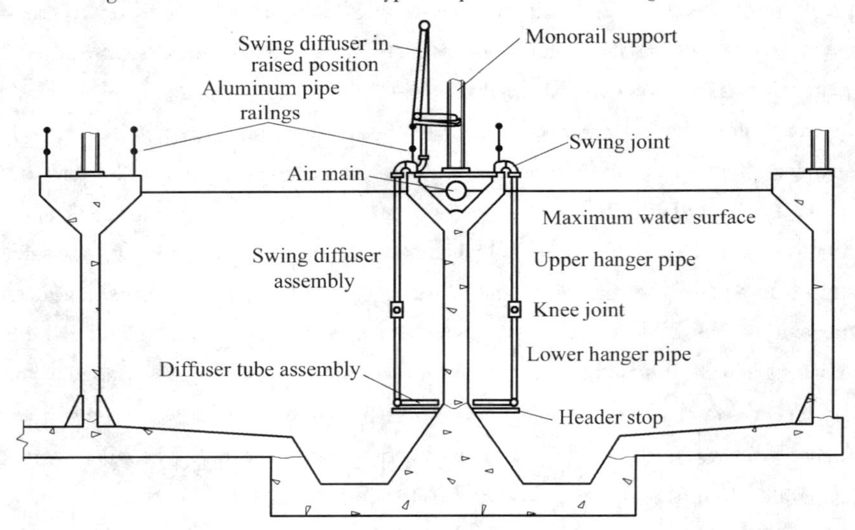

Figure 2.28 Typical section through an aerated grit chamber

For grit removal, aerated grit chambers are often provided with grab buckets traveling on monorails and centered over the grit collection and storage trough. An added advantage of a grab bucket grit-removal system is that dropping the grit from the bucket through the tank contents can further wash grit. Other installation are equipped with chain-and-bucket conveyors, running the full length of the storage troughs, which move the grit to one end of the trough and elevate it above the wastewater level in a continuous operation. Screw conveyors, tubular conveyors, jet pump and airlifts have also been used. In areas where industrial wastewater is discharged to the collection system, the release of volatile organic compounds (VOCs) by the air agitation in aerated grit chambers needs to be considered. The release of significant amounts of VOCs can be a health risk to the treatment plant operators. Where release of VOCs is an important consideration, covers may be required or nonaerated-type grit chambers used.

3. Vortex-Type Grit Chambers

Grit is also removed in devices that use a vortex flow pattern. Two types of devices are shown on Figure 2.29. In one type, illustrated on Figure 2.29(a), wastewater enters and exits tangentially. The rotating turbine maintains constant flow velocity, and its adjustable-pitch blades promote separation of organics from the grit. The action of the propeller produces a toroidal flow path for grit particles. The grit settles by gravity into the hopper in one revolution of the basin's contents. Solids are removed from the hopper by a grit pump or an airlift pump.

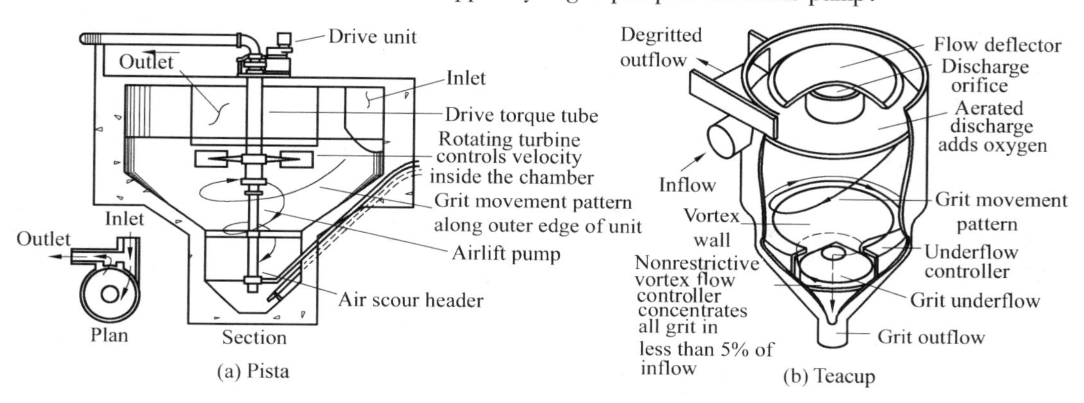

Figure 2.29 Vortex-type grit chambers

In the second type, illustrated on Figure 2.29(b), a vortex is generated by the flow entering tangentially at the top of the unit. Effluent exits the center of the top of the unit from a rotating cylinder, or "eye" of the fluid. Centrifugal and gravitational forces within this cylinder minimize the release of particles with densities greater than water. Grit settles by gravity to the bottom of the unit, while organics, including those separated from grit particles by centrifugal forces, exit principally with the effluent. Organics remaining with the settled grit are separated as the grit particles move along the unit floor.

4. Solids (Sludge) Degritting

Where grit chambers are not used and the grit is allowed to settle in the primary settling tanks, grit removal is accomplished by pumping dilute quantities of primary sludge to a cyclone degritter. The cyclone degritter acts as a centrifugal separator in which the heavy particles of grit and solids are separated by the action of a vortex and discharged separately from the lighter particles and the bulk of the liquid. The principal advantage of cyclone degritting is the elimination of the cost of building, operating, and maintaining grit chambers. The disadvantages are (1) pumping of dilute quantities of solids usually requires solids thickeners, and (2) pumping of grit with the liquid primary solids increases the cost of operating and maintaining solids collectors and the primary sludge pumps.

5. Grit Processing and Disposal

Grit consists of sand, gravel, cinders, or other heavy materials that have specific gravities or settling velocities considerably greater than those of organic particles. In addition to these materials, grit includes eggshells, bone chips, seeds, coffee grounds, and large organic particles. Grit separators and grit washers may accomplish removal of a major part of the organic material contained in grit. When some of the heavier organic matter remains with the grit, grit washers are commonly used to provide a second stage of volatile solids separation. Examples of grit separation and washing units are shown on Figure 2.30. Two principal types of grit washers are available. One type relies on an inclined submerged rake that provides the necessary agitation for separation of the grit from the organic materials and, at the same time, raises the washed grit to a point of discharge above water level. Another type of grit washer uses an inclined screw and moves the grit up the ramp. Both types can be equipped with water sprays to assist in the cleansing action.

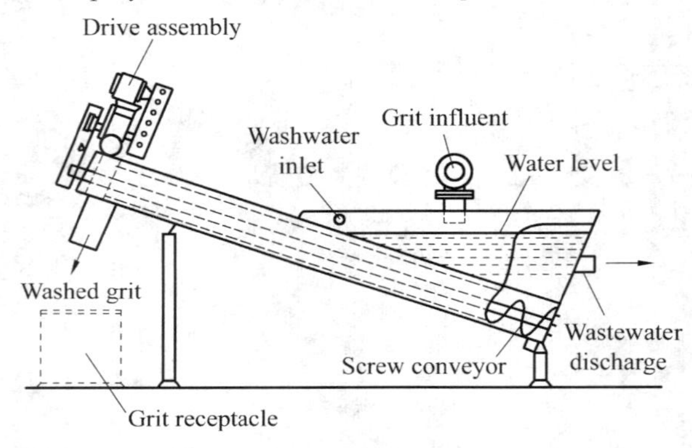

Figure 2.30 Schematic of grit separation and washing unit

The most common method of grit disposal is transport to a landfill. In some large plants, grit is incinerated with solids. As with screenings, some states require grit to be lime stabilized before disposal in a landfill. Disposal in all cases should be done in conformance with the appropriate environmental regulations.

2.4.5 Primary Sedimentation

The objective of treatment by sedimentation is to remove readily settleable solids and floating material and thus reduce the suspended solids content. Primary sedimentation is used as a preliminary step in the further processing of the wastewater. Efficiently designed and operated primary sedimentation tanks should remove from 50 to 70 percent of the suspended solids and from 25 to 40 percent of the BOD. Almost all treatment plants use mechanically cleaned sedimentation tanks of standardized circular or rectangular design. The selection of the type of sedimentation unit for a given application is governed by the size of the installation, by rules and regulations of local control authorities, by local site conditions, and by the experience and judgment of the engineer. Two or more tanks should be provided so that the process may remain in operation while one tank is out of service for maintenance and repair work. At large plants, the number of tanks is determined largely by size limitations.

1. Rectangular Tanks

Rectangular sedimentation tanks may use either chain-and-flight solids collectors or traveling-bridge-type collectors. A rectangular tank that uses a chain-and-flight-type collector is shown on Figure 2.31. Equipment for settled solids removal generally consists of a pair of endless conveyor chains, manufactured of alloy steel, cast iron, or thermoplastic. Attached to the chains at approximately 3 m intervals are scraper flights made of wood or fiberglass, extending the full width of the tank or bay. The solids settling in the tank are scraped to solids hoppers in small tanks and to transverse troughs in large tanks. Rectangular tanks may also be cleaned by a bridge-type mechanism that travels up and down the tank on rubber wheels or on rails supported on the sidewalls. One or more scraper blades are suspended from the bridge.

Because flow distribution in rectangular tanks is critical, one of the following inlet designs is used: (1) full-width inlet channels with inlet weirs, (2) inlet channels with submerged ports or orifices, (3) or inlet channels with wide gates and slotted baffles. Inlet weirs, while effective in spreading flow across the tank width, introduce a vertical velocity component into the solids hopper that may resuspend the solids particles.

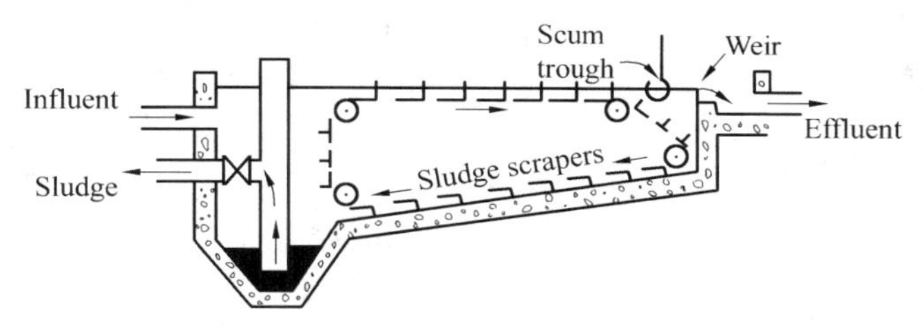

Figure 2.31 Schematic of Rectangular sedimentation tank

Scum is usually collected at the effluent end of rectangular tanks with the flights returning at the liquid surface. The scum can be scraped manually up an inclined apron, or it can be removed hydraulically or mechanically, and for scum removal a number of means have been developed. For small installations, the most common scum draw off facility consists of a horizontal slotted pipe that can be rotated by a lever or a screw. Except when drawing scum, the open slot is above the normal tank water level. Another method for removing scum is by a transverse rotating helical wiper attached to a shaft. Scum is removed from the water surface and moved over a short inclined apron for discharge to a cross-collecting scum trough.

2. Circular Tanks

In circular tanks the flow pattern is radial (as opposed to horizontal in rectangular tanks). To achieve a radial flow pattern, the wastewater to be settled can be introduced in the center or around the periphery of the tank, as shown on Figure 2.32. Both flow configurations have proved to be satisfactory generally, although the center-feed type is more commonly used, especially for primary treatment. Solids are usually withdrawn by sludge pumps for discharge to the solids processing and disposal units.

In the center-feed design (see Figure 2.32(a)), the wastewater is transported to the center of the tank in a pipe suspended from the bridge, or encased in concrete beneath the tank floor. At the center of the tank, the wastewater enters a circular well designed to distribute the flow equally in all directions. The center well has a diameter typically between 15 and 20 percent of ten total tank diameter and ranges from 1 to 2.5 m in depth and should have a tangential energy-dissipating inlet within the feedwell.

In the peripheral-feed design (see Figure 2.32(b)), a suspended circular baffle forms an annular space into which the inlet wastewater is discharged in a tangential direction. The wastewater flows spirally around the tank and underneath the baffle, and the clarified liquid is skimmed off over weirs on both sides of a centrally located weir trough. Grease and scum are confined to the surface of the annular space. Peripheral feed tanks are used generally for secondary clarification.

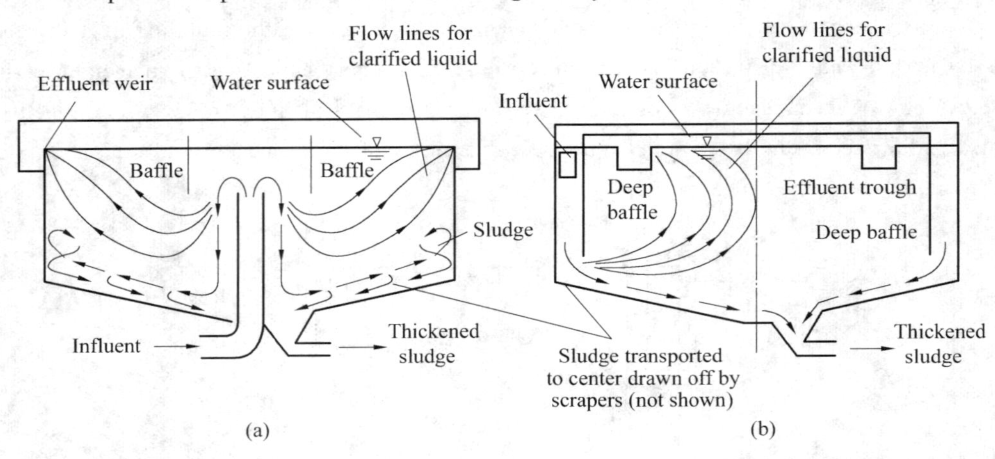

Figure 2.32 Typical circular primary sedimentation tanks: (a) center feed; (b) peripheral feed

The efficiency of sedimentation basins with respect to the removal of BOD and TSS is reduced by (1) eddy currents formed by the inertia of the incoming fluid, (2) wind-induced circulation cells formed in uncovered tanks, (3) thermal convection currents, (4) cold or warm water causing the formation of density currents that move along the bottom of the basin and warm water rising and flowing across the top of the tank, and (5) thermal stratification in hot arid climates.

2.5 Chemical Unit Processes of Wastewater Treatment

Chemical processes, in conjunction with various physical operations, have been developed for the complete secondary treatment of untreated (raw) wastewater, including the removal of either nitrogen or phosphorus or both. Chemical processes have also been developed to remove phosphorus by chemical precipitation, and are designed to be used in conjunction with biological treatment. Other chemical processes have been developed for the removal of heavy metals and for specific organic compounds and for the advanced treatment of wastewater. Currently the most important applications of chemical unit processes in wastewater treatment are for (1) the disinfection of wastewater, (2) the precipitation of phosphorus, and (3) the coagulation of particulate matter found in wastewater at various stages in the treatment process.

One of the inherent disadvantages associated with most chemical unit processes, as compared with the physical unit operations, is that they are additive processes (i.e., something is added to the wastewater to achieve the removal of something else). As a result, there is usually a net increase in the dissolved constituents in the wastewater. If the treated wastewater is to be reused, the increase in dissolved constituents can be a significant factor. This additive aspect is in contrast to the physical unit operations and the biological unit processes, which may be described as being subtractive, in that wastewater constituents are removed from the wastewater.

2.5.1 Chemical Precipitation For Improved Plant Performance

The term "chemical coagulation" as used in this text includes all of the reactions and mechanisms involved in the chemical destabilization of particles and in the formation of larger particles through perikinetic flocculation (aggregation of particles in the size range from 0.01 to 1μm). In general, a coagulant is the chemical that is added to destabilize the colloidal particles in wastewater so that floc formation can result. A flocculent is a chemical, typically organic, added to enhance the flocculation process. Typical coagulants and flocculants include natural and synthetic organic polymers, metal salts such as alum or ferric sulfate, and prehydrolized metal salts such as polyaluminum chloride (PACI) and polyiron chloride (PICI).

Chemical precipitation, as noted previously, involves the addition of chemicals to alter the physical state of dissolved and suspended solids and facilitate their removal by sedimentation. In current practice, chemical precipitation is used (1) as a means of improving the performance of primary settling facilities, (2) as a basic step in the independent physical-chemical treatment of wastewater, (3) for the removal of phosphorus, and (4) for the removal of heavy metals.

Over the years a number of different substances have been used as precipitants. The degree of clarification obtained depends on the quantity of chemicals used and the care with which the process is controlled. It is possible by chemical precipitation to obtain a clear effluent, substantially free from matter in suspension or in the colloidal state. The chemicals added to wastewater interact with substances that are either normally present in the wastewater or added for this purpose. The most common chemicals are listed in Table 2.8.

Table 2.8　Inorganic chemicals used most commonly for coagulation and precipitation processes in wastewater treatment

Chemical	Formula	Molecular weight	Equivalent weight	Availability	
				Form	Percent
Alum	$Al_2(SO_4)_3 \cdot 18H_2O$	666.5		Liquid Lump	8.5 (Al_2O_3) 17 (Al_2O_3)
	$Al_2(SO_4)_3 \cdot 14H_2O$	594.4	114	0Liquid Lump	8.5 (Al_2O_3) 17 (Al_2O_3)
Aluminum chloride	$AlCl_3$	133.3	44	Liquid	
Calcium hydroxide (lime)	$Ca(OH)_2$	56.1 as CaO	40	Lump Powder Slurry	63 ~ 73 as CaO 85 ~ 99 15 ~ 20
Ferric chloride	$FeCl_3$	162.2	91	Liquid Lump	20 (Fe) 20 (Fe)
Ferric sulfate	$Fe_2(SO_4)_3$	400	51.5	Granular	18.5 (Fe)
Ferrous sulfate (copperas)	$FeSO_4 \cdot 7H_2O$	278.1	139	Granular	20 (Fe)
Sodium aluminate	$Na_2Al_2O_4$	163.9	100	Flake	46 (Al_2O_3)

The degree of clarification obtained when chemicals are added to untreated wastewater depends on the quantity of chemicals used, mixing times, and the care with which the process is monitored and controlled. With chemical precipitation, it is possible to remove 80 to 90 percent of the total suspended solids (TSS) including some colloidal particles, 50 to 80 percent of the BOD, and 80 to 90 percent of the bacteria. Comparable removal values for well-designed and well-operated primary sedimentation tanks without the addition of chemicals are 50 to 70 percent of the TSS, 25 to 40 percent of the BOD, and 25 to 75 percent of the bacteria. Because of the variable characteristics of wastewater, the required chemical dosages should be determined from bench- or pilot-scale tests.

In some localities, industrial wastes have rendered municipal wastewater difficult to treat by biological means. In such situations, physical-chemical treatment may be an alternative approach. This method of treatment has met with limited success because of its lack of consistency in meeting discharge requirements, high costs for chemicals, handling and disposal of the great volumes of sludge resulting from the addition of chemicals, and numerous operating problems. Because of these reasons, new applications of physical-chemical treatment for municipal wastewater are rare. Physical-chemical treatment is used more extensively for the treatment of industrial wastewater. Depending on the treatment objectives, the required chemical dosages and application rates should be determined from bench- or pilot-scale tests. A flow diagram for the physical-chemical treatment of untreated wastewater is presented on Figure 2.33. As shown, after first-stage precipitation and pH adjustment by recarbonation (if required), the wastewater is passed through a granular-medium filter to remove any residual floc and then through carbon columns to remove dissolved organic compounds. The filter is shown as optional, but its use is recommended to reduce the blinding and headloss buildup in the carbon columns. The treated effluent from the carbon column is usually chlorinated before discharge to the receiving waters.

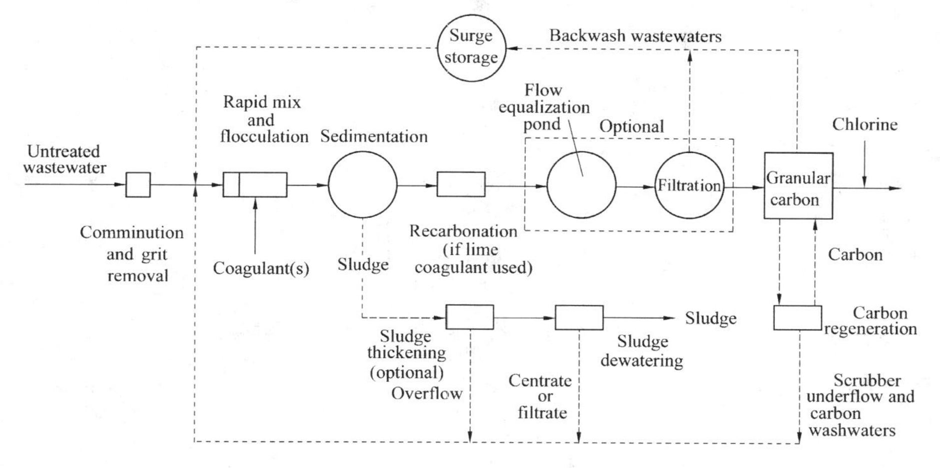

Figure 2.33 Typical flow diagram of an independent physical-chemical treatment plant

2.5.2 Chemical Precipitation For Phosphorus Removal

The removal of phosphorus from wastewater involves the incorporation of phosphate into TSS and the subsequent removal of those solids. Phosphorus can be incorporated into either biological solids (e.g., microorganisms) or chemical precipitates. The precipitation of phosphorus from wastewater can occur in a number of different locations within a process flow diagram (see Figure 2.34). The general locations where phosphorus can be removed may be classified as:

(1) Pre-precipitation. The addition of chemicals to raw wastewater for the precipitation of phosphorus in primary sedimentation facilities is termed "pre-precipitation." The precipitated phosphate is removed with the primary sludge.

(2) Coprecipitation. The addition of chemicals to form precipitates that are removed along with waste biological sludge is defined as "coprecipitation." Chemicals can be added to ① the effluent from primary sedimentation facilities, ② the mixed liquor (in the activated-sludge process), or ③ the effluent from a biological treatment process before secondary sedimentation.

(3) Postprecipitation. Postprecipitation involves the addition of chemicals to the effluent from secondary sedimentation facilities and the subsequent removal of chemical precipitates. In this process, the chemical precipitates are usually removed in separate sedimentation facilities or in effluent filters (see Figure 2.34).

Factors affecting the choice of chemical to use for phosphorus removal are reported in Table 2.9.

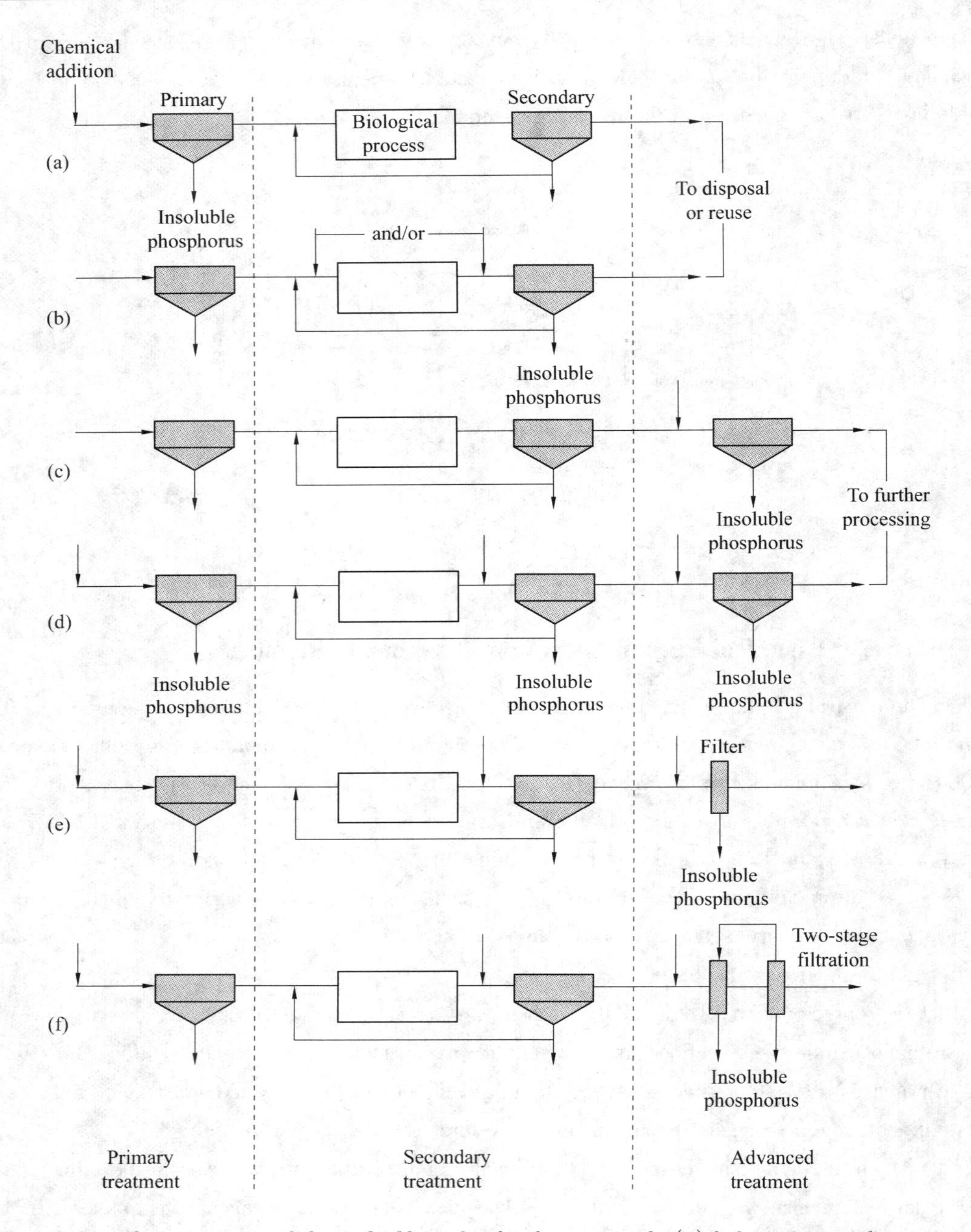

Figure 2.34 Alternative points of chemical addition for phosphorus removal: (a) before primary sedimentation, (b) before and/or following biological treatment, (c) following secondary treatment, and (d) ~ (f) at several locations in a process (known as "split treatment").

Table 2.9 Factors affecting the choice of chemical for phosphorus removal

Factors
1. Influent phosphorus level
2. Wastewater suspended solids
3. Alkalinity
4. Chemical cost (including transportation)
5. Reliability of chemical supply
6. Sludge handling facilities
7. Ultimate disposal methods
8. Compatibility with other treatment processes

1. Phosphorus Removal Using Metal Salts and Polymers

When aluminum or iron salts are added to untreated wastewater, they react with the soluble orthophosphate to produce a precipitate. Organic phosphorus and polyphosphate are removed by more complex reactions and by adsorption onto floc particles. The insolubilized phosphorus, as well as considerable quantities of BOD and TSS, are removed from the system as primary sludge. Adequate initial mixing and flocculation are necessary upstream of primary facilities, whether separate basins are provided or existing facilities are modified to provide these functions. Polymer addition may be required to aid in settling.

In trickling filter systems, the salts are added to the untreated wastewater or to the filter effluent. Multipoint additions have also been used. Phosphorus is removed from the liquid phase through a combination of precipitation, adsorption, exchange, and agglomeration, and removed from the process with either the primary or secondary sludges, or both. Theoretically, the minimum solubility of $AlPO_4$ occurs at about pH 6.3, and that of $FePO_4$ occurs at about pH 5.3; however, practical applications have yielded good phosphorus removal anywhere in the range of pH 6.5 to 7.0, which is compatible with most biological treatment processes.

In certain cases, such as trickling filtration and extended aeration activated-sludge processes, solids may not flocculate and settle well in the secondary clarifier. This settling problem may become acute in plants that are overloaded. The addition of aluminum or iron salts will cause the precipitation of metallic hydroxides or phosphates, or both. Aluminum and iron salts, along with certain organic polymers, can also be used to coagulate colloidal particles and to improve removals on filters. The resultant coagulated colloids and precipitates will settle readily in the secondary clarifier, reducing the TSS in the effluent and effecting phosphorus removal. Polymers may be added (1) to the mixing zone of a highly mixed or internally recirculated clarifier, (2) preceding a static or dynamic mixer, or (3) to an aerated channel.

2. Phosphorus Removal Using Lime

The use of lime for phosphorus removal is declining because of (1) the substantial increase in the mass of sludge to be handled compared to metal salts and (2) the operation and maintenance problems associated with the handling, storage, and feeding of lime. When lime is used, the principal variables controlling the dosage are the degree of removal required and the alkalinity of the wastewater. The operating dosage must usually be determined by onsite testing. Lime has been used customarily either as a precipitant in the primary sedimentation tanks or following secondary treatment clarification.

Both low and high lime treatment can be used to precipitate a portion of the phosphorus (usually about 65 to 80 percent). When lime is used, both the calcium and the hydroxide react with the orthophosphorus to form an insoluble hydroxyapatite [$Ca_5(OH)(PO_4)_3$]. A residual phosphorus level of 1.0 mg/L can be achieved with the addition of effluent filtration facilities to which chemicals can be added. In the high lime system, sufficient lime is added to raise the pH to about 11. After precipitation, the effluent must be recarbonated before biological treatment. In activated-sludge systems, the pH of the primary effluent should not exceed 9.5 or 10; higher pH values can result in biological process upsets. In the trickling filter process, the carbon dioxide generated during treatment is usually sufficient to lower the pH without recarbonation.

Lime can be added to the waste stream after biological treatment to reduce the level of phosphorus and TSS. Single-stage process and two-stage process flow diagrams for lime addition are shown on Figure 2.35. On Figure 2.35(a), a single-stage lime precipitation process is used for the treatment of secondary effluent. In the first-stage clarifier of the two-stage process shown on Figure 2.35(b), sufficient lime is added to raise the pH above 11 to precipitate the soluble phosphorus as basic calcium phosphate (apatite). The calcium carbonate precipitate formed in the process acts as a coagulant for TSS removal.

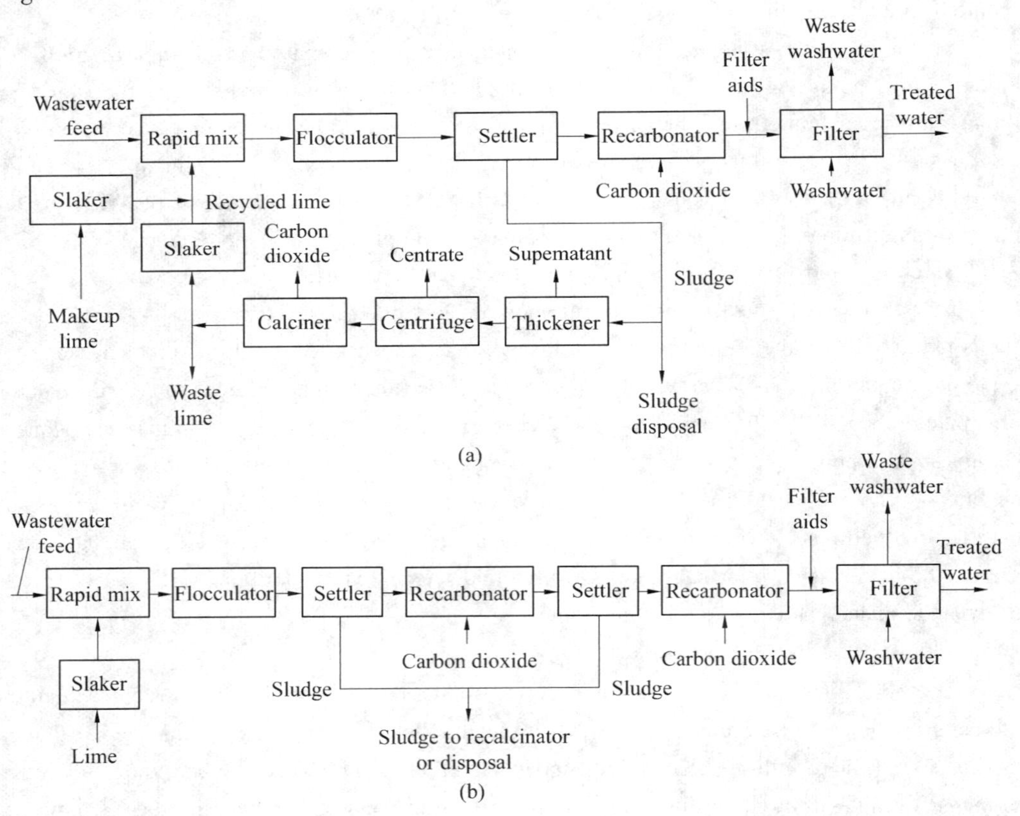

Figure 2.35 Typical lime treatment process flow diagram for phosphorus removal: (a) single-stage system; (b) two-stage system

2.5.3 Chemical Precipitation for Removal of Heavy Metals and Dissolved Inorganic Substances

The technologies available for the removal of heavy metals from wastewater include chemical precipitation, carbon adsorption, ion exchange, and reverse osmosis. Of these technologies, chemical precipitation is most commonly employed for most of the metals. Common precipitants include hydroxide (OH^-) and sulfide (S^{2-}). Carbonate (CO_3^{2-}) has also been used in some special cases. Metal may be removed separately or coprecipitated with phosphorus.

Metals of interest include arsenic (As), barium (Ba), cadmium (Cd), copper (Cu), mercury (Hg), nickel (Ni), selenium (Se), and zinc (Zn). Most of these metals can be precipitated as hydroxides or sulfides. In wastewater treatment facilities, metals are precipitated most commonly as metal hydroxides through the addition of lime or caustic to a pH of minimum solubility. However, several of these compounds are, as discussed previously, amphoteric (i.e., capable of either accepting or donating a proton) and exhibit a point of minimum solubility. The minimum effluent concentration levels that can be achieved in the chemical precipitation of heavy metals are reported in Table 2.10. In practice, the minimum achievable residual metal concentrations will also depend on the nature and concentration of the organic matter in the wastewater as well as the temperature.

Table 2.10 Practical effluent concentration levels achievable in heavy metals removal by precipitation

Metal	Achievable effluent concentration, mg/L	Type of precipitation and technology
Arsenic	0.05	Sulfide precipitation with filtration
	0.005	Ferric hydroxide coprecipitation
Barium	0.5	Sulfate precipitation
Cadmium	0.05	Hydroxide precipitation at pH 10-11
	0.05	Coprecipitation with ferric hydroxide
	0.008	Sulfide precipitation
Copper	0.02 ~ 0.07	Hydroxide precipitation
	0.01 ~ 0.02	Sulfide precipitation
Mercury	0.01 ~ 0.02	Sulfide precipitation
	0.001 ~ 0.01	Alum coprecipitation
	0.0005 ~ 0.005	Ferric hydroxide coprecipitation
	0.001 ~ 0.005	Ion exchange
Nickel	0.12	Hydroxide precipitation at pH 10
Selenium	0.05	Sulfide precipitation
Zinc	0.1	Hydroxide precipitation at pH 11

As discussed previously, precipitation of phosphorus in wastewater is usually accomplished by the addition of coagulants, such as alum, lime or iron salts, and polyelectrolytes. Coincidentally with the addition of these chemicals for the removal of phosphorus is the removal that occurs of various inorganic ions, principally some of the heavy metals. Where both industrial and domestic wastes are treated together, it may be necessary to add chemicals to the primary settling facilities, especially if onsite pretreatment measures prove to be ineffective. When chemical precipitation is used, anaerobic digestion for sludge stabilization may not be possible because of the toxicity of the precipitated heavy metals. As noted previously, one of the disadvantages of chemical precipitation is that it usually results in a net increase in the total dissolved solids of the wastewater that is being treated.

2.5.4 Chemical Oxidation

Chemical oxidation in wastewater treatment typically involves the use of oxidizing agents such as ozone (O_3), hydrogen peroxide (H_2O_2), permanganate (MnO_4), chloride dioxide (ClO_2), chlorine (Cl_2) or ($HOCl$), and oxygen (O_2), to bring about change in the chemical composition of a compound or a group of compounds.

Some of the more important applications of chemical oxidation in wastewater management are summarized in Table 2.11. In the past, chemical oxidation was used most commonly to (1) reduce the concentration of residual organics, (2) control odors, (3) remove ammonia, and (4) reduce the bacterial and viral content of wastewaters. Chemical oxidation is especially effective for the elimination of odorous compounds (e.g., oxidation of sulfides and mercaptans).

In addition to the applications reported in Table 2.11, chemical oxidation is now commonly used to (1) improve the treatability of nonbiodegradable (refractory) organic compounds, (2) eliminate the inhibitory effects of certain organic and inorganic compounds to microbial growth, and (3) reduce or eliminate the toxicity of certain organic and inorganic compounds to microbial growth and aquatic flora.

Table 2.11 Typical applications of chemical oxidation in wastewater collection, treatment, and disposal

Application	Chemicals used	Remarks
Collection		
Slime-growth control	Cl_2, H_2O_2	Control of fungi and slime-producing bacteria
Corrosion control (H_2S)	Cl_2, H_2O_2, O_3	Control brought about by oxidation of H_2S
Odor control	Cl_2, H_2O_2, O_3	Especially in pumping stations and long, flat sewers
Treatment		
Grease removal	Cl_2	Added before preaeration
BOD reduction	Cl_2, O_3	Oxidation of organic substances
Ferrous sulfate oxidation	Cl_2[b]	Production of ferric sulfate and ferric chloride
Filter-ponding control	Cl_2	Maintaining residual at filter nozzles
Filter-fly control	Cl_2	Maintaining residual at filter nozzles during fly season
Sludge-bulking control	Cl_2, H_2O_2, O_3	Temporary control measure

Continued Table 2.11

Control of filamentous microorganisms	Cl_2	Dilute chlorine solution sprayed on foam caused by filamentous organisms
Digester supernatant oxidation	Cl_2	
Digester foaming control	Cl_2	
Ammonia oxidation	Cl_2	Conversion of ammonia to nitrogen gas
Odor control	Cl_2, H_2O_2, O_3	
Oxidation of refractory organic compounds	O_3	
Dispersal		
Bacterial reduction	Cl_2, H_2O_2, O_3	Plant effluent, overflows, and stormwater
Odor control	Cl_2, H_2O_2, O_3	

Typical chemical dosages for both chlorine and ozone for the oxidation of the organics in wastewater increase with the degree of treatment. Because of the complexities associated with composition of wastewater, chemical dosages for the removal of refractory organic compounds cannot be derived from the chemical stoichiometry, assuming that it is known. Pilot-plant studies must be conducted when either chlorine, chlorine dioxide, or ozone is to be used for the oxidation of refractory organics to assess both the efficacy and required dosages.

The chemical process in which chlorine is used to oxidize the ammonia nitrogen in solution to nitrogen gas and other stable compounds is known as breakpoint chlorination. Perhaps the most important advantage of this process is that, with proper control, all the ammonia nitrogen in the wastewater can be oxidized. However, because the process has a number of disadvantages including the buildup of acid (HCl) which will react with the alkalinity, the buildup of total dissolved solids, and the formation of unwanted chloro-organic compounds, ammonia oxidation is seldom used today.

2.5.5 Chemical Neutralization, Scale Control, and Stabilization

The removal of excess acidity or alkalinity by treatment with a chemical of the opposite composition is termed neutralization. In general, all treated wastewaters with excessively low or high pH will require neutralization before they can be dispersed to the environment. Scaling control is required for nanofiltration and reverse osmosis treatment to control the formation of scale, which can severely impact performance. Chemical stabilization is often required for highly treated wastewaters to control their aggressiveness with respect to corrosion.

In a variety of wastewater-treatment operations and processes, there is often a need for pH adjustment. Because a number of chemicals are available that can be used, the choice will depend on the suitability of a given chemical for a particular application and prevailing economics. Sodium hydroxide (NaOH, also known as caustic soda) and sodium carbonate, although somewhat expensive, are convenient and are used widely by small plants or for treatment where small quantities are adequate. Lime, which is cheaper but somewhat less convenient, is the most widely used chemical. Lime can be purchased as quicklime or slaked hydrated lime, high-calcium or dolomitic lime, and in several physical forms. Limestone and dolomitic limestone are cheaper but less convenient to use and slower in reaction rate. Because they can become coated in certain waste-treatment applications, their use is limited. Calcium and magnesium chemicals often form sludges that require disposal. Alkaline wastes are less of a problem than acid wastes but nevertheless often require treatment. If acidic waste streams are not available or are not adequate to neutralize alkaline wastes, sulfuric acid is commonly employed.

With the increasing use that is being made of nanofiltration, reverse osmosis, and electrodialysis in wastewater reuse applications, adjustment of the scaling characteristics of the effluent to be treated is important to avoid calcium carbonate and sulfate scale formation. Depending on the recovery rate, the concentration of salts can increase by a factor of up to 10 within the treatment module. When such a salt concentration increase occurs, it is often possible to exceed the solubility product of calcium carbonate and other scale-forming compounds. The formation of scale within the treatment module will cause a deterioration in the performance, ultimately leading to the failure of the membrane module. Usually, $CaCO_3$ scale control can be achieved using one or more of the following methods: (1) acidifying to reduce pH and alkalinity, (2) reducing calcium concentration by ion exchange or lime softening, (3) adding a scale inhibitor chemical (antiscalant) to increase the apparent solubility of $CaCO_3$ in the concentrate stream, and (4) lowering the product recovery rate. Because it is not possible to predict a priori the value of pH in water treated with reverse osmosis, it is usually necessary to conduct pilot-scale studies using the same modules that will be used in the full-scale installation.

Wastewater effluent that is demineralized with reverse osmosis will generally require pH and calcium carbonate adjustment (stabilization) to prevent metallic corrosion, due to the contact of the demineralized water with metallic pipes and equipment. Corrosion occurs because material from the solid is removed (solubilized) to satisfy the various solubility products. Demineralized water typically is stabilized by adding lime, using the procedure outlined above.

2.6 Biological Treatment Processes of Wastewater Treatment

The overall objectives of the biological treatment of domestic wastewater are to (1) transform (i.e., oxidize) dissolved and particulate biodegradable constituents into acceptable end products, (2) capture and incorporate suspended and nonsettleable colloidal solids into a biological floc or biofilm, (3) transform or remove nutrients, such as nitrogen and phosphorus, and (4) in some cases, remove specific trace organic constituents and compounds. For industrial wastewater, the objective is to remove or reduce the concentration of organic and inorganic compounds. Because some of the constituents and compounds found in industrial wastewater are toxic to microorganisms, pretreatment may be required before the industrial wastewater can be discharged to a municipal collection system. For agricultural irrigation return wastewater, the objective is to remove nutrients, specifically nitrogen and phosphorus, that are capable of stimulating the growth of aquatic plants.

Schematic flow diagrams of various treatment processes for domestic wastewater incorporating biological processes are shown on Figure 2.36. Common terms used in the field of biological wastewater treatment and their definitions are presented in Table 2.12. The principal biological processes used for wastewater treatment can be divided into two main categories: suspended growth and attached growth (or biofilm) processes.

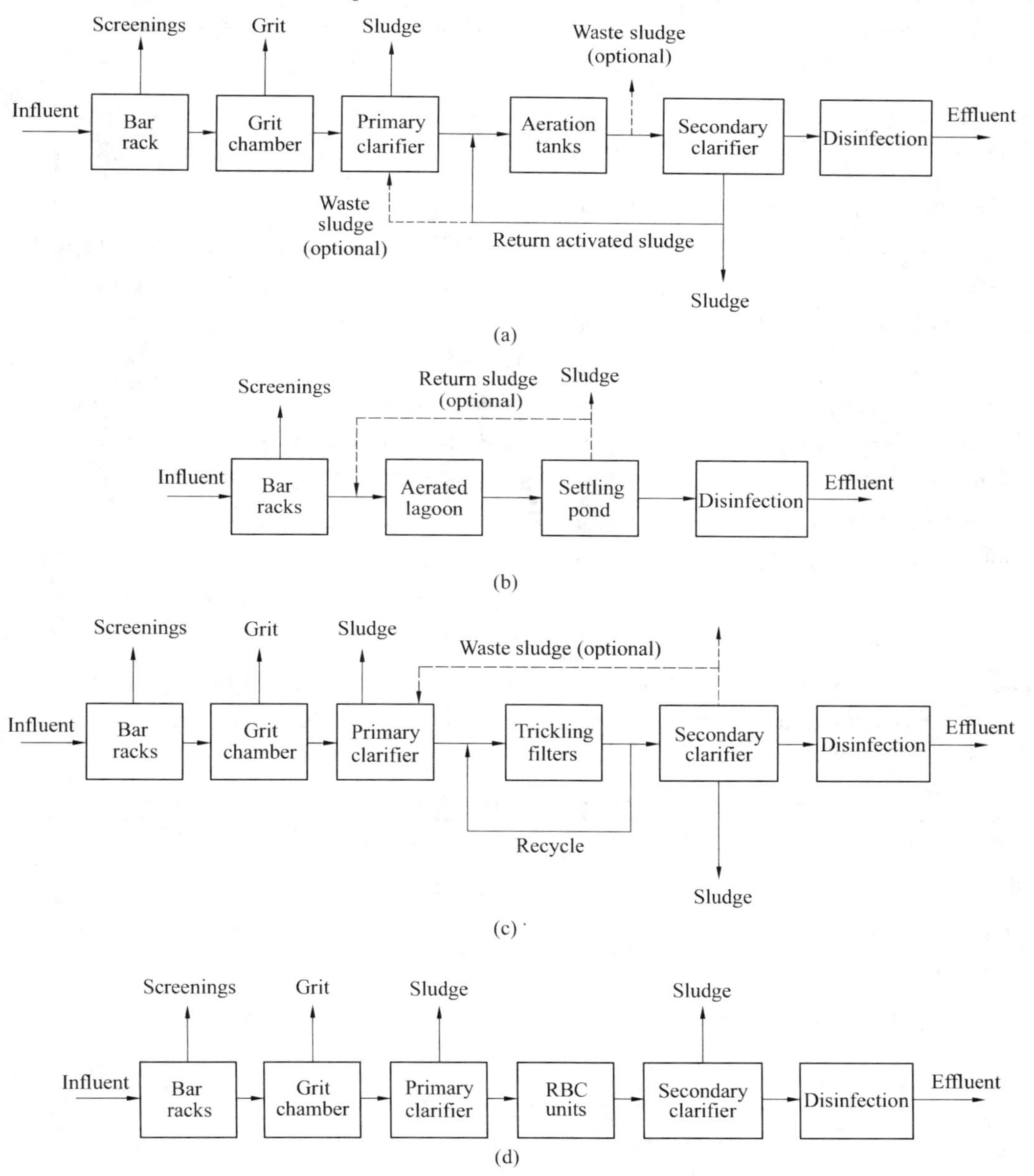

Figure 2.36 Typical (simplified) flow diagrams for biological processes used for wastewater treatment: (a) activated-sludge process, (b) aerated lagoons, (c) trickling filters, and (d) rotating biological contactors.

2.6.1 Role of Microorganisms in Wastewater Treatment

The removal of dissolved and particulate carbonaceous BOD and the stabilization of organic matter found in wastewater is accomplished biologically using a variety of microorganisms, principally bacteria. Microorganisms are used to oxidize (i. e., convert) the dissolved and particulate carbonaceous organic matter into simple end products and additional biomass as represented by the following equation for the aerobic biological oxidation of organic matter.

$$v_1(\text{organic material}) + v_2O_2 + v_3NH_3 + v_4PO_4^{3-} \xrightarrow{\text{microorganisms}} v_5(\text{new cells}) + v_6CO_2 + v_7H_2O \quad (2.9)$$

where v_i = the stoichiometric coefficient.

In Equation (2.9), oxygen (O_2), ammonia (NH_3), and phosphate (PO_4^{3-}) are used to represent the nutrients needed for the conversion of the organic matter to simple end products [i.e., carbon dioxide (CO_2) and water]. The term shown over the directional arrow is used to denote the fact that microorganisms are needed to carry out the oxidation process. The term new cells is used to represent the biomass produced as a result of the oxidation of the organic matter. Microorganisms are also used to remove nitrogen and phosphorus in wastewater treatment processes. Specific bacteria are capable of oxidizing ammonia (nitrification) to nitrite and nitrate, while other bacteria can reduce the oxidized nitrogen to gaseous nitrogen. For phosphorus removal, biological processes are configured to encourage the growth of bacteria with the ability to take up and store large amounts of inorganic phosphorus. Because the biomass has a specific gravity slightly greater than that of water, the biomass can be removed from the treated liquid by gravity settling. It is important to note that unless the biomass produced from the organic matter is removed on a periodic basis, complete treatment has not been accomplished because the biomass, which itself is organic, will be measured as BOD in the effluent.

Table 2.12 Definitions of common terminology used for biological wastewater treatment

Term	Definition
Metabolic function	
Aerobic (oxic) processes	Biological treatment processes that occur in the presence of oxygen
Anaerobic processes	Biological treatment processes that occur in the absence of oxygen
Anoxic processes	The process by which nitrate nitrogen is converted biologically to nitrogen gas in the absence of oxygen. This process is also known as denitrification
Facultative processes	Biological treatment processes in which the organisms can function in the presence or absence of molecular oxygen
Combined aerobic/anoxic/anaerobic processes	Various combinations of aerobic, anoxic, and anaerobic processes grouped together to achieve a specific treatment objective

Continued Table 2.12

Treatment processes	
Suspended-growth processes	Biological treatment processes in which the microorganisms responsible for the conversion of organic matter or other constituents in the wastewater to gases and cell tissue are maintained in suspension within the liquid
Attached-growth processes	Biological treatment processes in which the microorganisms responsible for the conversion of organic matter or other constituents in the wastewater to gases and cell tissue are attached to some inert medium, such as rocks, slag, or specially designated ceramic or plastic materials. Attached-growth treatment processes are also known as fixed-film processes
Combined processes	Term used to describe combined processes (e. g., combined suspended and attached growth processes)
Lagoon processes	A generic term applied to treatment processes that take place in ponds or lagoons with various aspect ratios and depths
Treatment functions	
Biological nutrient removal	The term applied to the biological removal of nitrogen and phosphorous in biological treatment processes
Biological phosphorous removal	The term applied to the biological removal of phosphorous by accumulation in biomass and subsequent solids separation
Carbonaceous BOD removal	Biological conversion of the carbonaceous organic matter in wastewater to cell tissue and various gaseous end products. In the conversion, it is assumed that the nitrogen present in the various compounds is converted to ammonia
Nitrification	The two-step biological process by which ammonia is converted first to nitrite and then to nitrate
Denitrification	The biological processes by which nitrate is reduced to nitrogen and other gaseous end products
Stabilization	The biological processes by which the organic matter in the sludges produced from the primary settling and biological treatment of wastewater is stabilized, usually by conversation to gases and cell tissue. Depending on whether this stabilization is carried out under aerobic or anaerobic conditions, the process is known as aerobic or anaerobic digestion
Substrate	The term used to denote the organic matter or nutrients that are converted during biological treatment or that may be limiting in biological treatment. For example, the carbonaceous organic matter in wastewater is referred to as the substrate that is during biological treatment

2.6.2 Suspended Growth Processes

Suspended growth systems comprise aggregates of microorganisms generally growing as flocs in intimate contact with the wastewater they are treating. The aggregates or flocs are responsible for the removal of polluting material and comprise a wide range of microbial species. The most prevalent and important of these microorganisms are the bacteria, the protozoa and the metazoa. Fungi and viruses are also found, but probably contribute little to the treatment of the wastewater. Suspended growth treatment systems permit the exploitation of the full range of microbial metabolic capabilities. The full spectrum of redox environments from aerobic, through anoxic to anaerobic can be found within the floc itself, but they can also be created by appropriate process reactor design. These environments allow the growth of both the organoheterotrophs (which oxidize organic carbon and remove BOD) and the lithoautotrophs (which are responsible for ammonia oxidation). Indeed, in waste stabilization pond systems, phototrophic organisms, which utilize a range of electron acceptors, can be exploited to achieve good treatment with negligible energy input.

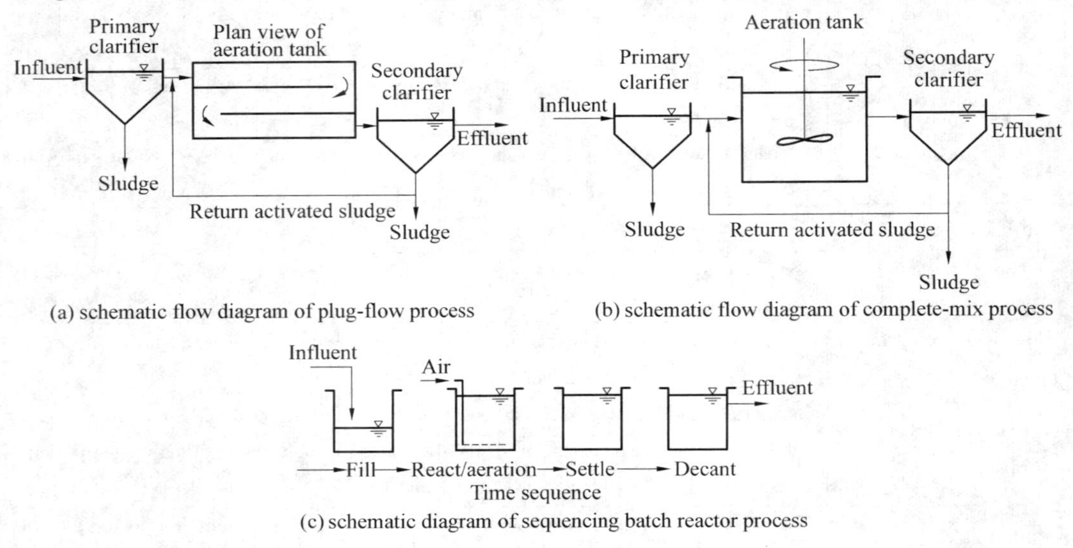

Figure 2.37 Typical activated-sludge processes with different types of reactors

Treatment of wastewaters in suspended growth environments offers many process advantages, which has led to the proliferation of this type of treatment system. Waste stabilization ponds, aerated lagoons and activated sludge (see Figure 2.37) are all examples of treatment options that rely on the actions of microorganisms growing in a suspension of the wastewater under treatment. In the aeration tank of activated sludge process, contact time is provided for mixing and aerating influent wastewater with the microbial suspension, generally referred to as the mixed liquor suspended solids (MLSS) or mixed liquor volatile suspended solids (MLVSS). Mechanical equipment is used to provide the mixing and transfer of oxygen into the process. The mixed liquor then flows to a clarifier where the microbial suspension is settled and thickened. The settled biomass, described as activated sludge because of the presence of active microorganisms, is returned to the aeration tank to continue biodegradation of the influent organic material. A portion of the thickened solids is removed daily or periodically as the process produces excess biomass that would accumulate along with the nonbiodegradable solids contained in the influent wastewater. If the accumulated solids are not removed, they will eventually find their way to the system effluent.

A suspended growth wastewater treatment processes is a biological reactor which has been engineered to encourage the growth of specific types of microorganisms that are able to undertake the reactions necessary to achieve purification of the influent wastewater. Their successful design requires the provision of: (1) a reactor (or series of reactors) of sufficient capacity to retain the wastewater long enough for the microorganisms to undertake the biological interconversions necessary to achieve the required effluent standard; (2) facilities to ensure that the microorganisms are retained in the reactor long enough to grow and divide, thus maintaining a stable population; (3) the correct redox environment to achieve the required biological reactions.

1. Effluent Quality Requirements

The sole role of a wastewater treatment process is to protect the receiving watercourse from environmental degradation and ensure that the impact of the effluent from the plant is minimized. This is normally achieved by means of standards, which are set by the appropriate regulatory authority, to describe a minimum quality that the effluent must attain. Ideally these standards should be evaluated based on the characteristics of the receiving watercourse, the likely benefits that will accrue from maintaining a healthy watercourse and the ability of the community to finance treatment. Standards always contain a requirement for the treatment plant to achieve a certain effluent quality in terms of the biochemical oxygen demand (BOD) and nearly always have a standard for suspended solids. Depending on the quality of the receiving watercourse they may then contain requirements for ammonia-nitrogen, total nitrogen and phosphorus. The minimum standards that apply throughout the European Community (EC) are summarized in Table 2.13, however, member states may impose stricter standards than this where it is thought appropriate.

Table 2.13 The minimum treatment standards for sewage effluents that must be achieved by EC member states

Parameter	Concentration /($mg \cdot L^{-1}$)	Minimum % of reduction[a]
(a) Requirements for discharges from urban waste water treatment plants		
BOD_5 at 20℃ without nitrification	25	70 ~ 90
COD	125	75
Total suspended solids	35[b] (> 10 000 pe)	90
	60 (2000 ~ 10 000 pe)	70
(b) Discharges to sensitive areas which are subject to eutrophication		
Total phosphorus	2 mg P/L (10 000 ~ 100 000 PE)	80
	1 mg P/L (> 100 000 PE)	
Total nitrogen[c]	15 mg N/L (10 000 100 000 PE)	70 ~ 80
	10 mg N/L (> 100 000 PE)	

[a]Reduction in relation to the load of the influent.

[b]Optional requirement.

[c] Organic and ammonical nitrogen and nitrite-nitrogen.

2. Microbial Reactions in Suspended Growth Systems

For each of the parameters listed in Table 2.13 (a) and (b) there is one or more species of bacteria which can be exploited to metabolize the parameter using it either as a source of electrons or as a terminal electron acceptor. These reactions are summarized briefly below.

(1) BOD Removal

The biochemical oxygen demand measures the amount of oxygen required to oxidize the organic carbon present in a wastewater according to the equation:

$$C_6H_{12}O_6 + 6O_6 \rightarrow 6CO_2 + 6H_2O \quad (2.10)$$

This oxidation reaction, referred to as respiration or catabolism, provides the energy necessary for bacterial growth and reproduction and there is a vast range of heterotrophic species which will carry it out in suspended growth processes. In addition to energy, the microorganisms need a source of carbon to build new cell material. This is also provided by the organic material measured in the BOD test according to Equation (2.11), where $C_5H_7O_2$ represents a new bacterial cell.

$$C_6H_{12}O_6 \rightarrow C_5H_7O_2 \quad (2.11)$$

All biological wastewater treatment systems are designed to remove BOD, both in its particulate and soluble form. However, as much as 40% of the total BOD of a wastewater is particulate, consequently, it is generally most cost effective to remove this fraction by sedimentation. That is the role of the primary sedimentation tank, septic tank or anaerobic pond. The fraction that remains after sedimentation is a mixture of colloidal and soluble BOD that can only be removed biologically. The soluble BOD is usually biodegraded very rapidly, generally in less than one hour. The colloidal fraction is entrapped in the sludge floc and is degraded more slowly.

(2) Ammonia Removal

Ammonia is removed by the action of two groups of bacteria, collectively termed the nitrifying bacteria, which catalyse the reactions of nitrification. Nitrification is the process of ammonia oxidation in which ammonia is oxidized ultimately to nitrate in two reactions carried out by distinct groups of obligately aerobic bacteria. The first intermediate is nitrite and this reaction is catalysed by the genus nitrosomonas:

$$NH_3 + 1.5O_2 \rightarrow NO_2^- + H^+ + H_2O \quad (2.12)$$

Nitrite is further oxidized to nitrate by nitrobacter:

$$NO_2^- + 0.5O_2 \rightarrow NO_3^- \quad (2.13)$$

Nitrosomonas and nitrobacter are both autotrophic genera which reduce carbon dioxide (in the form of bicarbonate or carbonate) as a source of cellular carbon. Assuming a gross cell composition for a typical nitrifying bacteria of $C_5H_7NO_2$, then the overall reaction for the oxidation of ammonia, coupled to the synthesis of new nitrifying bacteria, can be represented as:

$$NH_3 + 1.83O_2 + 1.98HCO_3 \rightarrow 0.021\ C_5H_7NO_2 + 1.041H_2O + 0.98NO_3 + 1.88H_2CO_3 \quad (2.14)$$

The energy expenditure required to achieve the reduction of bicarbonate is relatively high, yet the nitrifying bacteria can achieve only a low yield of energy from the oxidation of their chosen substrates, ammonia and nitrite. Consequently, these organisms demonstrate very low growth yields and require a long retention time in the aeration basin to ensure they can divide and maintain a stable population. They are also very susceptible to temperature changes and below 20℃ their reaction rate slows dramatically.

(3) Nitrogen Removal

Nitrate itself is able to act as a terminal electron acceptor. In the absence of a supply of dissolved oxygen, the utilization of oxygen for respiration cannot take place. However, certain chemo-organotrophs are capable of replacing O_2 with NO_3 as an oxidizing agent and respiration can proceed with the reduction of nitrate to nitrite, nitric oxide, nitrous oxide or nitrogen. Equation (2.15) demonstrates the stoichiometric reaction for the reduction of nitrate using methanol as a source of electrons, with the production of nitrogen gas and new cell material. This process is known as anaerobic or nitrate respiration and is carried out by a variety of bacteria such as *Alcaligenes*, *Achromobacter*, *Micrococcus* and *Pseudomonas*. Not all these genera are capable of complete oxidation to nitrogen and a variety of gaseous products can be produced.

$$NO_3 + 1.08CH_3OH + 0.24H_2CO_3 \rightarrow 0.06C_5H_7NO_2 + 0.47N_2 + 1.68H_2O + HCO_3 \quad (2.15)$$

As denitrification is an oxidation reaction, it needs an electron donor and an electron acceptor. Nitrate is the electron acceptor and organic carbon generally provides the source of electrons. This is usually from the BOD present in the sewage, or as a supplementary carbon source such as methanol or ethanol. It generally requires 8 mg COD to achieve the removal of 1 mg nitrate and it is advisable to aim for a nitrate concentration < 12 mg/L in the final effluent to prevent sludge settlement problems.

(4) Phosphate Removal

Of the nutrients that are capable of supporting luxuriant growths of algae in a receiving water, phosphorus is rate limiting, and a concentration of 10 mg/L is required before algal growth will occur. It has been argued, therefore, that control over phosphorus-containing compounds in aquatic ecosystems presents a means of controlling the deleterious effects of eutrophication. Consequently, if the small concentration of phosphorus that is present in sewage effluents can be removed, then algae will not be able to flourish, regardless of the nitrogen concentration. Phosphorus load control has been demonstrated as one of the most effective ways of dealing with cultural eutrophication. A typical phosphorus standard is 1 mg/L dissolved orthophosphate (as phosphorus). The major sources of phosphorus in domestic wastewaters are from human excreta (50% ~ 65%) and synthetic detergents (30% ~ 50%) and typical concentrations are in the range 10 ~ 30 mg/L as phosphorus.

3. Manipulating Redox Environments

It is quite straightforward in a suspended growth wastewater treatment process to vary the electron source and the terminal electron acceptor and in such a way control the redox potential within the reactor. By doing so this allows a wide range of reactions to take place (see Table 2.14). Up to three reactor types are exploited; these are defined, based on the source of electrons, as aerobic (or oxic), anoxic and anaerobic.

Table 2.14 The reactions that can be achieved by manipulating the redox environment in a suspended growth reactor

Reaction	Electron donor	Electron acceptor	Redox potential
BOD removal	Organic material	Oxygen	> +200 mV
Nitrification	Ammonia	Oxygen	> +300 mV
Denitrification	Organic material	Nitrate	−100 ~ +150 mV
Phosphorus release	Polyphosphate	phb	< −300 mV
Phosphorus uptake	Acetate	Phosphate	> −150 mV6

(1) Aerobic Reactors

An aerobic reactor exploits oxygen as the terminal electron acceptor and is able to sustain a large number of important reactions that use a number of different electron donors. Principal among these are: oxidation of organic material, which reduces the BOD; oxidation of ammonia to nitrate, which reduces the ammonia concentration; and the luxury uptake of phosphate with the synthesis of polyphosphate, which reduces effluent phosphate concentrations.

A variety of options are available to achieve aerobic conditions in wastewaters. The strength of the incoming wastewater dictates the amount of oxygen that must be introduced into the reactor and guideline values are around 1.2 mg oxygen for every mg of BOD to be removed and 4.8 mg of oxygen for every mg ammonia to be removed. Natural treatment options, such as waste stabilization ponds, have long hydraulic retention times, typically between 15 and 40 days, with large reactor volumes. By contrast as the rate of oxygen transfer is increased by mechanical means, the retention time of the wastewater can be reduced to as little as a few hours in the activated sludge process. However, as aeration intensity increases, so does the cost of the treatment process and high intensity aeration is both capital and operating cost intensive.

When wastewaters are aerated with a residual dissolved oxygen > 1 mg/L, BOD removal occurs at a rate that is proportional to the amount of microorganisms held in the reactor. The more microorganisms that can be retained in the system the more rapid the rate of BOD removal will be and, consequently, the smaller the reactor that is required. However, there are two physical limits to the mass of microorganisms that can be retained, the first of these is the ability to transfer oxygen fast enough to match the oxygen uptake of the microorganisms and the second is the ability to remove the microorganisms by flocculation and sedimentation.

(2) Anoxic Reactors

The major anoxic reaction of importance in wastewater treatment is the reduction of nitrate to nitrogen gas. This is important for two reasons. First of all, if the nitrate concentration is high in the final sedimentation tank of the activated sludge process, denitrification can occur in the solids that have settled to the base of the tank. The nitrogen gas generated will buoy the settled sludge carrying it to the surface of the sedimentation tank and over the weir into the final effluent. This problem, termed a rising sludge, can be eliminated by ensuring the nitrate concentration in the final effluent does not rise above 12 mg/L. A second reason is that some treatment plants have effluent discharge standards for total nitrogen (Table 2.13b), in which case nitrate removal is mandatory.

Anoxic conditions are achieved by eliminating all the residual dissolved oxygen and ensuring an adequate supply of electrons. The final effluent from a nitrifying process has high concentrations of nitrate but lacks a supply of electrons. Thus, if it is directed to a reactor to undergo denitrification, the electron supply needs augmenting. This is achieved either by adding external carbon (usually as methanol or a secondary industrial waste such as molasses) or directing a fraction of the settled sewage to the anoxic tanks. The most cost effective way of achieving denitrification is simply to remove the aeration from a small section at the head of the reactor and ensure that both the return sludge and the settled sewage are fed to this section. It requires a retention time of around 1 hour together with gentle mixing to ensure solids do not deposit in the basin. Figure 2.38 shows four patterns of anoxic reactors: (a) Carbonaceous removal only: an aerobic basin with sludge return direct to the basin; (b) Carbonaceous removal and nitrification: an aerobic basin with sludge return to an anoxic zone which also receives the incoming settled sewage; (c) Carbonaceous removal, nitrification, denitrification and phosphorus removal: an anaerobic zone to permit phosphate release, an anoxic zone which denitrifies the return sludge from the aeration basin and recycles it to the anaerobic zone, and an aerobic basin which achieves carbonaceous removal, nitrification and luxury phosphate uptake; (d) Carbonaceous removal and phosphorus removal. An anaerobic zone receives return sludge from the aeration basin and permits phosphate release, followed by an aeration basin which achieves carbonaceous removal and luxury phosphate uptake.

(3) Anaerobic Zones

An anaerobic reactor is necessary to promote the release of phosphate and uptake of acetate by the phosphate-accumulating bacteria. The more anaerobic the conditions then the more acetate is generated. This in turn leads to more phosphate release. Luxury phosphate uptake in the aerobic zone is always proportional to the amount of phosphate released and thus there is less phosphate in the final effluent. In order to ensure anaerobiosis within a reactor, it is essential that there is no aeration device present and that the reactor is highly loaded with BOD such that any dissolved oxygen is removed rapidly. Anaerobic reactors are thus always sited at the head of the reactor treatment train (see Figure 2.38). If the aerobic stage of the treatment plant removes only BOD with no nitrification, then both the return sludge and the settled sewage can be fed to the anaerobic reactor. However, if the aerobic stage achieves nitrification, the nitrate in the return sludge will reduce the extent of anaerobiosis. Under such conditions the return sludge is fed to a separate anoxic reactor to undergo denitrification, and the nitrate free mixed liquor is recycled to the anaerobic reactor (see Figure 2.38).

2.6.3 Attached Growth Processes

In attached growth processes, the microorganisms responsible for the conversion of organic material or nutrients are attached to an inert packing material. The organic material and nutrients are removed from the wastewater flowing past the attached grow also known as a biofilm. Packing

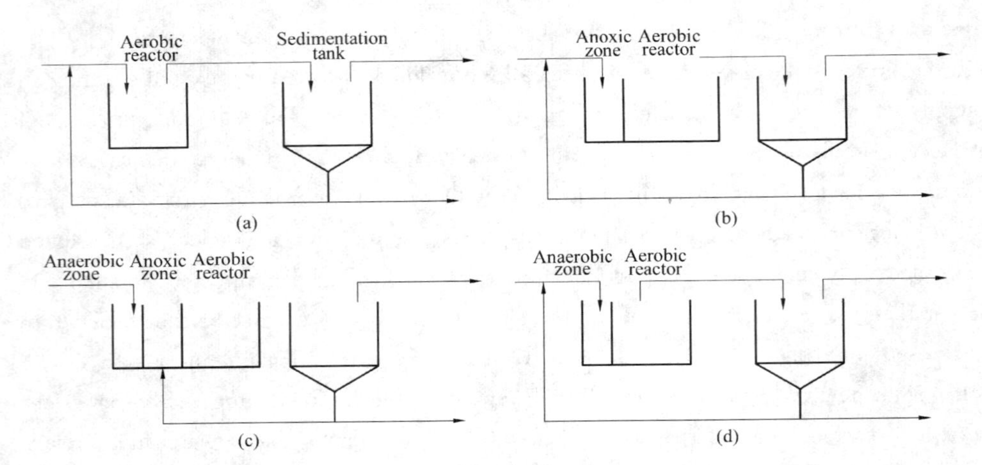

Figure 2.38 Reactor configurations for manipulating redox conditions

materials used in attached growth processes include rock, gravel, slag, sand, redwood, and a wide range of plastic and other synthetic materials. Attached growth processes can also be operated as aerobic or anaerobic processes. The packing can be submerged completely in liquid or not submerged, with air or gas space above the biofilm liquid layer.

1. Trickling Filters

The most common aerobic attached growth process used is the trickling filter in which wastewater is distributed over the top area of a vessel containing nonsubmerged packing material (see Figure 2.39). Historically, rock was used most commonly as the packing material for tricking filters with typical depths ranging from 1.25 to 2 m. Treatment occurs as the liquid flows over the attached biofilm. Rock filter beds are usually circular, and the liquid wastewater is distributed over the top of the bed by a rotary distributor. Many conventional trickling filters using rock as the packing material have been converted to plastic packing to increase treatment capacity. Virtually all new trickling filters are now constructed with plastic packing.

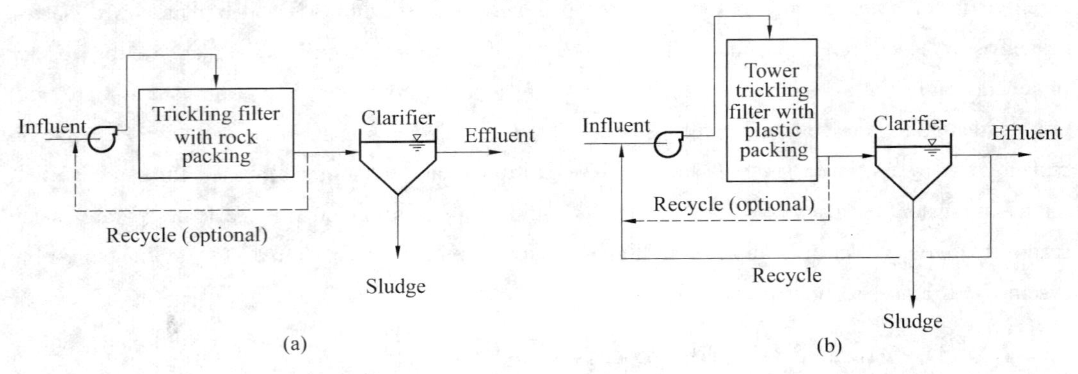

Figure 2.39 Attached growth biological treatment process (a) schematic of tricking filter with rock packing and (b) schematic of tower filter with plastic packing

A slime layer develops on the rock or plastic packing in the trickling filters and contains the microorganisms for biodegradation of the substrates to be removed from the liquid flowing over the packing. The biological community in the filter includes aerobic and facultative bacteria, fungi, algae, and protozoans. Higher animals, such as worms, insect larvae, and snails, are also present. The slime layer thickness can reach depths as much as 10 mm. Organic materials from the liquid is adsorbed onto the biological film or slime layer. In the outer portions of the biological slime layer (0.1 to 0.2 mm), the organic material is degraded by aerobic microorganisms. As the microorganisms grow and the slime layer thickness increases, oxygen is consumed before it can penetrate the full depth, and an anaerobic environment is established near the surface of the packing. As the slime layer increases in thickness, the substrate in the wastewater is used before it can penetrate the inner depths of the biofilm. Bacteria in the slime layer enter an endogenous respiration state and lose their ability to cling to the packing surface. The liquid then washes the slime off the packing, and a new slime layer starts to grow. The phenomenon of losing the slime layer is called sloughing and is primarily a function of the organic and hydraulic loading on the filter.

Trickling filter designs are classified by hydraulic or organic loading rates. Rock filter designs have been classified as low- or standard-rate, intermediate-rate, and high-rate. Plastic packing is used typically for high-rate designs; however, plastic packing has also been used at lower organic loadings, near the high end of those used for intermediate-rate rock filters. Much higher organic loadings have been used for rock or plastic packing designs in "roughing" applications where only partial BOD removal occurs.

2. Rotating Biological Contactors

An rotating biological contactors (RBCs) consists of a series of closely spaced circular disks of polystyrene or polyvinyl chloride that are submerged in wastewater and rotated through it [see Figure 2.40(a)]. The cylindrical plastic disks are attached to a horizontal shaft and are provided at standard unit sizes of approximately 3.5 m in diameter and 7.5 m in length. The surface area of the disks for a standard unit is about 9 300 m^2, and a unit with a higher density of disks is also available with approximately 13 900 m^2 of surface area. The RBC unit is partially submerged (typically 40 percent) in a tank containing the wastewater, and the disks rotate slowly at about 1.0 to 1.6 revolutions per minute [see Figure 2.40(a)]. Mechanical drives are normally used to rotate the units, but air-driven units have also been installed. In the air-driven units, an array of cups [see Figure 2.40(b)] is fixed to the periphery of the disks and diffused aeration is used to direct air to the cups to cause rotation. As the RBC disks rotate out of the wastewater, aeration is accomplished by exposure to the atmosphere. Wastewater flows down through the disks, and solids sloughing occurs. Similar to a trickling filter, RBC systems require pretreatment of primary clarification or fine screens and secondary clarification for liquid/solids separation.

Treatment systems with RBCs can be designed to provide secondary or advanced levels of treatment. Effluent BOD characteristics for secondary treatment are comparable to well-operated activated-sludge processes. Where a nitrified effluent is required, RBCs can be used to provide

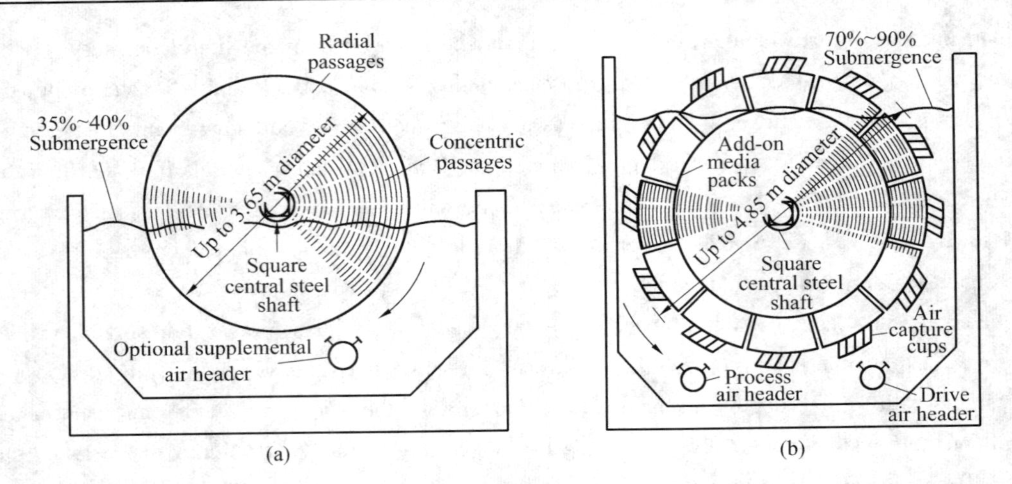

Figure 2.40　Typical RBC units: (a) conventional RBC with mechanical drive and optional air input, (b) submerged-type RBC equipped with air capture cups (air is used both to rotate and to aerate the biodisks)

combined treatment for BOD and ammonia nitrogen, or to provide separate nitrification of secondary effluent. An RBC process modification in which the disk support shaft is totally submerged has been used for denitrification of wastewater.

2.6.4　Combined Aerobic Treatment Processes

Several treatment process combinations have been developed that couple trickling filters with the activated-sludge process. The combined biological processes are known as dual processes or coupled trickling filter/activated-sludge systems. Combined processes have resulted as part of plant upgrading where either a trickling filter or activated-sludge process is added; they have also been incorporated into new treatment plant designs. Combined processes have the advantages of the two individual processes, which can include (1) the stability and resistance to shock loads of the attached growth process, (2) the volumetric efficiency and low energy requirement of attached growth process for partial BOD removal, (3) the role of attached growth pretreatment as a biological selector to improve activated-sludge settling characteristics, and (4) the high-quality effluent possible with activated-sludge treatment. The three principal types of combined processes are described below.

1. Trickling Filter/Solids Contact and Trickling Filter/Activated-Sludge Processes

The first group of the combined treatment processes, as illustrated on Figure 2.41, is commonly referred to as the trickling filter/solids contact (TF/SC) or trickling filter/activated-sludge (TF/AS) process. The principal difference between these processes is the shorter aeration period in the TF/SC process of minutes versus hours for the TF/AS process. Both processes use a trickling filter (with either rock or plastic packing), an activated-sludge aeration tank, and a final clarifier. In both processes, the trickling filter effluent is fed directly to the activated-sludge process without clarification and the return activated sludge from the secondary clarifier is fed to the activated-sludge aeration basin. A process modification that has been incorporated only into the TF/

SC process (see Figure 2.41(a)) is a return-sludge aeration tank and flocculating center-feed well for the clarifier. The TF/AS process is illustrated on Figure 2.41(b).

The most common application for the TF/AS process is where the trickling filter is designed as a roughing filter for 40 to 70 percent BOD removal and may be referred to as a roughing filter/activated-sludge (RF/AS) process. The trickling filter loading is about 4 times that used for the TF/SC process.

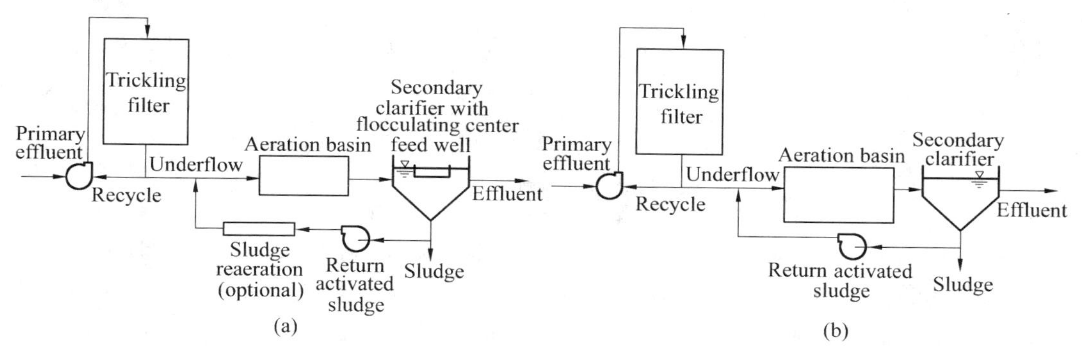

Figure 2.41 Combined trickling filter/ activated-sludge processes: (a) schematic flow diagram of tricking filter/ solids contact (TF/SC) process, (b) schematic flow diagram of tricking filter/ activated-sludge (TF/AS) process

2. Activated Biofilter and Biofilter Activated-Sludge Processes

The second group of combined processes is similar to the first, described above, with the exception that the return activated sludge (RAS) is returned directly to the trickling filter as illustrated on Figure 2.42 and an aeration basin may or may not be used. The two combined processes are termed the activated biofilter (ABF) and the biofilter/activated-sludge (BF/AS) process. The ABF and BF/AS processes are not used much today, in part because the original designs relied on redwood filter packing, which is available only at prohibitive cost.

In both cases, rock cannot be used because of the potential plugging and oxygen availability problems created by feeding return activated sludge to the trickling filter. High-rate plastic packing can be used in lieu of redwood packing. The ABF process can produce a secondary effluent quality at low organic loads to the trickling filter. At higher organic loading rates, an acceptable secondary effluent quality with the ABF process has been difficult to produce. To improve effluent quality, the ABF process is followed by a short Hydralic reteution time(HRT) aeration basin, which fits the description of the BF/AS process (see Figure 2.42(b)). In essence the BF/AS process is very similar to the RF/AS process with the exception of RAS return to the trickling filter instead of to the aeration basin.

3. Series Trickling Filter-Activated-Sludge Process

In the third approach employing combined processes, a trickling filter and an activated-sludge process are operated in series, with an intermediate clarifier between the trickling filter and activated-sludge process (see Figure 2.43). The combination of a trickling filter process followed by an activated-sludge process is often used (1) to upgrade an existing activated-sludge system, (2) to

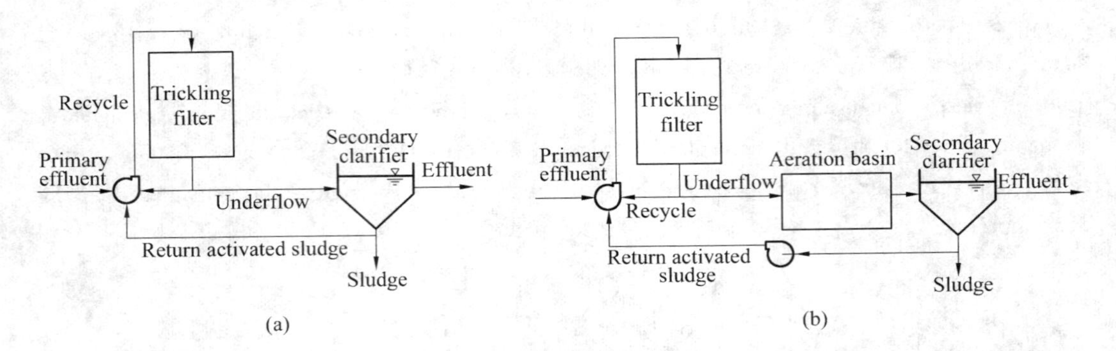

Figure 2.42　Combined trickling filter/ activated-sludge process with return sludge recycle to trickling filter: (a) schematic flow diagram of activated biofilter (ABF) and (b) schematic flow diagram of biofilter/ activated-sludge (BF/AS).

reduce the strength of wastewater where industrial and domestic wastewater is treated in common treatment facilities, and (3) to protect a nitrification activated-sludge process from toxic and inhibitory substances. In systems treating high-strength wastes, intermediate clarifiers are used between the trickling filters and the activated-sludge units to reduce the solids load to the activated-sludge system and to minimize the aeration volume required.

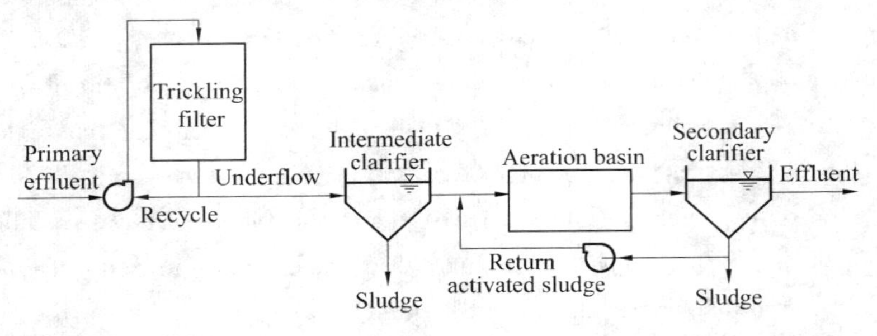

Figure 2.43　Schematic flow diagram of combined trickling filter activated-sludge process with intermediate clarifier

2.7　Sludge Treatment and Disposal

In the process of purifying the wastewater, another problem is created: sludge. The higher the degree of wastewater treatment, the larger the residue of sludge that must be handled. The exceptions to this rule are where land applications or polishing lagoons are used. The sludge is made of materials settled from the raw wastewater and of solids generated in the wastewater treatment processes. The quantities of sludge involved are significant. For primary treatment, they may be 0.25 to 0.35 percent by volume of wastewater treated. When treatment is upgraded to activated sludge, the quantities increase to 1.5 to 2.0 percent of this volume of water treated. Use of chemicals for phosphorus removal can add another 1.0 percent. The sludges withdrawn from the treatment processes are still largely water, as much as 97 percent. Sludge treatment processes, then, are concerned with separating the large amounts of water from the solid residues. The separated water is returned to the wastewater plant for processing.

The basic processes for sludge treatment are as follows: (1) Thickening: separating as much water as possible by gravity or flotation. (2) Stabilization: converting the organic solids to more refractory(inert) forms to that they can be handled or used as soil conditioners without causing a nuisance or health hazard through processes referred to as "digestion." (These are biochemical oxidation processes.) (3) Conditioning: treating the sludge with chemicals or heat, so that the water can be readily separated. (4) Dewatering: separating water by subjecting the sludge to vacuum, pressure or drying. (5) Reduction: converting the solids to a stable form by wet oxidation or incineration. (These are chemical oxidation processes; they decrease the volume of sludge, hence the term reduction.) Although a large number of alternative combinations of equipment and processes are used for treating sludges, the basic alternatives are fairly limited. The ultimate depository of the materials contained in the sludge must either be land, air, or water. Current policies discourage practices such as ocean dumping of sludge. Air pollution consideration necessitates air pollution control facilities as part of the sludge incineration process. The basic alternatives by which these processes may be employed are shown in Figure 2.44.

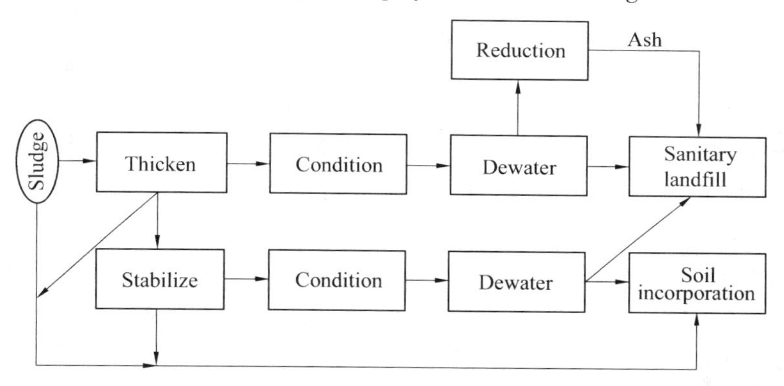

Figure 2.44 Basic sludge handling alternatives

2.7.1 Sources and Characteristics of Various Sludges

Before we begin the discussion of the various treatment processes, it is worthwhile to recapitulate the sources and nature of the sludges that must be treated.

Grit: The sand, broken glass, nuts, bolts, and other dense material that is collected in the grit chamber is not true sludge in the sense that it is not fluid. However, it still requires disposal. Because grit can be drained of water easily and is relatively stable in terms of biological activity (it is not biodegradable), it is generally trucked directly to without further treatment.

Primary or Raw Sludge: Sludge from the bottom of the primary clarifiers contains from 3 to 8 percent solids (1 percent solids $\approx$ 1 g solids/100 mL sludge volume), which is approximately 70 percent organic matter. This sludge rapidly becomes anaerobic, and is highly odiferous.

Secondary Sludge: This sludge consists of microorganisms and inert materials that have been wasted from the secondary treatment processes. Thus, the solids are about 90 percent organic matter. When the supply of air is removed, this sludge also becomes anaerobic, creating noxious conditions if not treated before disposal. The solids content depends on the source. Wasted activated sludge is typically 0.5 to 2 percent solids, while trickling filter sludge contains 2 to 5 percent solids. In some cases, secondary sludges contain large quantities of chemical precipitates because the aeration tank is used as the reaction basin for the addition of chemicals to remove phosphorus.

Tertiary Sludges: The characteristics of sludges from the tertiary treatment processes depend on the nature of the process. For example, phosphorus removal results in a chemical sludge that is difficult to handle and treat. When phosphorus removal occurs in the activated sludge process, the chemical sludge is combined will the biological sludge, making the latter more difficult to treat. Nitrogen removal by denitrification results in a biological sludge with properties very similar to those of waste activated sludge.

2.7.2 Thickening

Thickening is usually accomplished in one of two ways: the solids are floated to the top of the liquid (flotation) or are allowed to settle to the bottom (gravity thickening). The goal is to remove as much water as possible before final dewatering or digestion of the sludge. The processes involved offer a low-cost means of reducing sludge volumes by a factor of two or more. The costs of thickening are usually more than offset by the resulting savings in the size and cost of downstream sludge processing equipment.

1. Flotation

In the flotation thickening process (see Figure 2.45) air is injected into the sludge under pressure (275 to 550 kPa). Under this pressure, a large amount of air can be dissolved in the sludge. The sludge then flows into an open tank where, at atmospheric pressure, much of the air comes out of solution as minute bubbles. The bubbles attach themselves to sludge solids particles and float them to the surface. The sludge forms a layer at the top of the tank; this layer is removed by a skimming mechanism for further processing. The process typically increases the solids content of activated sludge from 0.5 ~ 1 percent to 3 ~ 6 percent. Flotation is especially effective on activated sludge. which is difficult to thicken by gravity.

2. Gravity Thickening

Gravity thickening is a simple and inexpensive process that has been used widely on primary sludges for many years. It is essentially a sedimentation process similar to that which occurs in all settling tanks. Sludge flows into a tank that is very similar in appearance to the circular clarifiers used in primary and secondary sedimentation (see Figure 2.46); the solids are allowed to settle to the bottom where a heavy-duty mechanism scrapes them to a hopper from which they are withdraw further processing. The type of sludge being thickened has a major effect on performance. The best results are obtained with purely primary sludges. As the proportion of activated sludge increases, the thickness of settled sludge solids decreases. Purely primary sludges can be thickened from 1 ~ 3

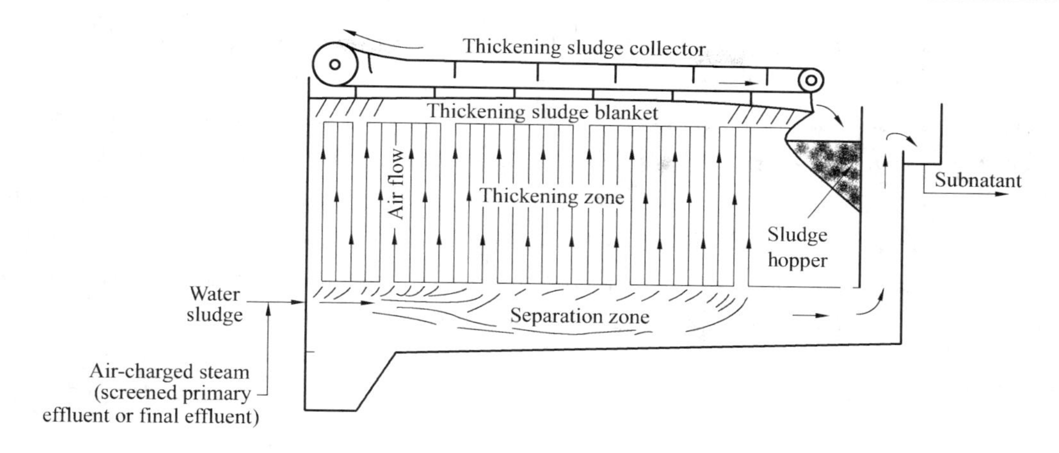

Figure 2.45 Air flotation thickener

percent to 10 percent solids. The current trend is toward using gravity thickening for primary sludges and flotation thickening for activated sludges, and then blending the thickened sludges for further processing.

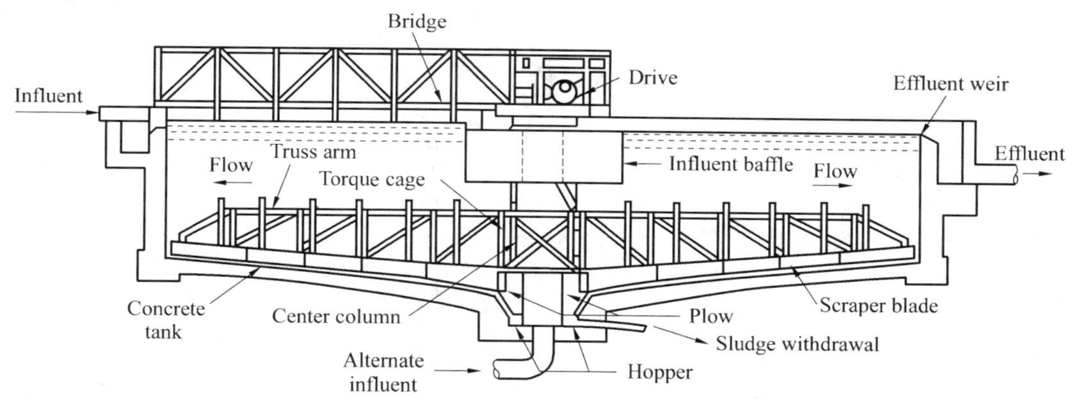

Figure 2.46 Gravity thickener

2.7.3 Stabilization

The principal purposes of sludge stabilization are to break down the organic solids biochemically so that they are more stable (less odorous and less putrescible) and more dewaterable, and to reduce the mass of sludge. If the sludge is to be dewatered and burned, stabilization is not used. There are two basic stabilization processes in use. One is carried out in closed tanks devoid of oxygen and is called anaerobic digestion. The other approach injects air into the sludge to accomplish aerobic digestion.

The aerobic digestion of biological sludges is nothing more than a continuation of the activated sludge process. When a culture of aerobic heterotrophs is placed in an environment containing a source of organic material, the microorganisms remove and utilize most of this material. A fraction of the organic material removed will be used for the synthesis of new biomass. The remaining material will be channeled into energy metabolism and oxidized to carbon dioxide, water, and soluble insert to provide energy for both synthesis and maintenance (life-support) function. Once the external source of organic material is exhausted, however, the microorganisms enter into endogenous respiration, where cellular material is oxidized to satisfy the energy of maintenance (that is, energy for life-support requirements). If this is continued over an extended period of time, the total quantity of biomass will be considerably reduced. Furthermore, that portion remaining will exist at such a low energy state that it can be considered biologically stable and suitable for disposal in the environment. This forms the basic principle of aerobic digestion.

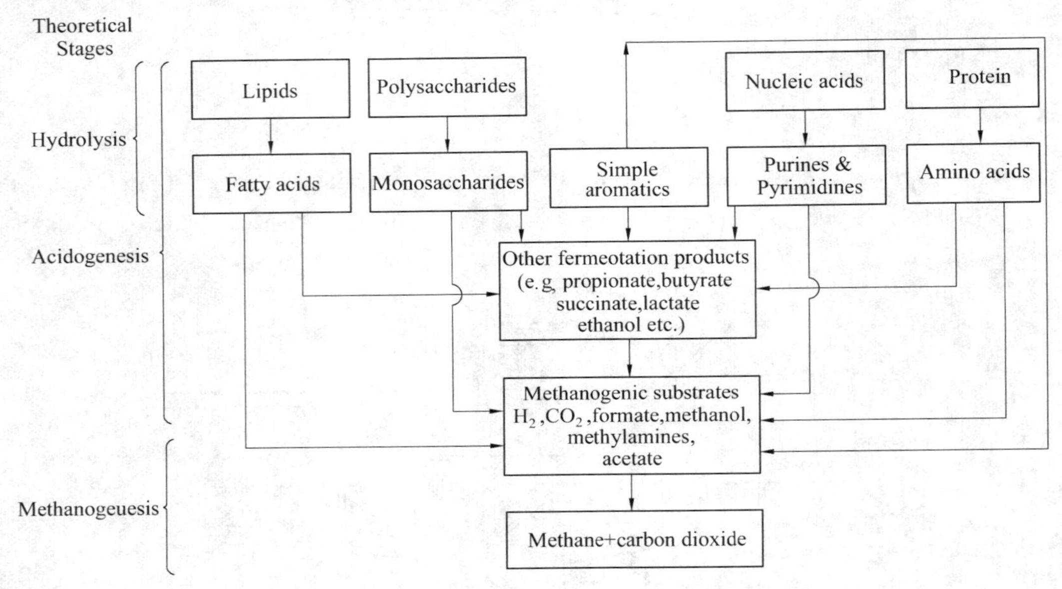

Figure 2.47　Schematic diagram of the patterns of carbon flow in anaerobic digestion

The anaerobic treatment of complex wastes involves three distinct stages. In the first stage, complex waste components, including fats, proteins, and polysaccharides, are hydrolyzed to their component subunits. This is accomplished by a heterogeneous group of facultative and anaerobic bacteria. These bacteria then subject the products of hydrolysis to fermentation and other metabolic processes leading to the formation of simple organic compounds and hydrogen in a process called acidogenesis or acetogenesis. The second stage is commonly referred to as acid fermentation. In this stage, organic material is simply converted to organic, alcohols, and new bacterial cells, so that little stabilization of BOD or COD is realized. In the third stage, the end products of the first stage are converted to gases (mainly methane and carbon dioxide) by several different species of strictly anaerobic bacteria. Thus, it is here that true stabilization of the organic material occurs. This stage is generally referred to as methane fermentation. The stages of anaerobic waste treatment are illustrated in Figures 2.47 and 2.48. Even though the anaerobic process is presented as being sequential in nature, all stages take place simultaneously and synergistically. The primary acid produced during acid fermentation is acetic acid. The significance of this acid as a precursor for methane formation is illustrated in Figure 2.48.

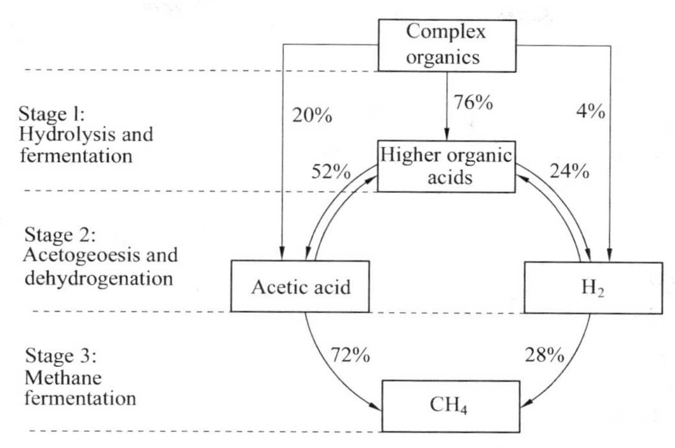

Figure 2.48 Steps in anaerobic digestion process with energy flow

2.7.4 Sludge Conditioning

Several methods of conditioning sludge to facilitate the separation of the liquid and solids are available. One of the most commonly used is the addition of coagulants such as ferric chloride, lime, or organic polymers. Ash from incinerated sludge has also found use as a conditioning agent. As happens when coagulants are added to turbid water, chemical coagulants act to clump the solids together so that they are more easily separated from the water. In recent years, organic polymers have become increasingly popular for sludge conditioning. Polymers are easy to handle, require little storage space, and are very effective. The conditioning chemicals injected into the sludge just before the dewatering process and are mixed with the sludge.

Another conditioning approach is to heat the sludge at high temperatures (175 to 230℃) and pressures (1 000 to 2 000 kPa). Under these conditions, much like those of a pressure cooker, water that is bound up in the solids is released, improving the dewatering characteristics of the sludge. Heat treatment has the advantage of producing a sludge that dewaters better than chemically conditioned sludge. The process has the disadvantages of relatively complex operation and maintenance and the creation of highly polluted cooking liquors that when recycled to the treatment plant impose a significant added treatment burden.

2.7.5 Sludge Dewatering

The most popular method of sludge dewatering in the past has been the use of sludge drying beds. These beds are especially popular in small plants because of their simplicity of operation and maintenance. Most of these plants are located in small and medium-sized communities, with an average flow rate of less than 0.10 m^3/s. Although the use of drying beds might be expected in the warmer, sunny regions, they are also used in several large facilities in northern climates.

A vacuum filter consists of a cylindrical drum covered with a filtering material or fabric, which rotates partially submerged in a vat of conditioned sludge. A vacuum is applied inside the drum to extract water, leaving the solids, or filter cake, on the filter medium. As the drum completes its rotational cycle, a blade scrapes the filter cake from the filter and the cycle begins again. In some systems, the filter fabric passes off the drum over small rollers to dislodge the cake. There is a wide variety of filter fabrics, ranging from Dacron to stainless-steel coils, each with its own advantages. The vacuum filter can be applied to digested sludge to produce a sludge cake dry enough (15 to 30 percent solids) to handle and dispose of by burial in a landfill or by application to the land as a relatively dry fertilizer. If the sludge is to be incinerated, it is not stabilized. In this case, the vacuum filter is applied to the raw sludge to dewater it. The sludge cake is then fed to the furnace to be incinerated. Vacuum filters are being replaced by continuous belt filter presses.

2.7.6 Reduction

If sludge use as a soil conditioner is not practical, or if a site is not available for landfill using dewatered sludge, cities may turn to the alternative of sludge reduction. Incineration completely evaporates the moisture in the sludge and combusts the organic solids to a sterile ash. To minimize the amount of fuel used, the sludge must be dewatered as completely as possible before incineration. The exhaust gas from an incinerator must be treated carefully to avoid air pollution.

2.7.7 Sludge Disposal

The wastewater treatment plant(WWTP) process residuals (leftover sludges, either treated or untreated) are the bane of design and operating personnel. Of the five possible disposal sites for residuals, two are feasible and only one is practical. Conceivably, one could ultimately dispose of residues in the following places: in the air, in the ocean, in "outer space," on the land, or in the marketplace. Disposal in the air by burning is in reality not ultimate disposal but only temporary storage until the residue falls to the ground. If you use air pollution control devices, then the residue from these devices must be disposed of. Disposal of sewage sludge at sea by barging is now prohibited in the United States. "Outer space" is not a suitable disposal site. Thus, we are left with land disposal and utilization of the sludge to produce a product.

The practice of applying WWTP residuals for the purposes of recovering nutrients, water, or reclaiming despoiled land such as strip mine spoils is called land spreading. In contrast to the other land disposal techniques, land spreading is land-use intensive. Application rates are governed by the character of the soil and the ability of the crops of forests on which the sludge is spread to accommodate it.

Sludge landfill can be defined as the planned burial of wastewater solids, including processed sludge, screenings, grit, and ash, at a designated site. The solids are placed into a prepare site or excavated trench and covered with a layer of soil. The soil cover must be deeper than the depth of the plow zone (about 0 20 to 0. 25 m). For the most part, landfilling of screenings, grit, and ash is accomplished with methods similar to those used for sludge landfilling.

Dedicated land disposal means the application of heavy sludge loadings to some finite land area that has limited public access and has been set aside or dedicated for all time to the disposal of wastewater sludge. Dedicated land disposal does not mean in place utilization No crops may be grown. Dedicated sites typically receive liquid sludges. While application of dewatered sludges is possible, it is not common. In addition, disposal of dewatered sludge in landfills is generally more cost-effective.

Wastewater solids may sometimes be used beneficially in ways other than as a soil nutrient. Of the several methods worthy of note, composting and co-firing with municipal solid waste are two which have received increasing amounts of interest in the last few years. The recovery of lime and the use of the sludge to form activated carbon have also been in practice to a lesser extent.

Review Questions

1. List the four categories of water quality for drinking water.

2. List the four categories of physical characteristics.

3. Differentiate between coagulation and flocculation.

4. Define hardness in term of the chemical constituents that cause it and in terms of the results as seen by the users of hard water.

5. Compare slow sand filters and rapid sand filter with respect to operation procedures and loading rates.

6. Explain why a disinfectant that has a residual is preferable to one that does not.

7. In raw domestic sewage, what are typical concentrations of BOD, TSS, nitrogen and phosphorus, and coliforms?

8. What minimum level of pollutant removal does secondary treatment accomplish?

9. What do the terms influent and effluent refer to?

10. What is the purpose of grit removal? How is it accomplished?

11. Give a brief description of a primary clarifier.

12. Sketch a flow diagram of an activated sludge treatment process.

13. Describe a method for removing phosphorus from sewage.

14. Define nitrified effluent and denitrification.

15. Brief discuss some general options for sludge disposal.

Chapter 3 Air Pollution and Control

3.1 The Atmosphere

The atmosphere (air) is composed of 78.1% nitrogen, 20.9% oxygen and a number of other gases such as argon, carbon dioxide, methane, and water vapor that total about 1%. Most of the atmosphere is held close to the Earth by the pull of gravitational force, so it gets less dense with increasing distance from the Earth. Throughout the various layers of the atmosphere, nitrogen and oxygen are the most common gases present, although the molecules are farther apart at higher attitudes. Figure 3.1 shows the various layers of the atmosphere and some of the characteristics of each layer. The atmosphere is divided into four sections. The troposphere extends from the Earth's surface to about 10 kilometers above the Earth. It actually varies from about 8 to 18 kilometers, depending on the position of the Earth and the season of the year. The temperature of the troposphere declines by about 6℃ for every kilometer above the surface. The troposphere contains most of the water of the atmosphere and is the layer which weather takes place. The stratosphere extends from the top of the troposphere to about 50 kilometers and contains most of the ozone. The ozone is in a band between 15 and 30 kilometers above the Earth's surface. Because the ozone layer absorbs sunlight, the upper layers of the stratosphere are warmer than the lower layers. The mesosphere is a layer with decreasing temperature from 50 to 80 kilometers above the Earth. The ther mosphere is a layer with increasing temperature that extends up to about 300 kilometers.

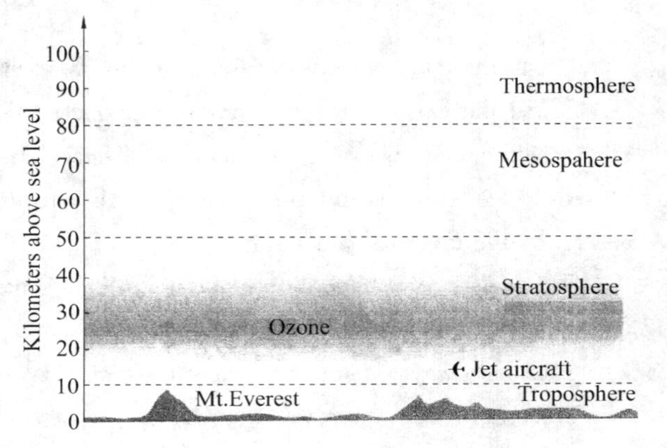

Figure 3.1 The atmosphere

There are several natural sources of gases and particles that degrade the quality of the air, including material emitted from volcanoes, dust from wind erosion, and gases from the decomposition of dead plants and animals. Since these events are not controlled by humans, they do not fit our definition of pollution. However, the pollutions from automobile emissions, chemical odors, factory smoke, and similar materials are directly related to the number of people living in an area and the kinds of activities in which they are involved. When a population is small and its energy use is low, the impact of people is minimal. The pollutants released into the air are diluted, carried away by the wind, washed from the air by rain, or react with oxygen in the air to form harmless materials. Thus, the overall negative effect is slight. However, our urbanized, industrialized civilization has dense concentrations of people that use large quantities of fossil fuels for manufacturing, transportation, and domestic purposes. These activities release large quantities of polluting by-products into our environment.

3.2 Categories of Air Pollutants

Around the world, five major types of substances are released directly into the atmosphere in their unmodified forms in sufficient quantities to pose a health risk and are called primary air pollutants. They are carbon monoxide, volatile organic compounds (hydrocarbons), particulate matter, sulfur dioxide, and oxides of nitrogen. Primary air pollutants may interact with one another in the presence of sunlight to form new compounds such as ozone that are known as secondary air pollutants. Secondary air pollutants also form from reactions with substances that occur naturally in the atmosphere. In addition, the U.S. Environmental Protection Agency(EPA) has a category of air pollutants known as criteria air pollutants. Criteria air pollutants are those for which specific are quality standards have been set (see Table 3.1). In addition, certain compounds with high toxicity are known as hazardous air pollutants or air toxics.

3.2.1 Carbon Monoxide

Carbon monoxide (CO) is produced when organic materials such as gasoline, coal, wood, and trash are burned with insufficient oxygen. When amount of oxygen is restricted, carbon monoxide is formed instead of carbon dioxide ($2C + O_2 \rightarrow 2CO$). Any process that involves the burning of fossil fuels has the potential to produce carbon monoxide. The single largest source of carbon monoxide is the automobile. About 60 percent of CO comes from vehicles driven on roads and 20 percent comes from vehicles not used on roads. The remainder comes from other processes that involve burning (power plants, industry, burning leaves, etc.). Although increased fuel efficiency and the use of catalytic converters have reduced carbon monoxide emissions per kilometer driven, carbon monoxide remains a problem because the number of automobiles on the road and the number of kilometers driven have risen.

Table 3.1 National Ambient Air Quality Standards

Pollutant	Standard
Carbon monoxide (CO)	8-hour average (9 ppm) 1-hour average (35 ppm)
Nitrogen dioxide (NO_2)	Annual mean (0.053 ppm)
Ozone (O_3)	8-hour average (0.08 ppm) 1-hour average (0.12 ppm)

Lead (Pb)	3-month average (1.5 $\mu g/m^3$)
Particulate matter (PM_{10})	Annual mean (50 $\mu g/m^3$) 24-hour average (150 $\mu g/m^3$)
Particulate matter ($PM_{2.5}$)	Annual mean (15 $\mu g/m^3$) 24-hour average (65 $\mu g/m^3$)
Sulfur dioxide (SO_2)	Annual mean (0.03 ppm) 24-hour average (0.14 ppm) 3-hour average (0.50 ppm)

Carbon monoxide is dangerous because it binds to the hemoglobin in blood and makes the hemoglobin less able to carry oxygen. Carbon monoxide is most dangerous in enclosed spaces, where it is not diluted by fresh air entering the space. Several hours of exposure to air containing only 0.001 percent of carbon monoxide can cause death. The amount of carbon monoxide produced in heavy traffic can cause headaches, drowsiness, and blurred vision. Fortunately, carbon monoxide is not a persistent pollutant. It readily combines with oxygen in the air to form carbon dioxide ($2CO + O_2 \rightarrow 2CO_2$). Therefore, the air can be cleared of its carbon monoxide if no new carbon monoxide is introduced into it. Catalytic converters reduce the amount of carbon monoxide released by vehicle engines, and specially formulated fuels that produce less carbon monoxide are used in many cities that have a carbon monoxide problem.

3.2.2 Volatile Organic Compounds

In addition to carbon monoxide, automobiles emit a variety of volatile organic compounds (VOCs). Volatile organic compounds are composed primarily of carbon and hydrogen, so they are often referred to as hydrocarbons. If they evaporate into the air, they are volatile; hence, airborne organic compounds are volatile organic compounds. They are either evaporated from fuel supplies or are remnants of fuel that did not burn completely. The use of internal combustion engines accounts for the majority of volatile organic compounds released into the air, although refineries and other industries add to the total atmospheric burden. Consumer products such as oil-based paint, charcoal lighter, and many other chemicals are also important. Some of these compounds are toxic and are known as hazardous air pollutants. They are also important because they contribute to the production of the secondary air pollutants found in smog.

Many modifications to automobiles have significantly reduced the amount of volatile organic compounds entering the atmosphere. Recycling some gases through the engine so they burn rather than escape, increasing the proportion of oxygen in the fuel-air mixture to obtain more complete burning of the fuel, and using devices to prevent the escape of gases from the fuel tank and crankcase are three of these modifications. In addition, catalytic converters allow unburned organic compounds in exhaust gases to be oxidized more completely so that fewer volatile organic compounds leave the tail pipe.

3.2.3 Particulate Matter

Particulate matter consists of minute (10 microns and smaller) solid particles and liquid droplets dispersed into the atmosphere. The U. S. Environmental Protection Agency has set standards for particles smaller than 10 microns (PM_{10}) and 2.5 microns ($PM_{2.5}$). Most of the coarse particles (greater than 2.5 microns) are primary pollutants that are released directly into the air. Fine particles (2.5 microns or less) are mostly secondary pollutants that form in the atmosphere from interactions of primary pollutants. Sulfates and nitrates formed from sulfur dioxide and nitrogen oxides are examples.

Particles can accumulate in the lungs and interfere with their ability to exchange gases. Such lung damage usually occurs in people who are repeatedly exposed to large amounts of particulate matter on the job. Miners and others who work in dusty conditions are most likely to be affected. Droplets and solid particles can also serve as centers for the deposition of other substances from the atmosphere. As we breathe air containing particulates, we come in contact with concentrations of other potentially more harmful materials that have accumulated on the particulates. Sulfuric, nitric, and carbonic acids, which irritate the lining of our respiratory system, frequently are associated with particulates.

3.2.4 Sulfur Dioxide

Sulfur dioxide (SO_2) is a compound of sulfur and oxygen that is produced when sulfur-containing fossil fuels are burned ($S + O_2 \rightarrow SO_2$). There is sulfur in coal and oil because they were produced from the bodies of organisms that had sulfur as a component of some of their molecules. The sulfur combines with oxygen to form sulfur dioxide when fossil fuels are burned. Today, the major source of sulfur dioxide is from the burning of coal in power plants. Sulfur dioxide has a sharp odor, irritates respiratory tissue, and aggravates asthmatic and other respiratory conditions. It also reacts with water, oxygen, and other materials in the air to form sulfur-containing acids. The acids can become attached to particles that, when inhaled, are very corrosive to lung tissue. These acid-containing particles are also involved in acid deposition.

3.2.5 Nitrogen Dioxide

The burning of fossil fuels produces a mixture of nitrogen-containing compounds commonly known as oxides of nitrogen (NO_x). These compounds are formed because the nitrogen and oxygen molecules in the air combine with one another when subjected to the high temperatures experienced during combustion. The primary molecule produced is nitrogen monoxide ($N_2 + O_2 \rightarrow 2NO$), but nitrogen monoxide can be converted to nitrogen dioxide in the air ($2NO + O_2 \rightarrow 2NO_2$) to produce a mixture of NO and NO_2. Thus, NO_2 is, for the most part, a secondary pollutant. Nitrogen dioxide is a reddish brown, highly reactive gas that is responsible for much of the haze seen over cities, causes respiratory problems, and is a component of acid precipitation. The primary source of nitrogen oxides is the automobile engine. Catalytic converters significantly reduce the amount of nitrogen monoxide released from the internal combustion engine. However, the increase in the numbers of cars and kilometers driven has resulted in an increase of NO_x levels, and NO_2 is a significant pollutant.

3.2.6 Lead

Lead (Pb) can enter the body through breathing airborne particles or consuming lead that was deposited on surfaces. Lead accumulates in the body and causes a variety of health effects, including mental retardation and kidney damage. At one time, the primary source of airborne lead was from additives in gasoline. Lead was added to gasoline to help engines run more effectively. Another major source of lead is paints. Many older homes have paints that contain lead, since various lead compounds are colorful pigments. Dust from flaking paint, remodeling, or demolition is released into the atmosphere. Although the amount of lead may be small, its presence in the home can result in significant exposure to inhabitants, particularly young children who chew on painted surfaces and often eat paint chips. Today, however, the major sources of lead are from industrial sources where metals are smelted or batteries manufactured.

3.2.7 Ground-Level Ozone and Photochemical Smog

Ozone (O_3) is an extremely reactive molecule that irritates respiratory tissues and can cause permanent lung damage. It also damages plants and reduces agricultural yields. Ozone is a secondary pollutant that is formed as a component of photochemical smog. Photochemical smog is a mixture of pollutants including ozone, aldehydes, and peroxyacetyl nitrates resulting from the interaction of nitrogen dioxide and volatile organic compounds with sunlight in a warm environment. The two most destructive components of photochemical smog are ozone and peroxyacetyl nitrates. Both of these secondary pollutants are excellent oxidizing agents that will react readily with many other compounds, including those found in living things, causing destructive changes. Ozone is particularly harmful because it destroys chlorophyll in plants and injures lung tissue in humans and other animals. Peroxyacetyl nitrates, in addition to being oxidizing agents, are eye irritants.

For photochemical smog to develop, several ingredients are required. Nitrogen monoxide, nitrogen dioxide, and volatile organic compounds must be present, and sunlight and warm temperatures are important to support the chemical reactions involved. During morning rush-hour traffic, the amounts of nitrogen monoxide and volatile organic compounds increase. This leads to an increase in the amount of NO_2, and NO levels fall as NO is converted to NO_2. During this same time, ozone levels rise as NO_2 is broken down by sunlight, and levels of peroxyacetyl nitrates rise as well. As the sun sets, production of ozone lessens. In addition, ozone and other smog components react with their surroundings and are destroyed, so smog conditions improve in the evening. Smog problems could be substantially decreased by reducing the NO_x and VOCs associated with the use of internal combustion engines (perhaps eliminating them completely) or by moving population centers away from the valleys where thermal inversions occur. Currently, major advances have been made in reducing the smog problem by reformulating gasoline and installing devices on automobiles that reduce the amount of NO_x and VOCs released.

3.3 Acid Deposition

Acid deposition is the accumulation of potential acid-forming particles on a surface. The acid-forming particles can be dissolved in rain, snow, or fog or can be deposited as dry particles. When dry particles are deposited, an acid does not actually form until these materials mix with water. Even though the acids are formed and deposited in different ways, all of these sources of acid-forming particles are commonly referred to as acid rain. Acids result from natural causes, such as vegetation, volcanoes, and lightning, and from human activities, such as the burning of coal and use of the internal combustion engine (see Figure 3.2). These combustion processes produce sulfur dioxide (SO_2) and oxides of nitrogen (NO_x). Oxidizing agents, such as ozone, hydroxide ions, or hydrogen peroxide, along with water, are necessary to convert the sulfur dioxide or nitrogen oxides to sulfuric or nitric acid. Acid rain is a worldwide problem. Reports of high acid-rain damage have come from Canada, England, Germany, France, Scandinavia, and the United States. Rain is normally slightly acidic, (pH between 5.6 and 5.7) since atmospheric carbon dioxide dissolves in water to produce carbonic acid. But acid rains sometimes have a concentration of acid a thousand times higher than normal.

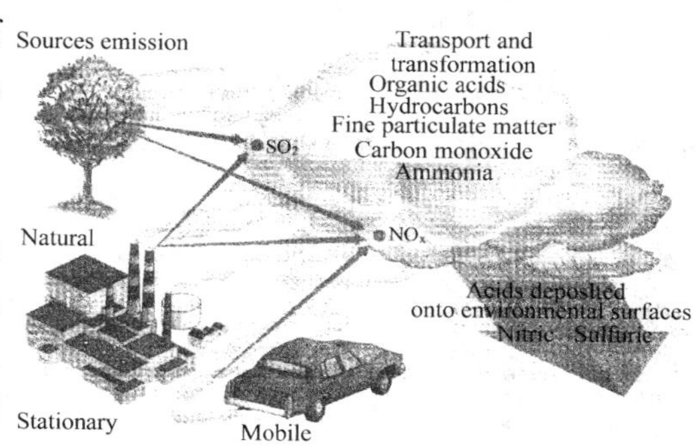

Figure 3.2 Acid deposition

Acid rain can cause damage in several ways. Buildings and monuments are often made from materials that contain limestone (calcium carbonate, $CaCO_3$), because limestone is relatively soft and easy to work. Sulfuric acid (H_2SO_4), a major component of acid rain, converts limestone to gypsum ($CaSO_4$), which is more soluble than calcium carbonate and is eroded over many years of contact with acid rain. Metal surfaces can also be attacked by acid rain.

The effects of acid rain on ecosystems are often subtle and difficult to quantify. However, in many parts of the world, acid rain is suspected of causing the death of many forests and reducing the vigor and rate of growth of others. In central Europe, many forests have declined significantly, resulting in the death of about 6 million hectares (14.8 million acres) of trees. Northeastern North America has been affected with significant tree death and reduction in vigor, particularly at higher elevations. Some areas have had 50 percent mortality of red spruce trees. A strong link can be established between the decline of the forests and acid rain. Sulfur dioxide and oxides of nitrogen are the primary molecules that contribute to acid rain. The deposition of acids causes major changes to the soils in areas where the soils are not able to buffer the additional acid. As soil becomes acidic, aluminum is released from binding sites and becomes part of the soil water, where it interferes with the ability of plant roots to absorb nutrients. A recent long-term study in New Hampshire strongly suggests that the many years of acid precipitation have reduced the amount of calcium and magnesium in the soil, which are essential for plant growth. Because there are no easy ways to replace the calcium even if acid rain were to stop, it would still take many years for the forests to return to health. Reduction in the pH of the soil may also change the kinds of bacteria in the soil and reduce the availability of nutrients for plants. While none of these factors alone would necessarily result in tree death, each could add to the stresses on the plant and may allow other factors, such as insect infestations, extreme weather conditions (particularly at high elevations), or drought, to further weaken trees and ultimately cause their death.

The effects of acid rain on aquatic ecosystems are much more clear-cut. In several experiments, lakes were purposely made acidic and the changes in the ecosystems recorded. The experiments showed that as lakes became more acidic, there is a progressive loss of many kinds of organisms. The food web becomes less complicated, many organisms fail to reproduce, and many others die. Most healthy lakes have a PH above 6. At a pH of 5.5, many desirable species of fish are eliminated; at a PH of 5, only a few starving fish may be found, and none are able to reproduce. Lakes with a pH of 4.5 are nearly sterile.

Several reasons account for these changes. Many of the early developmental stages of insects and fish are more sensitive to acid conditions than are the adults. In addition, the young often live in shallow water, which is most affected by a flood of acid into lakes and rivers during the spring snowmelt. The snow and its acids have accumulated over the winter, and the snowmelt releases large amounts of acid all at once. Crayfish and other crustaceans need calcium to form their external skeleton. As the pH of the water decreases, the crayfish are unable to form new exoskeletons and so they die. Reduced calcium availability also results in the development of some fish with malformed skeletons. As mentioned earlier, increased acidity also results in the release of aluminum, which impairs the function of a fish's gills.

About 14 000 lakes in Canada and 11 000 in the United States have been seriously altered by becoming acidic. Many lakes in Scandinavia are similarly affected. The extent to which acid deposition affects an ecosystem depends on the nature of the bedrock in the area and the ecosystem's proximity to acid-forming pollution sources (see Figure 3.3). In any aquatic ecosystem, the following factors increase the risk of damage from acid deposition:

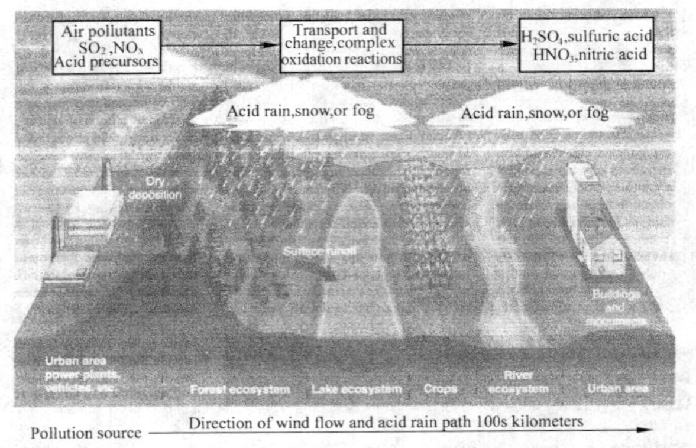

Figure 3.3 Factors that contribute to acid rain damage

Source: data from U.S. EPA, acid rain.

(1) location downwind from a major source of pollution; (2) hard, insoluble bedrock with a thin layer of infertile soil in the watershed; (3) low buffering capacity in the soil of the watershed; (4) a low lake surface area to watershed ratio. Soils derived from igneous rock are not capable of buffering the effects of acid deposition, while soils derived from sedimentary rocks such as limestone release bases that neutralize the effects of acids. Because of this, eastern Canada and the U.S. Northeast are particularly susceptible to acid rain. These areas have a high proportion of granite rock and are downwind from the major air-pollution sources of North America. Scandinavian countries have a similar geology and receive pollution from industrial areas in the United Kingdom and Europe. Thousands of kilometers of streams and up to 200 000 lakes in eastern Canada and the northeastern United States are estimated to be in danger of becoming acidified because of their location and geology.

3.4 Ozone Depletion

In the 1970s, various sectors of the scientific community became concerned about the possible reduction in the ozone layer in the Earth's upper atmosphere. Ozone is a molecule of three atoms of oxygen (O_3). In 1985, it was discovered that a significant thinning of the ozone layer over the Antarctic occurred during the Southern Hemisphere spring. Some regions of the ozone layer showed 95 percent depletion. Ozone depletion also has been found to be occurring farther north than previously. Measurements in Arctic regions suggest a thinning of the ozone layer there also. These findings caused several countries to become involved in efforts to protect the ozone layer.

The ozone in the outer layers of the atmosphere, approximately 15 to 35 kilometers from the Earth's surface, shields the Earth from the harmful effects of ultraviolet light radiation. Ozone absorbs ultraviolet light and is split into an oxygen molecule and an oxygen atom:

$$O_3 \xrightarrow{\text{Ultraviolet light}} O_2 + O$$

Oxygen molecules are also split by ultraviolet light to form oxygen atoms :

$$O_2 \xrightarrow{\text{Ultraviolet light}} 2O$$

Recombination of oxygen atoms and oxygen molecules allows ozone to be formed again and to be available to absorb more ultraviolet light ($O_2 + O \rightarrow O_3$). This series of reactions results in the absorption of 99 percent of the ultraviolet light energy coming from the sun and prevents it from reaching the Earth's surface. Less ozone in the upper atmosphere would result in more ultraviolet light reaching the Earth's surface, causing increased skin cancers and cataracts in humans and increased mutations all living things.

Chlorofluorocarbons are strongly implicated in the ozone reduction in the upper atmosphere. Chlorine reacts with ozone in the following way to reduce the quantity of ozone present: $Cl + O_3 \rightarrow ClO + O_2$; $ClO + O \rightarrow Cl + O_2$. These reactions both destroy ozone and reduce the likelihood that it will be formed because atomic oxygen (O) is removed as well. It is also important to note that it can take 10 to 20 years for chlorofluorocarbon molecules to get into the stratosphere, and then they can react with the ozone for up to 120 years. As a result, the World Meteorological Association predicts ozone depletion will worsen before improvements are seen.

Since the 1970s, when chlorofluorocarbons were linked to the depletion of the ozone layer in the upper atmosphere, their use as propellants in aerosol cans has been banned in the United States, Canada, Norway, and Sweden, and the European Union agreed to reduce use of chlorofluorocarbons in aerosol cans. In other parts of the world, however, chlorofluorocarbons are still widely used as aerosol propellants.

In 1987, several industrialized countries, including Canada, the United States, the United Kingdom, Sweden, Norway, Netherlands, the Soviet Union, and West Germany, agreed to freeze chlorofluorocarbon production at current levels and reduce production by 50 percent by the year 2000. This document, known as the Montreal Protocol, was ratified by the U.S. Senate in 1988. Although the initial concerns related to the problem of ozone depletion, efforts to reduce chlorofluorocarbons have been effective at removing a greenhouse gas as well. As a result of the 1987 Montreal Protocol, chlorofluorocarbon emissions dropped 87 percent from their peak in 1988. In 1990, in London, international agreements were reached to further reduce the use of chlorofluorocarbons. A major barrier to these negotiations was the reluctance of the developed countries of the world to establish a fund to help less-developed countries implement technologies that would allow them to obtain refrigeration and air conditioning without the use of chlorofluorocarbons. In 1991, DuPont announced the development of new refrigerants that would not harm the ozone layer. These and other alternative refrigerants are now used in refrigerators and air conditioners in many nations, including the United States. In 1996, the United States stopped producing chlorofluorocarbons. As a result of these international efforts and rapid changes in technology, the use of chlorofluorocarbons has dropped rapidly, and concentrations of chlorofluorocarbons in the atmosphere will slowly fall over the next few decades.

3.5 Global Warming and Climate Change

In recent years, scientists suspected that the average temperature of the Earth was increasing and looked for causes for the change. In 1996, Intergovernmental Panel on Climate Change (IPCC) published its Assessment and concluded that climate change is occurring and that it is highly probable that human activity is an important cause of the change. What actually causes global warming? An explanation is relatively straight forward. Several gases in the atmosphere are transparent to ultraviolet and visible light but absorb infrared radiation. These gases allow sunlight to penetrate the atmosphere and be absorbed by the Earth's surface. This sunlight energy is reradiated as infrared radiation (heat), which absorbed by the greenhouse gases in the atmosphere. Because the effect is similar to what happens in a greenhouse (the glass allows light to enter but retards the loss of heat), these gases are called greenhouse gases, and the warming thought to occur from their increase is called the greenhouse effect (see Figure 3.4). The most important greenhouse gases are carbon dioxide (CO_2), chlorofluorocarbons (primarily CCl_3F and CCl_2F_2), methane (CH_4), and nitrous oxide (N_2O).

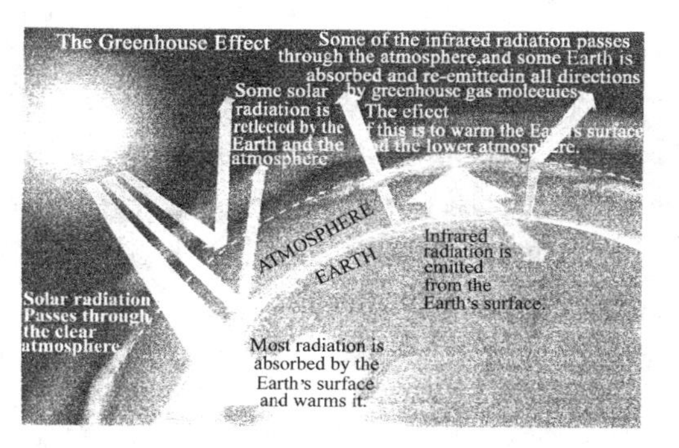

Figure 3.4 Greenhouse Effect

Carbon dioxide (CO_2) is the most abundant of the greenhouse gases. It occurs as a natural consequence of respiration. However, much larger quantities are put into the atmosphere as a waste product of energy production. Coal, oil, natural gas, and biomass are all burned to provide heat and electricity for industrial processes, home heating, and cooking. Another factor contributing to the increase in the concentration of carbon dioxide in the atmosphere is deforestation. Cutting down trees to convert forested land to other uses releases the carbon in their structure, and a reduction in the amount of forest lessens its ability to remove carbon dioxide from the atmosphere. The combination of these factors (fossil-fuel burning and deforestation) has resulted in an increase in the concentration of carbon dioxide in the atmosphere.

Chlorofluorocarbons (CFCs) are entirely the result of human activity. They were widely used as refrigerant gases in refrigerators and air conditioners, as cleaning solvents, as propellants in aerosol containers, and as expanders in foam products. Although they are present in the atmosphere in minute quantities, they are extremely efficient as greenhouse gases (about 15 000 times more efficient at retarding heat loss than is carbon dioxide). Some methane enters the atmosphere from fossil-fuel sources, but the majority comes primarily from biological sources. Several kinds of bacteria that are particularly abundant in wetlands and rice fields release methane into the atmosphere. Methane-releasing bacteria are also found in large numbers in the guts of termites and various kinds of ruminant animals such as cattle. Nitrous oxide, a minor component of the greenhouse gas picture, enters the atmosphere primarily from fossil fuels and fertilizers. It could be reduced by more careful use of nitrogen-containing fertilizers.

It is important to recognize that although a small increase in the average temperature of the Earth may seem trivial; this increase could set in motion changes that could significantly alter the climate of major regions of the world. Computer models suggest that rising temperature will lead to increased incidences of severe weather and changes in rainfall patterns that would result in more rain in some areas and drought in others. These models suggest that the magnitude and rate of change will differ from region to region. Furthermore, some natural ecosystems or human settlements will be able to withstand or adapt to the changes, while others will not.

Poorer nations are generally more vulnerable to the consequences of global warming. These nations tend to be more dependent on climate-sensitive sectors, such as subsistence agriculture, and lack the resources to buffer themselves against the changes that global warming may bring. The

Intergovernmental Panel on Climate Change has identified Africa as "the continent most vulnerable to the impacts of projected changes because widespread poverty limits adaptation capabilities." In addition to changes in weather, there are many other potential consequences of warmer temperatures and changes in climate (see Figure 3.5). These include rising sea levels, disruption of the water cycle, potential health concerns, changing forests and natural areas, and challenges to agriculture and the food supply.

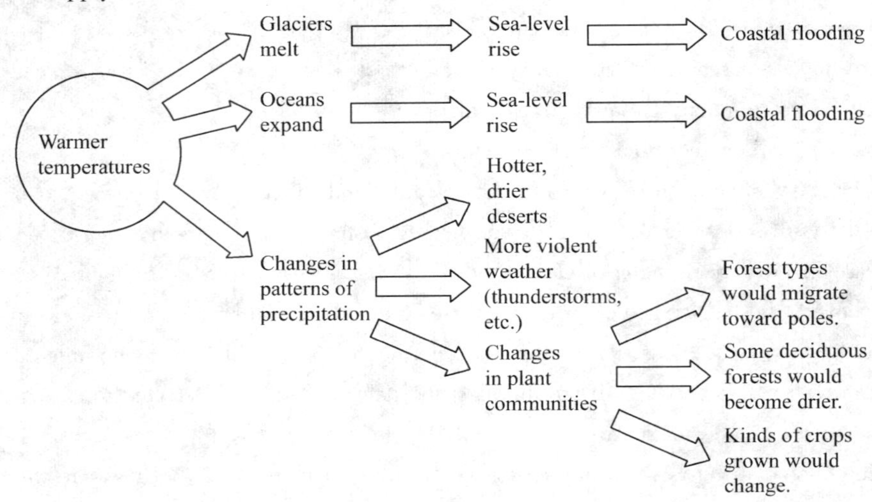

Figure 3.5　Effects of Global Warming

3.6　Air Pollution Control

There are several approaches or strategies for air pollution control. The most effective control would be to prevent the pollution from occurring in the first place. Complete source shutdown would accomplish this, but shutdown is only practical under emergency conditions, and even then it causes economic loss. Nevertheless, state public health officials can force industries to stop operations and can curtail highway traffic if an air pollution episode is imminent or occurring. A source shutdown can only offer, at best, a very temporary solution to local problems of air pollution.

Another option for air pollution control is source location in order to minimize the adverse impacts in a particular locality. Community air zoning may be included in municipal master plans, requiring power plants or industrial facilities to be located where fewer people will be affected by the pollutants. The location of these zones can be established on the basis of prevailing wind patterns and weather conditions. This option has a limitation; although local air quality may be somewhat protected, the pollutants can still be "air-mailed" to neighboring communities by the dispersion of the plumes emitted from tall chimneys or smoke stacks.

Tall smoke stacks take pollutants high into the atmosphere, allowing the processes of mixing and dispersion to dilute the contaminants, reducing pollution levels. However, "what goes up must come down," and when it comes down, the problem of air pollution returns. Acid deposition (or acid rain) is a notable example of an air pollution problem that transcends local boundaries.

An important approach for air pollution control is to encourage or require industries to make fuel substitutions or process changes. For example, making more use of solar, hydroelectric, and geothermal energy would eliminate much of the pollution caused by fossil fuel combustion at power-generating plants. Nuclear power would do the same, but other problems related to high-level radioactive waste disposal and safety remain to be solved. Also, using natural low-sulfur coal and oil would reduce SO_2 emissions from existing power-generating stations. Technology is available for treating and desulfurizing dirty fossil fuels prior to their combustion, but it is expensive. A complete change of some industrial manufacturing processes can also reduce air pollution; one example is the use of electric furnaces instead of open-hearth furnaces in the steel industry.

Fuel substitutions are also effective in reducing pollution from mobile sources. For example, the use of reformulated gasolines or alternative fuels such as liquefied petroleum gas, compressed natural gas, or methanol for highway vehicles would help to clear the air. Ultimately, complete replacement of gasoline-powered vehicles with electric-powered vehicles may eliminate one of the major sources of ambient air pollution. California was the first state to require that, by the year 1998, 2 percent of new car sales be zero-emission vehicles (that is, electric cars). The requirement increases to 10 percent by the year 2003. Other states, including Massachusetts and New York, have adopted similar quotas and timetables.

The use of correct operation and maintenance practices is important for minimizing air pollution and should not be overlooked as an effective control strategy. For example, if a power plant operator allows too much air into the boiler furnace, fly ash emissions will increase. Adding too much sulfur at a sulfuric acid manufacturing plant, without providing enough air, can cause excessive SO_2 emissions. Even a failure to properly lubricate a fan motor at an incinerator can lead to unnecessary pollution. Finally, one of the most important strategies for controlling emissions from mobile sources is an effective motor vehicle inspection and maintenance program.

3.6.1 Control of Primary Particles

The optimum strategy for air quality protection is to reduce the amount of pollution at its source, primarily by fuel substitutions or process changes. When this is not possible or is simply insufficient to accomplish the goal, some type of air cleaning equipment must be installed at the source. Several types of air cleaning devices can collect or trap air pollutants before they are emitted into the atmosphere. Some of these devices serve to control only suspended particulates and others control only gaseous pollutants. The design or selection of a particular type of air cleaning apparatus depends on the physical and chemical properties of the pollutant to be removed as well as on temperature, corrosivity, and other characteristics of the pollutant and the carrier gas.

1. Wall Collection Devices

The first three types of control devices we consider—gravity settlers, cyclone separators, and electrostatic precipitators—all function by driving the particles to a solid wall, where they adhere to each other to form agglomerates that can be removed from the collection device and disposed of. Although these devices look different from one another, they all use the same general idea and are described by the same general design equations.

(1) Gravity Settlers

A gravity settler is simply a long chamber through which the contaminated gas passes slowly, allowing time for the particles to settle by gravity to the bottom. It is an old, unsophisticated device that must be cleaned manually at regular intervals. But it is simple to construct, requires little maintenance, and has some use in industries treating very dirty gases, e. g., some smelters and metallurgical processes. Furthermore, the mathematical analysis for gravity settlers is very easy; it will reappear in modified form for cyclones and electrostatic precipitators.

(2) Centrifugal Separators

At even modest velocities and common radii, the centrifugal forces acting on particles can be two orders of magnitude larger than the gravity forces. For this reason centrifugal particle separators are much more useful than gravity settlers. The most successful type is sketched in Figure 3.6 called a cyclone separator, or simply a cyclone. It is probably the most widely used particle collection device in the world. In any industrial district of any city, a sharp-eyed student can find at least a dozen of these outside various industrial plants.

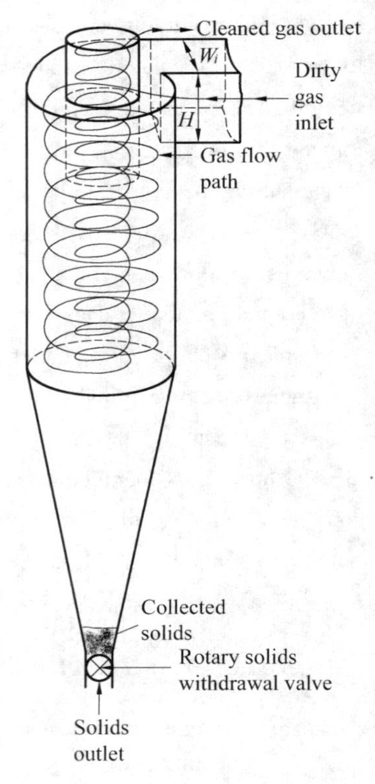

Figure 3.6　Schematic of a cyclone separator

A cyclone consists of a vertical cylindrical body, with a dust outlet at the conical bottom. The gas enters through a rectangular inlet, normally twice as high as it is wide, arranged tangentially to the circular body of the cyclone, so that the entering gas flows around the circumference of the cylindrical body, not radially inward. The gas spirals around the outer part of the cylindrical body with a downward component, then turns and spirals upward, leaving through the outlet at the top of the device. During the outer spiral of the gas the particles are driven to the wall by centrifugal force, where they collect, attach to each other, and form larger agglomerates that slide down the wall by gravity and collect in the dust hopper in the bottom.

There are many other variants on the centrifugal collector idea, but none approaches the cyclone in breadth of application. These devices are simple and almost maintenance-free. Because any medium-sized welding shop can make one, the big suppliers of pollution control equipment, who have test data on the effects of small changes in the internal geometry, have been unwilling to make these data public. The same basic device as the cyclone separator is used in other industrial settings

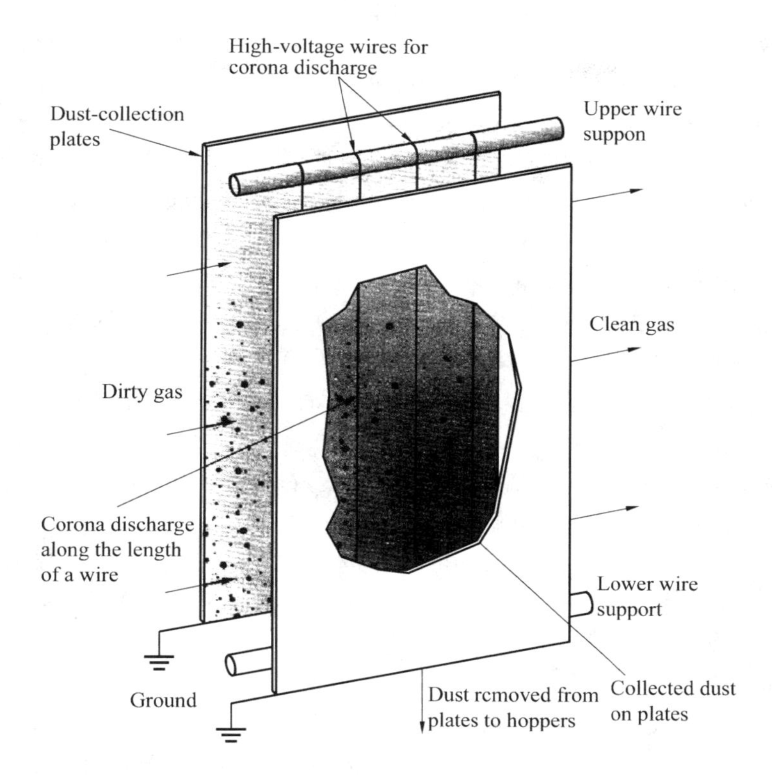

Figure 3.7 Diagrammatic sketch of a simplified ESP with two plates, four wires, and one flow channel

where the goal is not air pollution control, but some other kind of separation. When it is used to separate solids from liquids it is generally called a hydroclone. A cyclone called an air-swept classifier is attached to many industrial grinders. It passes those particles ground fine enough, and collects those that are too coarse, returning them to the grinder.

(3) Electrostatic Precipitators (ESP)

The electrostatic precipitator (ESP) is like a gravity settler or centrifugal separator, but electrostatic force drives the particles to the wall. It is effective on much smaller particles than the previous two devices. The basic idea of all ESPs is to give the particles an electrostatic charge and then put them in an electrostatic field that drives them to a collecting wall. In one type of ESP, called a two-stage precipitator, charging and collecting are carried out in separate parts of the ESP This type, widely used in building air conditioners, is sometimes called an electronic air filter. However, for most industrial applications the two separate steps are carried out simultaneously in the same part of the ESP. The charging function is done much more quickly than the collecting function, and the size of the ESP is largely determined by the collecting function.

Figure 3.7 shows in simplified form a wire-and-plate ESP with two plates. The gas passes between the plates, which are electrically grounded (i.e., voltage = 0). Between the plates are rows of wires, held at a voltage of typically -40 000 volts. The power is obtained by transforming ordinary alternating current to a high voltage and then rectifying it through some kind of solid-state rectifier. This combination of charged wires and grounded plates produces both the free electrons to charge the particles and the field to drive them against the plates. On the plates the particles lose their charge and adhere to each other and the plate, forming a "cake." The cleaned gas then passes out the far side of the precipitator. Solid cakes are removed by rapping the plates at regular time intervals with a mechanical or electromagnetic rapper that strikes a vertical or horizontal blow on the edge of the plate. Through science, art, and experience designers have learned to make rappers that cause most of the collected cake to fall into hoppers below the plates. Some of the cake is always re-entrained, thereby lowering the efficiency of the system. If the collected particles are liquid, e.g., sulfuric acid mist, they run down the plate and drip off. For liquid droplets the plate is often replaced by a circular pipe with the wire down its center. Some ESPs (mostly the circular pipe variety) have a film of water flowing down the collecting surface, to carry the collected particles to the bottom without rapping.

The ESP industry is now well established. Standard package units are available for small flows (down to the size of home air conditioners), and large power plants have precipitators costing up to $30 million.

2. Dividing Collection Devices

Gravity settlers, cyclones, and ESPs collect particles by driving them against a solid wall. Filters and scrubbers do not drive the particles to a wall, but rather divide the flow into smaller parts where they can collect the particles. The public often refers to any kind of pollution control device as a filter, giving the word filter the meaning "cleaning device." Technically, a filter is one of the devices described in this section. Other devices are not truly filters.

(1) Surface Filters

Most of us have personal experience with surface filters, which is a membrane (sheet steel, cloth, wire mesh, or filter paper) with holes smaller than the dimensions of the particles to be retained. Although this kind of filter is sometimes used for air pollution control purposes, it is not common because constructing a filter with holes as small as many of the particles we wish to collect is very difficult. The industrial air filters rarely have holes smaller than the smallest particles captured; they often act as if they did. The reason is that, as fine particles are caught on the sides of the holes of a filter, they tend to bridge over the holes and make them smaller. Thus as the amount of collected particles increases, the cake of collected material becomes the filter, and the filter medium (usually a cloth) that originally served as a filter to collect the cake now serves only to support the cake, and no longer as a filter. This cake of collected particles will have average pore sizes smaller than the diameter of the particles in the oncoming gas stream, and thus will act as a sieve for them. The particles collect on the front surface of the growing cake. For that reason this is called a surface filter.

The theory of cake accumulation and pressure drop for this type of device is well-known from industrial filtration. The two most widely used designs of industrial surface filters are shown in

Figure 3.8 and 3.9. Because the enclosing sheet metal structure in both figures is normally the size and roughly the shape of a house, this type of gas filter is generally called a baghouse. The design in Figure 3.8, most often called a shake-deflate filter, consists of a large number of cylindrical cloth bags that are closed at the top like a giant stocking, toe upward. These are hung from a support. Their lower ends slip over and are clamped onto cylindrical sleeves that project upward from a plate at the bottom. The dirty gas flows into the space below this plate and up inside the bags. The gas flows outward through the bags, leaving its solids behind. The clean gas then flows into the space outside the bags and is ducted to the exhaust stack or to some further processing.

For the baghouse in Figure 3.8 there must be some way of removing the cake of particles that accumulates on the filters. When the gas flow has been switched off, the bags are shaken by the support to loosen the collected cake. A weak flow of gas in the reverse direction may also be added to help dislodge the cake, thus deflating the bags. The cake falls into the hopper at the bottom of the baghouse and is collected or disposed of in some way. Typically, for a major continuous source like a power plant, about five baghouses will be used in parallel, with four operating as gas cleaners during the time that the other one is being shaken and cleaned. Each baghouse might operate for two hours and then be cleaned for 10 minutes; at all times one baghouse would be out of service for cleaning or waiting to be put back into service. Thus the baghouse must be sized so that four of them operating together provide adequate capacity for the expected gas flow rate.

The other widely used baghouse design, called a pulse-jet filter, is shown in Figure 3.9. In it the flow during filtration is inward through the bags, which are similar to the bags in Figure 3.8 except their ends open at the top. The bags are supported by internal wire cages to prevent their collapse. The bags are cleaned by intermittent jets of compressed air that flow into the inside of the bag to blow the cake off. Often these baghouses are cleaned while they are in service; the internal pulse causes much of the collected solids to fall to the hopper, but some are drawn back to the filter cloth. Just after the cleaning the control efficiency will be less than just before the next cleaning, but the average efficiency meets the legal control requirements.

The flow velocities through such filters are very low, typically a few feet per minute. In contrast, in devices like cyclones the flow is about 60 feet per second. A wind velocity equal to the typical flow through such a filter is so low that someone standing in it could not tell in which direction it was blowing and would report that there was no wind at all.

(2) Depth Filters

Another class of filters, widely used for air pollution control, collects particles throughout the entire filter body, are called depth filtters, in which a mass of randomly oriented fibers (not woven to form a single surface) collects particles as the gas passes through it. The filters are often used where the particles to be caught are fine drops of liquids that are only moderately viscous. Such drops will coalesce on the fibers and then run off as larger drops, leaving the fibers ready to catch more fine drops. If the particles were solid, then this type of filter would require regular cleaning; for the liquid application it does not. The most widespread air pollution control use of depth filters is

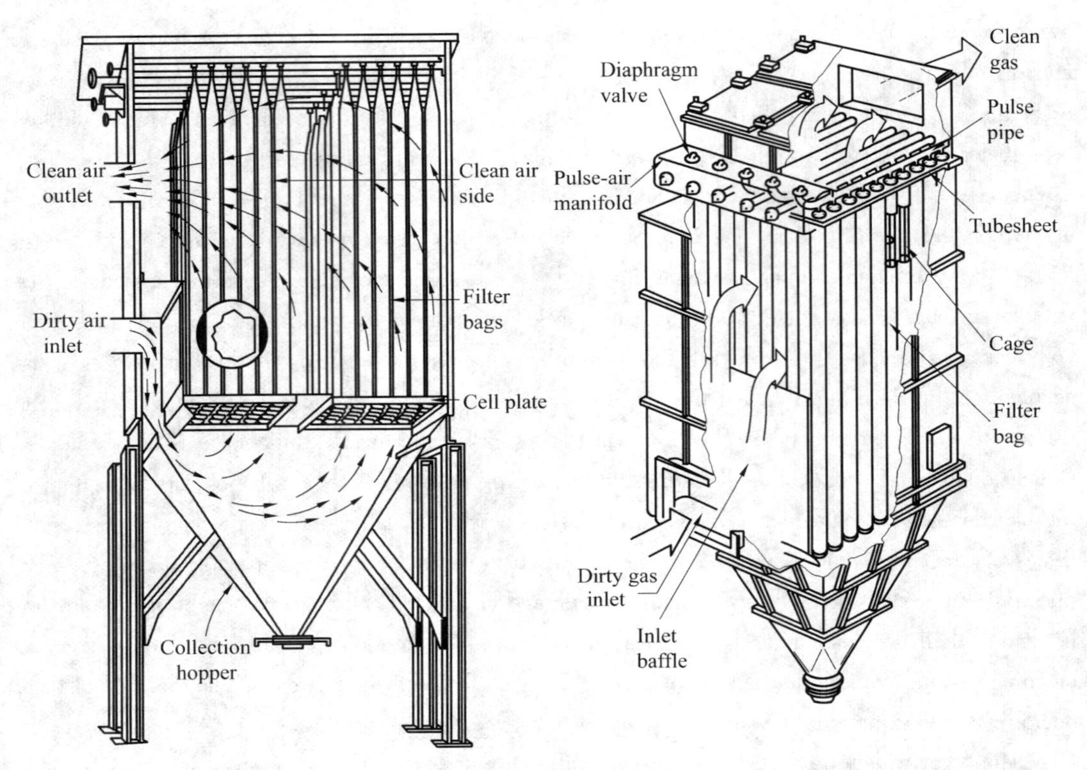

Figure 3.8 Typical industrial baghouse of the shake-deflate design

Figure 3.9 Typical industrial baghouse of the pulse-jet design

in the collection of very fine liquid drops, sulfuric acid mist, produced in sulfuric acid plants. This kind of device is also used for removing solid particles from gas streams that contain few of them, e.g., for cleaning the air of industrial clean rooms or hospital surgical suites and in personal protection dust masks. The filters are thrown away when they have collected enough particles that their pressure drop begins to increase. The depth filters used in those applications are normally called high-efficiency, particle-arresting (HEPA) or absolute filters. The air filters on household furnaces operate this way as well; typically the fibers are coated with a sticky substance to improve the retention of the collected dust and lint.

Depth filters collect particles mostly by impaction. Some older types of particle collectors also used impaction, to catch particles on solid walls, but they are seldom used now. Some size-specific particle analyzers (impactors or cascade impactors) use impaction on collecting surfaces to collect specific sizes of particles.

(3) Filter Media

Whether a filter behaves as a surface or a depth filter depends on the type of filter medium used. For shake-deflate baghouses the filter bags are made of tightly woven fibers, much like those in a pair of jeans. Pulse-jet baghouses use high-strength felted fabrics, so that they act partly as depth filters and partly as surface filters. This allows them to operate at superficial velocities (air-to-cloth ratios) two to four times those of shake-deflate baghouses; in recent years this higher capacity per unit size has allowed them to take market share away from the previously dominant shake-deflate type baghouses.

Filter fabrics are made of cotton, wool, glass fibers, and a variety of synthetic fibers. The choice depends on price and suitability for the expected service. Cotton and wool cannot be used above 180 and 200 °F, respectively, without rapid deterioration, whereas glass can be used to 500 °F (and short-term excursions to 550 °F). The synthetics have intermediate service temperatures. In addition the fibers must be resistant to acids or alkalis if these are present in the gas stream or the particles as well as to flexing wear caused by the repeated cleaning. Typical bag service life is 3 to 5 years. Generally fibers that have many small microfibers sticking out their sides form better cakes than those that do not.

3.6.2 Control of Volatile Organic Compounds (VOCs)

Volatile organic compounds (VOCs) are liquids or solids that contain organic carbon (carbon bonded to carbon, hydrogen, nitrogen, or sulfur, but not carbonate carbon as in $CaCO_3$ nor carbide carbon as in CaC_2 or CO or CO_2), which vaporize at significant rates. VOCs are probably the second-most widespread and diverse class of emissions after particulates. VOCs are a large family of compounds. Some (e.g., benzene) are toxic and carcinogenic, and are regulated individually as hazardous pollutants. Most VOCs are believed not to be toxic (or not very toxic) to humans. Our principal concern with VOCs is that they participate in the "smog" reaction and also in the formation of secondary particles in the atmosphere. Some VOCs are powerful infrared absorbers and thus contribute to the problem of global warning.

1. Control by Prevention

If possible, we prevent the formation of a VOC-containing air or gas stream, which we must treat by some kind of tailpipe control device. The ways of doing this for VOCs are substitution, process modification, and leakage control.

(1) Substitution and Process Modification

Oil-based paints, coatings, and inks harden by the evaporation of VOC solvents such as paint thinner into the atmosphere. Water-based paints are concentrated oil-based paints, emulsified in water. After the water evaporates, the small amount of organic solvent in the remaining paint must also evaporate for the paint to harden. Switching from oil- to water-based paints, coatings, and inks greatly reduces but does not totally eliminate the emissions of VOCs from painting, coating, or printing. For many applications, e.g., house paint, the water-based paints seem just as good as oil-based paints. But water-based paints have not yet been developed that can produce auto body finishes as bright, smooth, and durable as the high-performance oil-based paints and coatings now used. There are numerous other examples where a less volatile or nonvolatile solvent can be substituted for the more volatile one. This replacement normally reduces but does not eliminate the emission of VOCs. In addition, a less toxic solvent can often be substituted for a more toxic one, although the more toxic solvents often have special solvent properties that are hard to replace.

Replacing gasoline as a motor fuel with compressed natural gas or propane is also a form of substitution that reduces the emissions of VOCs, because those fuels can be handled, metered, and burned with fewer VOC emissions than can gasoline. The petroleum industry is working hard to improve the burning properties, handling, and use of gasoline, to make it as low-emission a fuel as compressed natural gas and propane, so that gasoline can keep its dominant position in the auto fuel market.

Process modification to prevent or reduce the formation of the VOC stream may be more economical than applying the control options discussed below. Often substitution and process modification are indistinguishable. Replacing gasoline-powered vehicles with electric-powered vehicles is a form of process modification that reduces the emissions of VOCs, as well as emission of carbon monoxide and nitrogen oxides, in the place the vehicle is. On the other hand, it causes other emissions where the electricity is generated. If we consider the process as "get workers from their homes to their place of employment", then improved public transport, mandatory ride pools, etc., are modifications of the process that reduce emissions of VOCs (and of CO and NO_x).

Many coating, finishing, and decoration processes that at one time depended on evaporating solvents have been replaced by others that do not, e.g., fluidized-bed powder coating and ultraviolet lithography. Finding alternatives to VOC solvents and fuels can be difficult, but it is often the most cost-effective way to reduce VOC emissions.

(2) Leakage Control

Many small emissions of VOCs occur as leaks at seals. In recent years these have come under regulatory control because, as the larger sources are controlled, these become a more significant part of the remaining problem.

Figure 3.10 shows three kinds of seals. Figure 3.10(a) shows a static seal, as exists

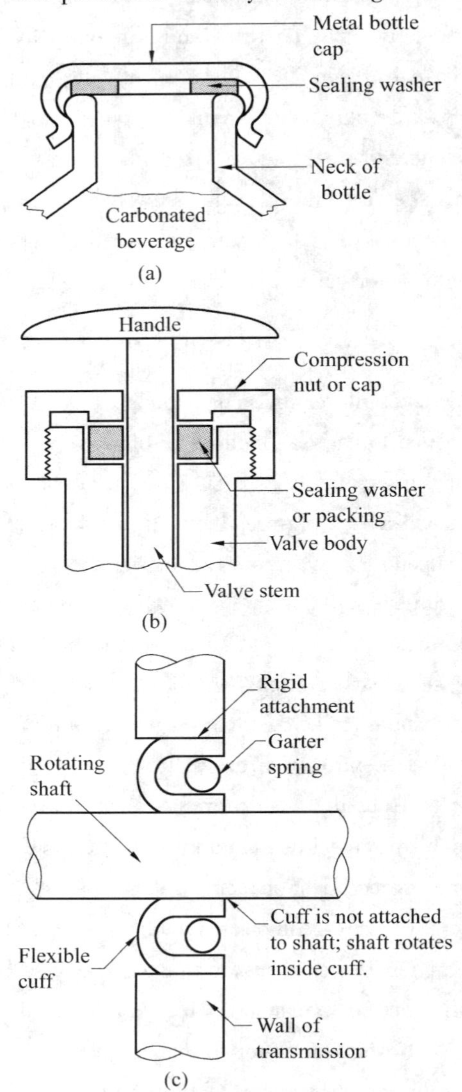

Figure 3.10 Three kinds of seals: (a) a static seal, as exists between a carbonated beverage bottle and its bottle caps; (b) a packed seal, as exists between the valve stem and valve body of simple faucet, and as also exists on many simple pumps; (c) a rotary seal of the type common on the drive shafts of automobiles and some pumps

between the bottle cap and the top of a bottle of carbonated beverage. A thin washer of elastomeric material is compressed between the metal cap and the glass bottle top. This compressed material forms a seal that prevents the escape of CO_2 (carbonation), often for many years. Leaks through this kind of seal are generally unimportant. Sealing is more difficult when one of the two surfaces involved in the seal moves relative to the other. Figure 3.10(b) shows a simple compression seal between a housing and a shaft. The example shown is a water faucet, in which a nut screws down over the body of the faucet to compress an elastomeric seal that is trapped between the body of the faucet and the stem of the valve. The compressed seal must be tight enough to prevent leakage of the high-pressure water inside the valve out along the edge of the stem, but not so tight that the valve cannot be easily rotated by hand. Students are probably aware from personal experience that this type of seal often leaks. If the leak is a small amount of water into the bathroom sink, that causes little problem; tightening the nut normally reduces the leak to a rate low enough that it becomes invisible (but does not become zero). Figure 3.10(c) shows in greatly simplified form the seal that surrounds the drive shaft of an automobile at the point where the shaft exits from the transmission. The inside of the transmission is filled with oil. The elastomeric seal is like a shirt cuff turned back on itself with the outside held solidly to the wall of the transmission and the inside held loosely against the rotating shaft by a garter spring. If we set that spring loosely, then there will be a great deal of leakage. If we set it tightly, then the friction and wear between the cuff and the shaft that rotates inside it will be excessive. Setting the tension on that spring requires a compromise between the desire for low leakage and the desire for low friction and wear. That compromise normally leads to a low, but not a zero, leakage rate; a small amount of oil is always dripping out, and accumulating on the floor of our garages. Valves and pumps also have shafts that must rotate, and hence they have the same kind of leakage problem.

All of the pumps and valves in facilities that process VOCs have this same kind of leakage problem. The seals regularly used are more complex versions of types b and c in Figure 3.10. There is considerable regulatory pressure for the seals to be made more and more leak-tight. Mostly this goal will be accomplished by replacing simple, low-quality seals on pumps and valves with more complex and expensive, higher-quality seals.

2. Control by Concentration and Recovery

Adsorption means the attachment of molecules to the surface of a solid. In contrast, absorption means the dissolution of molecules within a collecting medium, which may be liquid or solid. Generally, absorbed materials are dissolved into the absorbent, like sugar dissolved in water, whereas adsorbed materials are attached onto the surface of a material, like dust on a wall. Absorption mostly occurs into liquids, adsorption mostly onto solids. The most widely used adsorbent for VOCs is activated carbon.

(1) Adsorption

Adsorption is mostly used in air pollution control to concentrate a pollutant that is present in dilute form in an air or gas stream. The material collected is most often a VOC like gasoline or various paint thinners and solvents. The solid is most often some kind of activated carbon. For large-scale air pollution applications, like collecting the solvent vapors coming off a large paint-drying oven or a large printing press, the normal procedure is to use several adsorption beds. As shown in Figure 3.11, the contaminated air stream passes through two vessels in series. Inside each of the vessels is a bed of adsorbent that removes the VOCs. From the second vessel the cleaned air, normally containing at most a few parts per million of VOCs, passes to the atmosphere. Meanwhile, a third vessel is being regenerated. Steam passes through it, removing the adsorbed VOCs from the adsorbent. The mixture of steam and VOCs coming from the top of the vessel passes to a water-cooled condenser that condenses both the VOCs and the steam. Both pass in liquid form to a separator, where the VOCs, which are normally much less dense than water and have little solubility in water, float on top and are decanted and sent to solvent recovery.

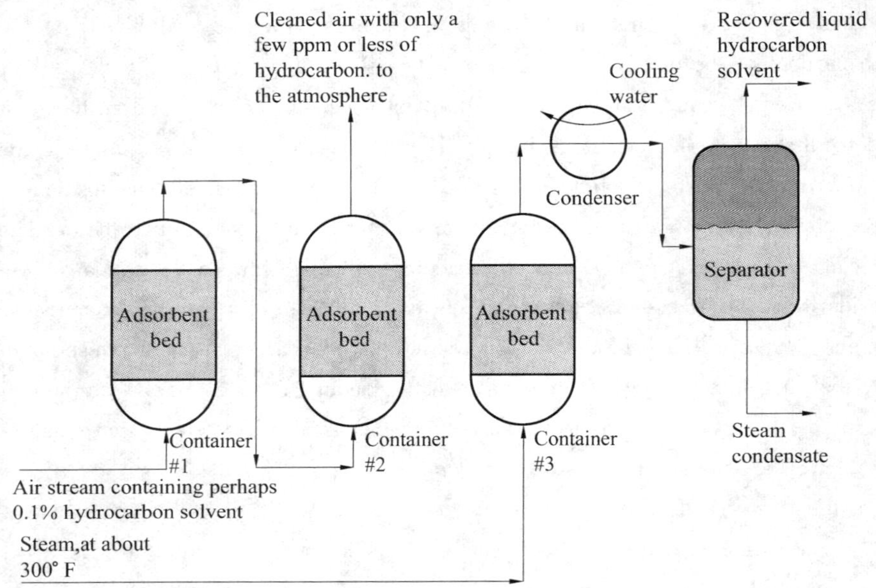

Figure 3.11 The typical arrangement for adsorption of a VOC from a gas stream, using three adsorbed beds; automatic switching valves; and steam desorption, condensation, and gravity separation

After a suitable time period a set of automatically programmed valves changes the position of the containers in the flow sheet. (The containers do not move; their place in the piping arrangement changes.) Container 1, which is most heavily loaded, goes to the regeneration position. Container 2, which is lightly loaded with VOCs, goes to the position where container 1 was; and container 3, which is now regenerated and very clean, goes to the position previously held by container 2, making the final cleanup on the air stream. Figure 3.11 shows the steam condensate leaving the separator, without specifying where it goes. As discussed above, this condensate will be saturated with dissolved VOC. The VOC concentration may be high enough to prevent its being sent back to the steam boiler, or for it to be discharged to a sewer. If there is no good way to deal with this stream, then the absorber solves a large air pollution problem but creates a small water pollution problem.

(2) Absorption (Scrubbing)

If we can find a liquid solvent in which the VOC is soluble and in which the remainder of the contaminated gas stream is insoluble, then we can use absorption to remove and concentrate the VOC for recovery and re-use, or destruction. The standard chemical engineering method of removing any component from a gas stream—absorption and stripping—is sketched in Figure 3.12. If we can find a liquid solvent in which the gaseous component we wish to selectively remove is much more soluble than are the other components in the gas stream, the procedure is quite straightforward. The feed gas enters the absorber, which is a vertical column in which the gas passes upward and the liquid solvent passes downward. Normally, bubble caps, sieve trays, or packing is used in the interior of the column to promote good countercurrent contact between the solvent and the gas. The stripped solvent enters the top of the column and flows countercurrent to the gas. By the time the gas has reached the top of the column, most of the component we wish to remove has been dissolved into the solvent; the cleaned gas passes on to the atmosphere or to its further uses. The loaded solvent, which now contains most of the component we are removing from the gas, passes to the stripper, which normally is operated at a higher temperature and/or a lower pressure than the absorber. At this higher temperature and/or lower pressure, the solubility of the gas in the selective solvent is greatly reduced so the gas comes out of solution. In Figure 3.12 the separated component is shown leaving as a gas for use, sale, or destruction. In some cases it is condensed and leaves as a liquid. The stripped or lean solvent is sent back to the absorber column. Very large absorption-stripping systems often use tray columns, but the small ones used in most air pollution control applications use internal packings.

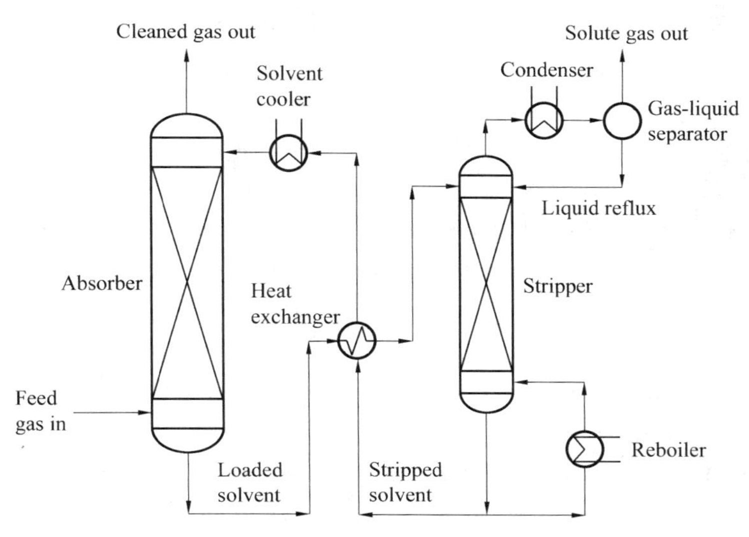

Figure 3.12 The flow diagram for the most common method for removing one component from a gas stream

Functionally, this is the same as the adsorption process sketched in Figure 3.11. The chosen component is selectively removed from the gas stream onto an adsorbent or into an absorbent in one vessel and is subsequently removed at much higher concentration (often practically pure) in another vessel at a higher temperature and/or lower pressure. The absorption-stripping scheme in Figure 3.12 is mechanically simpler because it is easy to move liquids with pumps and pipes. It is much harder to move solids the same way. The adsorption equivalents of Figure 3.12 have been tried, but the mechanical difficulties have been severe enough that most adsorption is done with the solids remaining in place as shown in Figure 3.11, using a semisteady-state operation.

The absorption solvent must have the following properties: ① It must afford reasonable solubility for the material to be removed, and, if this material is to be recovered at reasonable purity, it must not dissolve and thus carry along any of the other components of the gas stream. ② In the absorber, the gas being treated will come to equilibrium with the stripped solvent. The vapor pressure of the solvent, at absorber temperature, must be low enough that if the cleaned gas is to be discharged to the atmosphere, the emission of solvent is small enough to be permissible. Some solvent is lost this way; the cost of replacing it must be acceptable. If the solvent is water this is not a problem (unless we need the gas to be dry for its next use), but for other solvents this can be a problem. ③ At the higher temperature (or lower pressure) of the stripping column, the absorbed material must come out of solution easily, and the vapor pressure of the solvent must be low enough that it does not contaminate the recovered VOC. If the solvent vapor pressure in the stripper is too large, one may replace the stripper by a standard distillation column (combination stripper and rectifier) to recover the transferred material at adequate purity. ④ The solvent must be stable at the conditions in the absorber and stripper, and be usable for a considerable time before replacement. ⑤ The solvent molecular weight should be as low as possible, to maximize its ability to absorb. This requirement conflicts with the low solvent vapor pressure requirement, so that a compromise must be made.

3. Control by Oxidation

The final fate of VOCs is mostly to be oxidized to CO_2 and H_2O, as a fuel either in our engines or furnaces, in an incinerator, in a biological treatment device, or in the atmosphere (forming ozone and fine particles). VOC-containing gas streams that are too concentrated to be discharged to the atmosphere but not large enough to be concentrated and recovered are oxidized before discharge, either at high temperatures in an incinerator or at low temperatures by biological oxidation.

(1) Combustion

One of the first major undertakings in the history of air pollution control was the control of emissions from coal-burning boilers, furnaces, etc. Unburned coal or products of incomplete combustion formed a substantial part of these emissions. These were one of the easiest pollutants to control; all that is required for good control is sufficient excess air and adequate mixing between the burning coal, its decomposition products, and the air.

To get complete combustion with imperfect mixing, one must supply excess air in addition to that needed for stoichiometric combustion. The amount of excess air to be used is determined by economics. At zero excess air, some valuable fuel escapes unburned to pollute the atmosphere. Large amounts of excess air lower the combustion temperature by diluting the combustion products, and carry away more heat in the exhaust gas. This lowers the furnace's efficiency (fraction of heating value of the fuel transferred to whatever is being heated). Large industrial furnaces operate with 5 to 30 percent excess air. Autos have variable excess air, depending on engine load. The optimum amount of excess air for VOC destruction is generally higher than the optimum for fuel efficiency; air pollution control officers try to induce furnace operators to use the optimum amount for VOC destruction.

The mixing problem is especially difficult in flares. These are safety devices used in oil refineries and many other processing plants. All vessels containing fluids under pressure have high-pressure relief valves that open if the internal pressure of the vessel exceeds its safe operating value. All household water heaters have such a valve to prevent tank rapture in some unlikely but not impossible circumstances. With a hot water heater, if the valve opens, hot water drops onto the floor. In the case of a large petroleum-processing vessel (distillation column, cracker, isomerizer, etc.) the material released is an inflammable VOC, which cannot safely be dropped on the floor. The outlets of a refinery's relief valves are piped to a flare (or "flare stack"), which is an elevated pipe with pilot lights to ignite any released VOCs. Many have steam jets running constantly to mix air into the gas being released. These flares handle significant amounts of VOCs only during process upsets and emergencies at the facilities they serve. When there is a small release, the steam jets can often mix the gas and air well enough that there is practically complete combustion. For a large release the mixing is inadequate, and the large, bright orange, smoky flame from the flare indicates a significant release of unburned or partly burned VOC.

In the coal combustion process one difficulty, even in well-designed modern furnaces, is that some particles of coal and some hydrocarbons pass out of the flame zone before they can be combusted. These are called soot. In modem steam boilers this soot will collect in parts of the furnace where it is too cold for soot to burn, typically on the tubes in which the water is boiled or the steam superheated. If soot is allowed to collect there, it will impede heat transfer and make the boiler less efficient. The cure for this problem is a soot blower, which is typically a fixed or moving steam jet that blows high-pressure steam onto the surface of the tubes to remove this soot. Normally, soot blowing is required only a few minutes per day. Soot dislodged in this way exits the furnace as short-period emissions of black smoke. Most public relations officers ask plant engineers to do all soot blowing at night.

(2) Biological Oxidation (Biofiltration)

As discussed above, the ultimate fate of VOCs is to be oxidized to CO_2 and H_2O, either in our engines or furnaces, or incinerators, or in the environment. Many microorganisms will carry out these reactions fairly quickly at room temperature. They form the basis of most sewage treatment plants (oxidizing more complex organic materials than the simple VOCs of air pollution interest). Microorganisms can also oxidize the VOCs contained in gas or air streams. The typical biofilter (not truly a filter but commonly called one; better called a highly porous biochemical reactor) consists of the equivalent of a swimming pool, with a set of gas distributor pipes at the bottom, covered with several feet of soil or compost or loam in which the microorganisms live. The contaminated gas enters through the distributor pipes and flows slowly up through soil, allowing time for the VOC to dissolve in the water contained in the soil, and then to be oxidized by the microorganisms that live there.

Typically these devices have soil depths of 3 to 4 ft, void volumes of 50%, upward gas velocities of 0.005 to 0.5 ft/s, and gas residence times of 15 to 60 s. They work much better with polar VOCs, which are fairly soluble in water than with HCs whose solubility is much less. The microorganisms must be kept moist, protected from conditions that could injure them, and in some cases given nutrients. Because of the long time the gases must spend in them, these devices are much larger and take up much more ground surface than any of the other devices discussed in this section. In spite of these drawbacks, there are some applications for which they are economical, and for which they are used industrially.

3.6.3 Control of Sulfur Oxides

The control of particulates and VOCs is mostly accomplished by physical processes (cyclones, ESPs, filters, leakage control, vapor capture, condensation) that do not involve changing the chemical nature of the pollutant. Some particles and VOCs are chemically changed into harmless materials by combustion. The sulfur oxides and nitrogen oxides that cannot be economically collected by physical means nor rendered harmless by combustion. Their control is largely chemical rather than physical. Sulfur and nitrogen oxides are ubiquitous pollutants, which have many sources. SO_2, SO_3, and NO_2 are strong respiratory irritants that can cause health damage at high concentrations. Gases of NO_2 and SO_2 also form secondary particles in the atmosphere, contributing to our PM_{10} and $PM_{2.5}$ problems and impairing visibility. They are the principal causes of acid rain.

1. The Elementary Oxidation-Reduction Chemistry of Sulfur and Nitrogen

Both sulfur and nitrogen in the elemental state are relatively inert and harmless to humans. Both are needed for life; all animals require some N and S in their bodies. However, the oxides of sulfur and nitrogen are widely recognized air pollutants. The reduced products also are, in some cases, air pollutants. Reduction means the addition of hydrogen or the removal of oxygen. If we reduce nitrogen, we produce ammonia. Similarly, if we reduce sulfur, we produce hydrogen sulfide. Both hydrogen sulfide and ammonia are very strong-smelling substances, gaseous at room temperature (−60 ℃ and −33 ℃ boiling points, respectively), and toxic in high concentrations. Neither ammonia nor hydrogen sulfide has been shown to be toxic in the low concentrations that normally exist in the atmosphere.

When nitrogen is oxidized, nitric oxide (NO) and then nitrogen dioxide (NO_2) form; likewise, sulfur forms sulfur dioxide (SO_2) and then sulfur trioxide (SO_3). These are all gases at room temperature or slightly above room temperature (boiling points 21℃, 34℃, −10℃, and 45℃, respectively). The oxides have higher boiling points than the hydrides. Both nitrogen and sulfur can also form other oxides, but these are the ones of principal air pollution interest. In the atmosphere NO_2 and SO_3 react with water to form nitric and sulfuric acids, which then react with ammonia or any other available cation to form particles of ammonium nitrate or sulfate or some other nitrate or sulfate. They are significant contributors to urban PM_{10} and $PM_{2.5}$ problems. They are the principal causes of acid deposition and of visibility impairment in our national parks. NO and NO_2 also play a significant role in the formation of O_3.

2. The Removal of Reduced Sulfur Compounds from Petroleum and Natural Gas Streams

When the sulfur is present in reduced form, it can be removed from gas streams which occur in many natural gas deposits and in many by-product gases produced in oil refining and in the fuel gases produced by coal gasification. To make the system practical, one must find a solvent that can absorb much H_2S. Fortunately, for many of the gases of air pollution and industrial interest, we can do that. H_2S, SO_2, SO_3, NO_2, HCI, and CO_2 are acid gases, which form acids by dissolving in water. For H_2S the process is

$$H_2S\ (\text{gas}) \rightleftharpoons H_2S\ (\text{dissolved in water}) \rightleftharpoons H^+ + HS^- \tag{3.1}$$

If we can add something to the scrubbing solution that will consume either the H^+ or the HS^-, then more H_2S can dissolve in the water, and much less water is needed. For acid gases, the obvious choice is some alkali, a source of OH^- that can remove the H^+ by

$$H^+ + OH^- \rightleftharpoons H_2O \tag{3.2}$$

Removing the H^+ on the fight side of Equation(3.1) drives the equilibrium to the fight, greatly increasing the amount of H_2S absorbed.

(1) The Uses and Limitations of Absorbers and Strippers for Air Pollution Control

Absorber-stripper combinations are widely used to remove HCs from exhaust gas streams. This example shows that the removal of H_2S from natural gas and similar streams is simple and straightforward. The system also works extremely well for removing ammonia from a gas stream, because NH_3 is very soluble in water or in weak acids, forming a weak alkali by the following reaction:

$$NH_3 + H_2O \rightarrow NH_4^+ + OH^- \tag{3.3}$$

It is possible to make practically complete removal of NH_3 from gas streams with water or weak acids. The solubility of ammonia is so high that generally the simplest possible forms of this arrangement are satisfactory.

To remove SO_2 from gas streams by this method is also relatively easy if there are no other acid gases present. For example, SO_2 could be easily removed from N_2 by the scheme using any weak alkali (for example, ammonium hydroxide), and the solution would be easily regenerated to produce pure SO_2. The problem of removing sulfur dioxide from combustion gases is much more complex and difficult.

NO and NO_2 are not readily removed from gas streams by the process shown in Figure 3.11. Although NO_2 is an acid gas that produces nitric acid by reaction with water,

$$3NO_2 + H_2O \rightarrow 2HNO_3 + NO \tag{3.4}$$

the reaction rate is slow. NO is not an acid gas, so that although we can remove NO_2 from a gas stream with an alkaline solvent, we cannot remove NO with the same solvent. For this reason, weak alkali solvents are not successful for the joint removal of NO and NO_2 or for the rapid removal of NO_2 alone. No other solvent is known that serves well for this task. The scheme in Figure 3.12 is widely used in the chemical and petroleum industries to make separations not directly related to pollution control, e.g., the separation of CO_2 from H_2. The absorption column can also be used without regenerating the absorbent solution if the amount of material to be collected is small and there is some acceptable way of disposing of the loaded absorbent.

(2) Sulfur Removal from Hydrocarbons

Once H_2S has been separated from the other components of the gas, it is normally reacted with oxygen from the air in controlled amounts to oxidize it only as far as elemental sulfur,

$$H_2S + \frac{1}{2}O_2 \rightarrow S + H_2O \tag{3.5}$$

and not as far as SO_2,

$$H_2S + \frac{3}{2}O_2 \rightarrow SO_2 + H_2O \tag{3.6}$$

The elemental sulfur is either sold for use in the production of sulfuric acid or land filled if them is no nearby market for it. Although the chemical reaction in Equation (3.5) for production of sulfur is simple enough, there are a variety of ways of carrying it out, and the details can be complex. Hundreds of such plants operate successfully throughout the world; every major petroleum refinery has at least one.

Because elemental sulfur is inert and harmless and because reduced sulfur in the form of hydrogen sulfide or related compounds can be easily oxidized to sulfur or sulfur oxides, the entire strategy of the petroleum and natural gas industries in dealing with reduced sulfur in petroleum, natural gas, and other process gases is to keep the sulfur in the form of elemental sulfur or reduced sulfur (for example, H_2S). Oxygen from the air is virtually free, so we can always move in the oxidation direction at low cost. In contrast, hydrogen is an expensive raw material, so that moving in the reduction direction is expensive.

Sulfur in hydrocarbon fuels (natural gas, propane, gasoline, jet fuel, diesel fuel, furnace oil) is normally converted to SO_2 during combustion and then emitted to the atmosphere. Large oil-burning facilities can have equipment to capture that SO_2, but autos, trucks, and airplanes do not. The only way to limit the SO_2 emissions from these sources is to limit the amount of sulfur in the fuel. For this reason the Clean Air Act of 1990 limits the amount of sulfur in diesel fuel to 0.05 percent by weight. Crude oils vary in their sulfur contents: low-sulfur crudes are called "sweet"; high-sulfur crudes, "sour." If the fraction of the crude oil going to gasoline or diesel fuel has too high a sulfur content (which many do under current regulations), most of that sulfur is removed by catalytic hydrodesulfurization,

$$\left(\begin{matrix}\text{Hydrocarbon} \\ \text{containing S}\end{matrix}\right) + H_2 \xrightarrow{\text{Ni or Co catalyst promoted with Mo or W}} \text{hydrocarbon} + H_2S \quad (3.7)$$

The mixture leaving the reactor is cooled, condensing most of the hydrocarbons. The remaining gas stream, a mixture of H_2 and H_2S, is one of the streams treated in a refinery for H_2S removal by the process shown in Figure 3.11. Some petroleum streams in refineries are treated over these catalysts to remove both sulfur and nitrogen because those elements interfere with the catalysts used for subsequent processing. The resulting gas streams contain both H_2S and NH_3.

Whether the treatment of gases with high concentrations of H_2S and NH_3 should be considered as air pollution control is an open question. For natural gas fields with H_2S, treatment is a market requirement, because the typical purchase specification for natural gas in the United States is $H_2S \leqslant$ 4 ppm. However, at one time in oil refineries H_2S – containing gases were customarily burned for internal heat sources in the refineries if the H_2S content was modest. Current U. S. EPA air pollution regulations forbid the burning of such refinery waste gases if they contain more than 230 mg/dscm (dry standard cubic meter) of H_2S, so the removal of H_2S down to that concentration in oil refinery gases is done by the method shown in Figure 3. 11 to meet air pollution control regulations.

3. Removal of SO_2 From Lean Waste Gases

The major source of SO_2, except near uncontrolled copper, lead, zinc, and nickel smelters, which no longer exist in the United States, but do in some developing countries, is the stacks of large coal- or oil-burning facilities. Most of the largest ones are coal-burning electric power plants. For them, the typical SO_2 content of the exhaust gas is about 0. 1 percent SO_2, or 1 000 ppm, which is much too low for profitable recovery as H_2SO_4.

The most widely used procedure for controlling SO_2 emissions from these sources is scrubbing with water containing finely ground limestone; the whole process is called flue gas desulfurization. There are several drawbacks to this procedure for dealing with the SO_2 from an electric power plant. First, it requires a large amount of water. Second, the waste water stream, which is 80 percent saturated with SO_2, would emit this SO_2 back into the atmosphere at ground level (river level); causing an SO_2 problem that might be more troublesome than the emission of the same amount of SO_2 from the power plant's stack. Third, in aqueous solution SO_2 undergoes Reaction(3.8) bellows,

$$SO_2 + \frac{1}{2}O_2 \xrightarrow{\text{vanadium catalyst}} SO_3 \quad (3.8)$$

which would remove most of the dissolved O_2 in the river, making it impossible for fish to live in it. For this reason alone, simple dissolution of large quantities of SO_2 in most rivers is prohibited.

However, the first large power plant to treat its stack gas for SO_2 removal did remove SO_2 with river water. The Battersea Plant of the London Power Company is located on the banks of the Thames River, which is large enough to supply the water it needed. Furthermore, the water of the Thames is naturally alkaline because its course passes through many limestone formations, so that it will absorb substantially more SO_2 than would pure water. To prevent the dissolved SO_2 from consuming O_2 in the river, the effluent from the gas washers was held in oxidizing tanks, where air was bubbled through it until the dissolved SO_2 was mostly oxidized to sulfate (SO_4^{2-}), before being discharged to the Thames. In this form the sulfur has a low vapor pressure and does not reenter the air nor kill the fish by consuming the river's dissolved oxygen. Although this pioneering plant had its problems, it was a technical success—removing over 90 percent of the SO_2—and operated from 1933 to 1940.

In the previous examples we said little about the internal features of the absorbing column. For the high-pressure treatment of H_2S, either plate or packed towers are used, with little problem. For the SO_2 problem, three plausible arrangements are sketched in Figure 3.13. The first of these is a simple bubbler, in which the gas is forced under pressure through perforated pipes submerged in the scrubbing liquid. As the bubbles rise through the liquid, they approach chemical equilibrium with it. If the liquid is deep enough and the bubbles are small enough, this kind of device will bring the gas close to chemical equilibrium with the liquid. However, it has a high pressure drop. The gas pressure must at least equal the hydrostatic head of the liquid. If, for example, the liquid is a foot deep, then the hydrostatic head will be 12 inches of liquid, which is large enough to be quite expensive. Plate-type distillation and absorption columns are, in effect, a series of such bubblers, stacked one above the other, with the gas flowing up from one to the next and the liquid flowing down from one to the next through pipes called downcomers. At high pressures, where pressure drops are unimportant, they are the most widely used device.

The second arrangement is a spray chamber. In it the gas flows up through an open chamber while the scrubbing liquid falls from spray nozzles, much like the heads in bathroom showers, through the gas. In this arrangement the gas pressure drop is small, but it is difficult to approach equilibrium because the gas does not contact the liquid as well as it does in the bubbler. Nonetheless, it is widely used because of its simplicity, low pressure drop, and resistance to scale deposition and plugging.

The third arrangement is a packed column, which is similar to the spray chamber except that the open space is filled with some kind of solid material that allows the liquid to coat its surface and run down over it in a thin film. The gas passes between pieces of solid material and comes in good contact with the liquid films, in the most primitive of these; the solid materials were gravel or crushed rocks. More advanced ones use special shapes of ceramic, plastic, or metal that are fabricated to provide the optimum distribution of liquid surface for contact with the gas. This third kind of contactor can be designed to have a better mass transfer per unit of gas pressure drop than either of the other two kinds. All three of these arrangements, plus combinations of them, plus some other arrangements are in current use for removal of SO_2 from power plant stack gases.

The gas velocities in such devices range from about 1 ft/s in a packed tower to 10 ft/s for a spray chamber. Such devices are almost always cylindrical, because that shape is easier and cheaper to fabricate than, for example, a rectangular vessel of equal cross-sectional area. A typical length in the flow direction would be 50 ft. That is a very large diameter for any piece of chemical plant

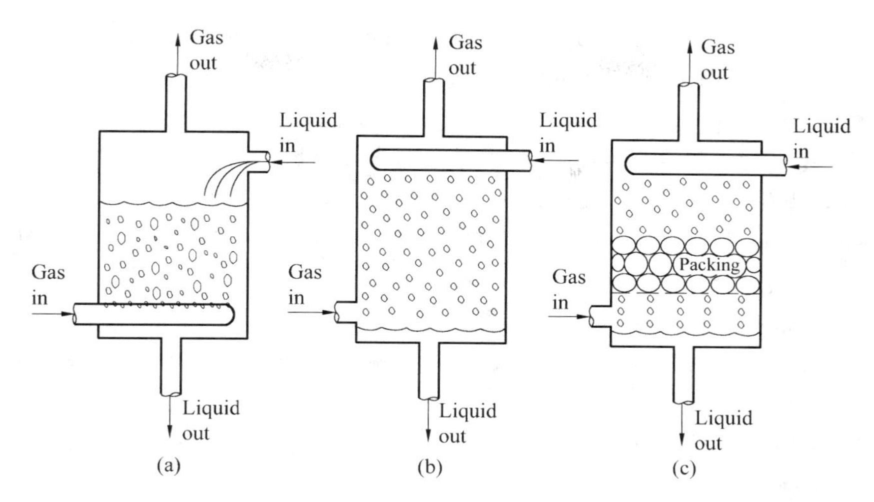

Figure 3.13 Three plausible arrangements for scrubbing a gas with a liquid: (a) bubbler, (b) spray chamber, (c) packed column.

equipment, but not for a power plant. Often the flow will be divided into several smaller scrubbers in parallel. This choice avoids having to ship or fabricate too large a vessel and ensures that one of the vessels can be taken out of service for maintenance while the rest are in operation. Thus the power plant can continue to operate while one part of the scrubber is out of service.

What problems might power plant operators encounter? First, there is the question of what to do with the sodium sulfate produced. Sodium sulfate (also called "salt cake") is used in detergent manufacture and in paper making, as well as in some miscellaneous uses. However, for those uses it must be quite pure. The sodium sulfate produced in this process would be contaminated with fly ash from the coal. Thus if we wished to sell the sodium sulfate, we would have to get it out of solution (by evaporation and crystallization) and then purify it. If we did, we would find that the total amount produced in a few power plants would glut the current market, so that although a few power plants might sell their sodium sulfate, most could not. Because of its water solubility, it is not generally acceptable in landfills unless they are well protected from water infiltration.

But the real difficulty is with carbon dioxide. Here we assumed that we could treat the exhaust gas with dilute alkaline solutions and remove the SO_2, which is an acid gas. However, the exhaust gas from combustion sources contains another acid gas, CO_2. Normally its concentration is about 12 percent, or 120 times that of the SO_2. We are not generally concerned with the fate of CO_2, but if it gets into solution it will use up sodium hydroxide by the reaction

$$2\ NaOH + CO_2 \rightarrow Na_2CO_3 + H_2O \tag{3.9}$$

Any sodium hydroxide used up this way is not available to participate in the reaction

$$2\ NaOH + SO_2 + \frac{1}{2}O_2 \rightarrow Na_2SO_4 + H_2O \tag{3.10}$$

The real problem is how to absorb one acid gas while not absorbing another acid gas that is present in much higher concentration.

If the problem were to use NaOH to remove SO_2 from a gas stream that contained no other acid gases, this would be a simple problem for which ordinary chemical engineering techniques would be satisfactory. The real problem is different from this one for the following reasons: ① There is another acid gas, CO_2, present that will use up our alkali unless we keep the solution acid enough to exclude it. ② The amount of alkali needed is high, and the cost of sodium hydroxide is enough that we would prefer to use a cheaper alkali if possible. ③ We have to do something with the waste product, either sell it or permanently dispose of it. ④ Because the volume of gas to be handled is very large, we must be very careful to keep the gas pressure drop in the scrubber low. The pressure drops that are normally used in the chemical and petroleum industry in gas absorbers are much too large to be acceptable here.

4. Alternatives to "Burn and Then Scrub"

When the electric power industry firstly faced regulations requiring it to reduce the SO_2 emissions from power plants, it decided for the most part to leave the power plant alone and to scrub the gas leaving the power plant. This approach is still the most common, using either wet limestone scrubbers or lime spray dryers. But the industry never entirely abandoned the investigation of alternative approaches. With strong pressure from the Clean Air Amendments of 1990 to reduce emissions of acid rain precursors, the electric power industry has renewed interest in these other possibilities.

(1) Change to a Lower Sulfur Content Fuel

If the management of a power plant can replace a high-sulfur coal with a low-sulfur coal, it reduces the SO_2 emissions quickly, simply, and without having to install expensive SO_2 control devices or to deal with their solid effluent. (Switching coals can cause some problems in the plant, which was presumably designed for the coal originally used, but such problems are generally manageable.) Many power plants that burned high-sulfur eastern coals switched to lower-sulfur coals from the Rocky Mountain states. This decision was a boon to the economies of Wyoming and Montana and a blow to the economies of the midwestern and eastern coal-producing states. This approach has been vigorously attacked, mostly on the grounds of job losses, by the midwestern and eastern coal miners and their elected representatives; it is a continuing political struggle.

(2) Remove Sulfur from the Fuel

Another alternative is to remove the sulfur from the fuel before it is burned. Pyritic sulfur can be removed by grinding the coal to a small enough size that the pyrites are mostly present as free pyrite particles. Gravity methods are then used to separate the low-density coal (s.g. = 1.1 to 1.3) from the high-density pyrites (s.g. = 5.0). This approach is particularly suited for coals in which a substantial fraction of the sulfur is present as pyrites. Unfortunately, pyrite particles are generally quite small, so that very fine grinding is needed to separate them from the rest of the coal.

It is also possible to dissolve coal in strong enough solvents and then to treat the solution by the same kind of catalytic hydrogenation processes that are used to remove sulfur from petroleum products. The mineral (ash-forming) materials do not dissolve, so they are rejected by filtration or settling. When the solvent is then removed for reuse, the remaining product is a very clean-burning combustible solid, free of ash and sulfur, called solvent-refined coal. Considerable development work on this process showed that it can be done, but so far not at a price comparable to "burn and then scrub."

(3) Modify the Combustion Process

The standard way of burning large amounts of coal (pulverized-coal furnace) is to grind the coal to about 50- to 150-μ size and blow it with hot air into a large combustion chamber. There the small coal particles decompose and burn in the one to four seconds that they spend in the furnace, transferring most of the heat generated to the walls of the furnace as radiant heat. The furnace walls are made of steel tubes in which fluid (most often water turning to steam) is heated. The hot gases leaving the furnace then pass over banks of tubes and transfer much of their remaining sensible heat to the fluid being heated.

Fluidized bed combustion is an alternative way to burn coal that is currently in the demonstration plant stage. In it, coal is burned in gravel-sized pieces by injecting them into a hot fluidized bed of limestone particles instead of as a finely dispersed powder in air. A fluidized bed is a dense bed of solid particles suspended in air; such beds are widely used in chemical engineering, e.g., in fluidized bed catalytic cracking. The coal spends much longer in the bed than it would in a pulverized coal furnace, because more time is needed to get complete combustion of the much larger particles.

In such a fluidized bed combustor SO_2 is formed in the presence of a large number of limestone particles and has a high probability of reacting with one of them in the combustion bed. Here the temperatures are much higher than in the dry processes, and most of the limestone has been converted to CaO, so that the reaction of SO_2 with CaO is rapid enough to provide adequate SO_2 control. The limestone in the bed is steadily replaced, and the material withdrawn has largely been converted to $CaSO_4$. Here again, a dry powder waste is produced instead of a wet scrubber sludge.

The fluidized bed has tubes full of water and steam projecting into the bed. The heat transfer between the hot bed in which the coal is burned and the tubes is much better than that between the flames and the walls of an ordinary coal-fired boiler. For this reason fluidized bed combustors are smaller and operate at lower temperatures than ordinary coal-fired boilers. This saves on some costs and greatly reduces the formation of nitrogen oxides. These boilers, however, have other problems, so that they are not yet a clear winner over conventional boilers.

A second combustion modification alternative, also in the demonstration plant stage, is to convert the coal to a synthetic fuel gas and then burn that in combination gas-turbine steam-turbine power plants, Figure 3.14. This seems complex and costly, but it has the advantage that in the synthetic fuel gas the sulfur is present as H_2S; since no other acid gas is present, the sulfur can be easily removed from the gas by the methods described. The second, and more important, advantage is that modern gas-turbine steam-turbine plants have a much higher thermal efficiency than typical coal-fired steam plants (perhaps 45 percent vs. 33 percent). If the problems with this technology can be solved, it may offer a more efficient and economical way of converting coal to electricity than the systems currently used even though it is much more complex.

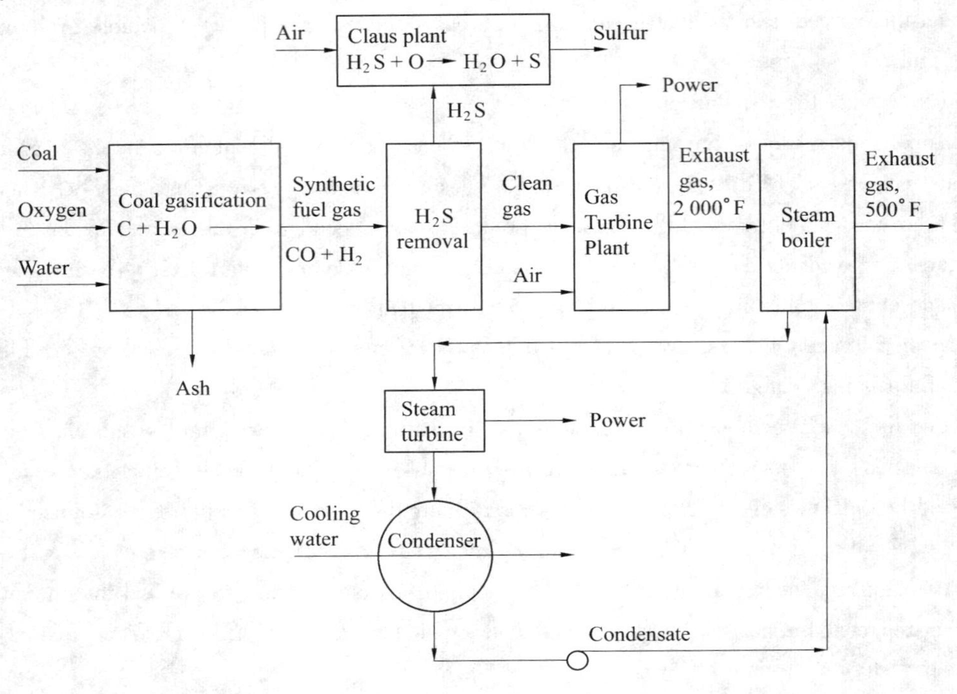

Figure 3.14 Schematic flow diagram of synthetic fuel gas, gas-turbine steam-turbine power plant

(4) Don't Burn at All

The majority of the SO_2 derived from human activities comes from coal and oil combustion in electric power plants. If we can produce electricity in some other way or reduce our use of electricity, we will consequently reduce our emissions of SO_2. For this reason more efficient electric devices (lights, refrigerators, motors) are, in effect, SO_2 control devices. So also are nuclear, wind, solar, tidal, geothermal, and hydroelectric power plants. There is currently a serious effort by the U.S. EPA and by the electric utility industry to improve the efficiency of electricity usage, and to encourage production of electricity from alternative energy sources for a variety of reasons, including reduction of SO_2 emissions.

3.6.4 Emission Controls for Mobile Sources

Highway vehicles-cars, trucks, buses, and motorcycles-are the principal moving sources of air pollution, although boats, trains, and airplanes also contribute to the problem. Pollutant emissions from these mobile sources are significant. In 1990, for example, about two thirds of CO emissions in the United States came from mobile sources, mostly from automobiles; in some urban areas, as much as 95 percent of CO emissions came from these sources. It is the mobility of huge numbers of these diverse, decentralized sources that makes their regulation and control even more of a challenge than control of stationary sources. Also, mobile sources such as highway vehicles are typically very close to the receptors-people living in cities.

In addition to carbon monoxide, the mobile emissions of most concern in the United States include nitrogen oxides and volatile organic compounds. Sulfur dioxide and particulate emissions are less of a problem largely because of fuel desulfurization, fewer diesel vehicles, and the generally good condition of motor vehicles. In many other countries, though, sulfur dioxide and particulate emissions are of concern, as are emissions of lead. Vehicular emissions of lead have effectively been eliminated in the United States because of the ban on the use of leaded gasoline.

1. Internal Combustion Engines

The four-stroke, gasoline-burning internal combustion engine is the prime mover of most highway vehicles, although motorcycles and small three-wheeled vehicles are also used, particularly in developing countries. There are four major emission points from internal combustion engines, including the exhaust or tailpipe, the engine crankcase vent, the carburetor, and the fuel tank. Hydrocarbons, carbon monoxide, and nitrogen oxides come from the exhaust, unburned gasoline and hydrocarbons come from the crankcase, and hydrocarbons evaporate from both the carburetor and fuel tank.

Evaporative losses from the fuel tank and carburetor are controlled by the use of an activated carbon canister that stores the vapors emitted when the engine is turned off. When the motor is on, the vapors are purged from the canister and burned in the engine. Crankcase vent emissions are controlled by positive crankcase ventilation (PCV) system, which server to recycle gasses that slip by the piston rings back to the engine intake manifold; a PCV check valve is used to prevent the buildup of excessive pressure in the engine crankcase.

Emissions from the tailpipe of an automobile, which account for roughly 60 percent of the hydrocarbons and almost all the carbon monoxide and nitrogen oxide emissions, depend on the mode of operation of the vehicle. Acceleration results in less carbon monoxide and hydrocarbon emissions, but more nitrogen oxides; deceleration results in less nitrogen oxides, but much higher hydrocarbon levels from the partially burned fuel. Because of these variations, exhaust emissions testing procedures are standardized so that test results will be meaningful.

Exhaust emissions from an automobile tailpipe are controlled by use of a catalytic converter, a device installed in the tailpipe ahead of the muffler. It allows almost complete oxidation of combustible gases in the exhaust to occur at relatively low temperatures; in effect, a catalytic converter allows "flameless combustion" to take place. The carbon monoxide and unburned hydrocarbons in the exhaust are converted to carbon dioxide and water. But any sulfur in the gasoline is oxidized to particulate sulfur trioxide, thereby increasing sulfur levels in the air.

To further reduce pollutant emissions from gasoline-powered motor vehicles, certain engine operating or design features can be controlled or modified. For example, increasing the air-to-fuel ratio helps combustion, thereby reducing carbon monoxide and hydrocarbon emissions. Carburetors are not very effective in accurately controlling the air-to-fuel ratio; new cars are equipped with computer-controlled fuel injectors that can optimize the air and fuel mixture for different driving conditions.

Other key engine design and operation features that affect emissions include the compression ratio and timing of the spark plugs. High compression ratios produce greater power, but the accompanying higher temperatures produce more nitrogen oxide emissions can be reduced by retarding the spark, but this reduces the performance of the engine. Clearly, control of air pollutant emissions from internal combustion engines is a complex problem.

As a result of emission controls required for gasoline-powered internal combustion engines, today's cars typically emit about 80 percent less pollution than did cars in the 1970s. On the other hand, total vehicle miles traveled have more than doubled since then, and many challenges remain in the effort to further reduce vehicle-related air pollution.

(1) Changes in Fuels

Fuel composition is yet another factor in mobile source emissions, and changing or modifying the fuel can play an important role in controlling air pollution from these sources. Removing impurities in the fuel is an obvious way to reduce pollution. Lead, a substance that was blended with gasoline to increase octane ratings and engine performance, is one substance no longer allowed in gasoline. Airborne lead is not only a health hazard, it also damages catalytic converters. The elimination of leaded gasoline in the United States has significantly reduced lead levels in urban air.

Another example of an undesirable substance in gasoline is sulfur, which not only leads to increased sulfur dioxide emissions, but interferes with the operation of catalytic converters, thereby contributing to increased carbon monoxide and hydrocarbon emissions. Because of more strictly controlled refining processes compared to other countries, gasolines produced in the United States contain less sulfur than fuels produced elsewhere and therefore cause less sulfur dioxide emissions.

Alternative fuels that can replace gasoline include compressed natural gas, liquefied petroleum gas, methanol, ethanol, propane, and hydrogen. Each of these possible replacements has a variety of advantages as well as disadvantages compared to gasoline. For example, compressed natural gas produces about 50 percent less carbon monoxide and volatile organic emissions and no air toxics such as benzene, but it has significant problems related to on-vehicle fuel handling, storage, and refueling. Hydrogen gas produces no carbon oxides or hydrocarbons and has the highest energy content of any combustible fuel, but there are numerous technological and safety problems to solve before hydrogen becomes a practical alternative fuel for automobiles

(2) Reformulated Gasoline (RFG)

As required by the 1990 Clean Air Act Amendments, a Reformulated Gasoline Program went into effect in 1995 in areas of the United States designated as having severe ozone pollution and smog, as well as in other areas of the country that opt to participate in the program. RFG is an oxygenated fuel, containing at least 2 percent of oxygenates by weight. The oxygenates replace or dilute other, less desirable compounds and enhances cleaner burning, especially in cold climates. This blend of gasoline also has limits on the amounts of benzene and other toxics, as well as hydrocarbons, it may contain.

RFG is expected to reduce emissions of hydrocarbons and toxics by at least 15 percent as compared to 1990 levels; emission of other criteria pollutants will also be reduced. Although the price of RFG is expected to raise fuel costs by 2 to 4 cents per gallon, no additional investments are necessary to retrofit highway vehicles or gasoline filling stations to accommodate the new fuel.

(3) Diesel Standards

In 2001, the EPA implemented standards to control emissions from diesel engines, a significant source of air pollution. Under these new standards, diesel producers will be required to virtually eliminate sulfur from the fuel. Sulfur produces soot and clogs up a vehicle's catalytic converter. With the removal of sulfur, manufacturers of diesel engines will be required to incorporate the sophisticated pollution control devices that are now standard equipment in all gasoline-burning cars. These changes will be phased in over time—80 percent of all diesel fuel must be virtually sulfur-free by year 2006, and the rest by 2010. Diesel engine manufacturers will have until 2010 to complete the required engine modifications. These new standards will cut air pollution from trucks and buses and other commercial vehicles by 95 percent, the equivalent of removing 13 million trucks from the road.

2. Zero-Emission Vehicles

Alternatives to internal combustion engines include solar-powered cars and electric vehicles. Solar-powered cars are still very much in the experimental stage, but electric cars are of particular interest at the present time. Some states (for example, New York and California) require that, by the year 2003, 10 percent of all motor vehicles sold in the state must be zero-emission vehicles. The only zero-emission vehicle presently capable of mass production is the electric-powered car, which uses lead-acid batteries to store the electricity.

Most electric-powered cars and vans have a range of only about 130 km before the batteries need recharging, and they are considerably more expensive than conventional gasoline-powered vehicles ($ 10 000 to $ 25 000 more by some estimates). Some researchers think that emissions from the mining, smelting, and recycling of lead needed to operate a large fleet of electric cars and vans could pose serious threats to public health. In addition, the extra electricity needed to recharge ear batteries will come from power plants that also generate pollution.

Clearly, there are trade-offs that must be considered by environmental engineers and policy makers in regard to zero-emission vehicles. Although electric-powered vehicles operate adequately for certain light-duty applications, more research and improvement in battery technology is needed before they are likely to become a significant alternative to internal combustion engines.

3. Hybrid Vehicles

A hybrid vehicle is one that can use two sources of power to turn the wheels. Some automobile makers, for example, have begun to make prototypes of hybrids that use a gasoline-powered internal combustion engine with an electric motor. The electric motor powers the car up to a speed of about 20 km/h (12 mph), below which gasoline engine emissions are highest. Above 20 km/h, the vehicle's relatively small (1.5-Li) gasoline engine starts up and becomes the main power source.

The electric motor can also operate when peak power is needed as when climbing a hill. When the gasoline engine operates, it also recharges the batteries for the electric motor, so the vehicle does not require any external recharger. It is very possible that hybrid vehicles such as this will be used until a technical breakthrough makes zero-emission vehicles fully competitive with other types of vehicles.

Review Questions

1. What are the major gaseous constituents of fresh (clean) air?

2. Give a brief definition of air pollution. What is meant by the term anthropogenic air pollution?

3. List three sources and types of natural air pollutants.

4. List the primary air pollutants commonly released into the atmosphere and their sources.

5. Define secondary air pollutants and give an example.

6. Describe three actions that can be taken to control air pollution.

7. What causes acid rain? List three probable detrimental consequences of acid rain.

8. Why is carbon dioxide called a "greenhouse gas"?

9. What would the consequences be if the ozone layer surrounding the Earth were destroyed?

Chapter 4 Solid Waste Management

4.1 Integrated Solid Water Management

Solid wastes comprise all the wastes arising from human and animal activities that are normally solid and that are discarded as useless or unwanted. The term solid waste as used in this text is all-inclusive, encompassing the heterogeneous mass of throwaways from the urban community as well as the more homogeneous accumulation of agricultural, industrial, and mineral wastes.

Solid waste management may be defined as the discipline associated with the control of generation, storage, collection, transfer and transport, processing, and disposal of solid wastes in a manner that is in accord with the best principles of public health, economics, engineering, conservation, aesthetics, and other environmental considerations, and that is also responsive to public attitudes. In its scope, solid waste management includes all administrative, financial, legal, planning, and engineering functions involved in solutions to all problems of solid wastes. The solutions may involve complex interdisciplinary relationships among such fields as political science, city and regional planning, geography, economics, public health, sociology, demography, communications, and conservation, as well as engineering and materials science.

The problems associated with the management of solid wastes in today's society are complex because of the quantity and diverse nature of the wastes, the development of sprawling urban areas, the funding limitations for public services in many large cities, the impacts of technology, and the emerging limitations in both energy and raw materials. As a consequence, if solid waste management is to be accomplished in an efficient and orderly manner, the fundamental aspects and relationships involved must be identified, adjusted for uniformity of data, and understood clearly.

In general, the activities associated with the management of solid wastes, from the point of generation to final disposal have been grouped into the six functional elements: (1) waste generation; (2) waste handling and separation, storage, and processing at the source; (3) collection; (4) separation and processing and transformation of solid wastes; (5) transfer and transport; and (6) disposal. When all of the functional elements have been evaluated for use, and all of the interfaces and connections between elements have been matched for effectiveness and economy, the community has developed an integrated waste management system, In this context integrated solid waste management (ISWM) can be defined as the selection and application of suitable techniques, technologies, and management programs to achieve specific waste management objectives and goals.

4.1.1 Hierarchy of Integrated Solid Waste Management

A hierarchy (arrangement in order of rank) in waste management can be used to rank actions to implement programs within the community. The ISWM hierarchy adopted by the U.S. Environmental Protection Agency (EPA) is composed of the following elements: source reduction, recycling, waste combustion, and landfilling. The ISWM hierarchy used in this book is source reduction, recycling, waste transformation, and landfilling. The term waste transformation is substituted for the U.S. EPA's term combustion, which is too limiting. In the broadest interpretation of the ISWM hierarchy, ISWM programs and systems should be developed in which the elements of the hierarchy are interrelated and are selected to complement each other. For example, the separate collection of yard wastes can be used to effect positively the operation of a waste-to-energy combustion facility.

1. Source Reduction

The highest rank of the ISWM hierarchy, source reduction, involves reducing the amount and/or toxicity of the wastes that are now generated. Source reduction is first in the hierarchy because it is the most effective way to reduce the quantity of waste, the cost associated with its handling, and its environmental impacts. Waste reduction may occur through the design, manufacture, and packaging of products with minimum toxic content, minimum volume of material, or a longer useful life. Waste reduction may also occur at the household, commercial, or industrial facility through selective buying patterns and the reuse of products and materials.

2. Recycling

The second highest rank in the hierarchy is recycling, which involves (1) the separation and collection of waste materials; (2) the preparation of these materials for reuse, reprocessing, and remanufacture; and (3) the reuse, reprocessing, and remanufacture of these materials. Recycling is an important factor in helping to reduce the demand on resources and the amount of waste requiring disposal by landfilling.

3. Waste Transformation

The third rank in the ISWM hierarchy, waste transformation, involves the physical, chemical, or biological alteration of wastes. Typically, the physical, chemical, and biological transformations that can be applied to municipal soild waste(MSW) are used (1) to improve the efficiency of solid waste management operations and systems, (2) to recover reusable and recyclable materials, and (3) to recover conversion products (e.g., compost) and energy in the form of heat and combustible biogas. The transformation of waste materials usually results in the reduced use of landfill capacity. The reduction in waste volume through combustion is a well-known example.

4. Landfilling

Ultimately, something must be done with (1) the solid wastes that cannot be recycled and are of no further use; (2) the residual matter remaining after solid wastes have been separated at a materials recovery facility; and (3) the residual matter remaining after the recovery of conversion products or energy. There are only two alternatives available for the long-term handling of solid wastes and residual matter: disposal on or in the earth's mantle, and disposal at the bottom of the ocean. Landfilling, the fourth rank of the ISWM hierarchy, involves the controlled disposal of wastes on or in the earth's mantle, and it is by far the most common method of ultimate disposal for waste residuals. Landfilling is the lowest rank in the ISWM hierarchy because it represents the least desirable means of dealing with society's wastes.

4.1.2 Planning for Integrated Waste Management

Developing and implementing an ISWM plan is essentially a local activity that involves the selection of the proper mix of alternatives and technologies to meet changing local waste management needs while meeting legislative mandates.

1. Proper Mix of Alternatives and Technologies.

A wide variety of alternative programs and technologies are now available for the management of solid wastes. Several questions arise from this variety: What is the proper mix between (1) the amount of waste separated for reuse and recycling, (2) the amount of waste that is composted, (3) the amount of waste that is combusted, and (4) the amount of waste to be disposed of in landfills? What technology should be used for collecting wastes separated at the source, for separating waste components at materials recovery facilities (MRFs), for composting the organic fraction of MSW, and for compacting wastes at a landfill? What is the proper timing for the application of various technologies in an ISWM system and how should decisions be made?

Because of the wide range of participants in the decision-making process for the implementation of solid waste management systems, the selection of the proper mix of alternatives and technologies for the effective management of wastes has become a difficult, if not impossible, task. The development of effective ISWM systems will depend on the availability of reliable data on the characteristics of the waste stream, performance specifications for alternative technologies, and adequate cost information.

2. Flexibility in Meeting Future Changes

The ability to adapt waste management practices to changing conditions is of critical importance in the development of an ISWM system. Some important factors to consider include (1) changes in the quantities and composition of the waste stream, (2) changes in the specifications and markets for recyclable materials, and (3) rapid developments in technology. If the ISWM system is planned and designed on the basis of a detailed analysis of the range of possible outcomes related to these factors, the local community will be protected from unexpended changes in local, regional, and larger-scale conditions.

3. Monitoring and Evaluation

Integrated solid waste management is an ongoing activity that requires continual monitoring and evaluation to determine if program objectives and goals (e.g., waste diversion goals) are being met. Only by developing and implementing ongoing monitoring and evaluation programs can timely changes be made to the ISWM system that reflect changes in waste characteristics, changing specifications and markets for recovered materials, and new and improved waste management technologies.

4.2 Sources, Types, and Composition of Municipal Solid Wastes

Solid wastes include all solid or semisolid materials that the possessor no longer considers of sufficient value to retain. The management of these waste materials is the fundamental concern of all the activities encompassed in solid waste management—whether the planning level is local, regional or subregional, or state and federal. For this reason, it is important to know as much about municipal solid waste (MSW) as possible. Knowledge of the sources and types of solid wastes, along with data on the composition and rates of generation, is basic to the design and operation of the functional elements associated with the management of solid wastes. To avoid confusion, the term refuse, often used interchangeably with the term solid wastes, is not used in this text.

4.2.1 Sources and Types of Solid Wastes

Sources of solid wastes in a community are, in general, related to land use and zoning. Although any number of source classifications can be developed, the following categories are useful: (1) residential, (2) commercial, (3) institutional, (4) construction and demolition, (5) municipal services, (6) treatment plant sites, (7) industrial, and (8) agricultural.

As a basis for subsequent discussions, it will be helpful to define the various types of solid wastes that are generated. It is important to be aware that the definitions of solid waste terms and the classifications vary greatly in the literature and in the profession. Consequently, the use of published data requires considerable care, judgment, and common sense. The following definitions are intended to serve as a guide and are not meant to be precise in a scientific sense.

1. Residential and Commercial

Residential and commercial solid wastes, excluding special and hazardous wastes discussed below, consist of the organic (combustible) and inorganic (noncombustible) solid wastes from residential areas and commercial establishments. Typically, the organic fraction of residential and commercial solid waste consists of materials such as food waste (also called garbage), paper of all types, corrugated cardboard (also known as paperboard and corrugated paper), plastics of all types, textiles, rubber, leather, wood, and yard wastes. The inorganic fraction consists of items such as glass, crockery, tin cans, aluminum, ferrous metals, and dirt. If the waste components are not separated when discarded, then the mixture of these wastes is also known as commingled residential and commercial MSW.

Wastes that will decompose rapidly, especially in warm weather, are also known as putrescible waste. The principal source of putrescible wastes is the handling, preparation, cooking, and eating of foods. Often, decomposition will lead to the development of offensive odors and the breeding of flies. In many locations, the putrescible nature of these wastes will influence the design and operation of the solid waste collection system. Although there are more than 50 classifications for paper, the waste paper found in MSW is typically composed of newspaper, books and magazines, commercial printing, office paper, other paperboard, paper packaging, other non-packaging paper, tissue paper and towels, and corrugated cardboard. The plastic materials found in MSW fall into the following seven categories: ① Polyethylene terephthalate (PETE/1); ②High-density polyethylene (HDPE/2); ③ Polyvinyl chloride (PVC/3); ④ Low-density polyethylene (LDPE/4); ⑤ Polypropylene (PP/5); ⑥Polystyrene (PS/6); ⑦ Other multilayered plastic materials (OTHER/7). The type of plastic container can be identified by number code (1 through 7) molded into the bottom of the container (see Figure 4.1). Mixed plastic is the term used for the mixture of the individual types of plastic found in MSW.

Figure 4.1 Code designation used for various types of plastics

Special wastes from residential and commercial sources include bulky items, consumer electronics, white goods, yard wastes that are collected separately, batteries, oil, and tires. These wastes are usually handled separately from other residential and commercial wastes. Wastes or combinations of wastes that pose a substantial present or potential hazard to human health or living organisms have been defined as hazardous wastes. The U. S. EPA has defined Resource Conservation and Recovery Act (RCRA) hazardous wastes in three general categories: (1) listed wastes, (2) characteristic hazardous wastes, and (3) other hazardous wastes.

2. Institutional

Institutional sources of solid waste include government centers, schools, prisons, and hospitals. Excluding manufacturing wastes from prisons and medical wastes from hospitals, the solid wastes generated at these facilities are quite similar commingled MSW. In most hospitals medical wastes are handled and process separately from other solid wastes.

3. Construction and Demolition

Wastes from the construction, remodeling, and repairing of individual residences, commercial buildings, and other structures are classified as construction wastes. The quantities produced are difficult to estimate. The composition is variable but may include dirt; stones; concrete; bricks; plaster; lumber; shingles; and plumbing, heating, and electrical parts. Wastes from razed buildings, broken-out streets, sidewalks, bridges, and other structures are classified as demolition wastes. The composition of demolition wastes is similar to construction wastes, but may include broken glass, plastics, and reinforcing steel.

4. Municipal Services

Other community wastes, resulting from the operation and maintenance of municipal facilities and the provision of other municipal services, include street sweepings, road side litter, wastes from municipal litter containers, landscape and tree trimmings, catch-basin debris, dead animals, and abandoned vehicles. Because it is impossible to predict where dead animals and abandoned automobiles will be found, these wastes are often identified as originating from nonspecific diffuse sources. Wastes from nonspecific diffuse sources can be contrasted to that of the residential sources, which are also diffuse but specific in that the generation of the wastes is a recurring event.

5. Treatment Plant Wastes and Other Residues

The solid and semisolid wastes from water, wastewater, and industrial waste treatment facilities are termed treatment plant wastes. The specific characteristics of these materials vary, depending on the nature of the treatment process. At present their collection is not the charge of most municipal agencies responsible for solid waste management. However, wastewater treatment plant sludges are commonly co-disposed with MSW in municipal landfills. In the future, the disposal of treatment plant sludges will likely become a major factor in any solid waste management plan.

Materials remaining from the combustion of wood, coal, coke, and other combustible wastes are categorized as ashes and residues. (Residues from power plants normally are not included in this category because they are handled am processed separately.) These residues are normally composed of fine, powder materials, cinders, clinkers, and small amounts of burned and partially burned materials. Glass, crockery, and various metals are also found in the residues from municipal incinerators.

6. Industrial Solid Waste Excluding Process Wastes

Sources and types of solid waste generated at industrial sites, grouped according to their Standard Industrial Classification (SIC), are reported in Table 4.1. This list excludes industrial process wastes and any hazardous wastes that may be generated.

7. Agricultural Wastes

Wastes and residues resulting from diverse agricultural activities—such as the planting and harvesting of row, field, tree and vine crops; the production of milk; the production of animals for slaughter; and the operation of feedlots—are collectively called agricultural wastes. At present, the disposal of these wastes is not the responsibility of most municipal and county solid waste management agencies. However, in many areas the disposal of animal manure has become a critical problem, especially from feedlots and dairies.

Table 4.1　Sources and types of industrial wastes

Code	SIC group classification[a]	Waste-generating processes	Expected specific wastes
19	Ordnance and accessories	Manufacturing and assembling	Metals, plastic, rubber, paper, wood, cloth, chemical residues
20	Food and kindred products	Processing, packaging, shipping	Meats, fats, oils, bones, offal, vegetables, fruits, nuts and shells, cereals

Continued Table 4.1

22	Textile mill products	Weaving, processing, dyeing, and shipping	Cloth and fiber residues
23	Apparel and other finished products	Cutting, sewing, sizing, pressing	Cloth, fibers, metals, plastics, rubber
24	Lumber and wood products	Sawmills, millwork plants, wooden container, miscellaneous wood products, manufacturing	Scrap wood, shaving, sawdust; in some instances metals, plastic, fibers, glues, sealers, paints, solvents
25a	Furniture, wood	Manufacture of household and office furniture, partitions, office and store fixtures, mattresses	Those listed under code 24; in addition, cloth and padding residues
25b	Furniture, metal	Manufacture of household and office furniture, lockers, bed-springs, frames	Metals, plastics, resins, glass, wood, rubber, adhesives, cloth, paper
26	Paper and allied products	Paper manufacture, conversion of paper and paperboard, manufacture of paperboard boxes and containers	Paper and fiber residues, chemicals, paper coatings and fillers, inks, glues, fasteners
27	Printing and publishing	Newspaper publishing, printing, lithography, engraving, and bookbinding	Paper, newsprint, cardboard, metals, chemicals, cloth, inks, glues
28	Chemical and related products	Manufacture and preparation of inorganic chemicals (ranges from drugs and soaps to paints and varnishes, and explosives)	Organic and inorganic chemicals, metals, plastics, rubber, glass, oils, paints, solvents, pigments
29	Petroleum refining and related industries	Manufacture of paving and roofing materials	Asphalt and tars, felts, asbestos, paper, cloth, fiber
30	Rubber and miscellaneous plastic products	Manufacture of fabricated rubber and plastic products	Scrap rubber and plastics, curing compounds, dyes
31	Leather and leather products	Leather tanning and fishing; manufacture of leather belting and packing	Scrap leather, thread, dyes, oils, processing and curing compounds
32	Stone, clay, and glass products	Manufacture of flat glass, fabrication and forming of glass; manufacture of concrete, gypsum, and plaster products; forming and processing of stone and stone products, abrasives, asbestos, and miscellaneous nonmineral products	Glass, cement, clay, ceramics, gypsum, asbestos, stone, paper, abrasives

Continued Table 4.1

33	Primary metal industries	Melting, casting, forging, drawing, rolling, forming, extruding operations	Ferrous and nonferrous metals scrap, slag, sand, cores, patterns, bonding agents
34	Fabricated metal products	Manufacture of metal cans, hand tools, general hardware, nonelectric heating apparatus, plumbing fixtures, fabricated structural products, wire, farm machinery and equipment, coating and engraving of metal	Metals, ceramics, sand, slag, scale, coatings, solvents, lubricants, pickling liquor
35	Machinery (except electrical)	Manufacture of equipment for construction, mining, elevators, moving stairways, conveyors, industrial trucks, trailers, stackers, machine tools, etc.	Slag, sand, cores, metal scrap, wood, plastics, resins, rubber, cloth, paints, solvents, petroleum products
36	Electronic and other electrical equipment and components	Manufacture of electric equipment, appliances, and communication apparatus; machining, drawing, forming, welding, stamping, winding, painting, planting, baking firing operations	Metal scrap, carbon black, glass, exotic metals, rubber, plastics, resins, fibers, cloth residues
37	Transportation equipment	Manufacture of motor vehicles, truck and bus bodies, motor vehicles parts and accessories, aircraft and parts, ship and boat building and repairing, motorcycles and bicycles and parts, etc.	Metal scrap, glass, fiber, wood, rubber, plastic, cloth, paints, solvents, petroleum products
38	Measuring, analyzing and controlling instruments	Manufacture of engineering, laboratory and research instruments and associated equipment	Metals, plastics, resins, wood, rubber, fibers, abrasives
39	Miscellaneous manufacturing	Manufacture of jewelry, silverware, plated ware, toys, amusement, sporting and athletic goods, costume novelties, buttons, brooms, brushes, signs, advertising displays	Metals, glass, plastic, resins, leather, rubber, bone, cloth, straw, adhesives, paints, solvents, composite materials

[a]Source: Standard Industrial Classification Manual (SIC) 1987, Executive Office of the President, Office of Management and Budget, U.S. Government Printing Office, Washington, DC.

4.2.2 Composition of Solid Wastes

Composition is the term used to describe the individual components that make up a solid waste stream and their relative distribution, usually based on percent by weight. Information on the composition of solid wastes is important in evaluating equipment needs, systems, and management programs and plans. For example, if the solid wastes generated at a commercial facility consist of only paper products, the use of special processing equipment, such as shredders and balers, may be appropriate. Separate collection may also be considered if the city or collection agency is involved in a paper-products recycling program.

Typical data on the distribution of MSW are presented in Table 4.2. As noted in Table 4.2, the residential and commercial portion makes up about 50 to 75 percent of the total MSW generated in a community. The actual percentage distribution will depend on (1) the extent of the construction and demolition activities, (2) the extent of the municipal services provided, and (3) the types of water and wastewater treatment processes that are used. The wide variation in the special wastes category (3 to 12 percent) is due to the fact that in many communities yard wastes are collected separately. The percentage of construction and demolition wastes varies widely depending on the part of the country and the general health of the local, state, and national economy. The percentage of treatment, plant sludges will also vary widely depending on the extent and type of water and wastewater treatment provided.

Information and data on the physical composition of solid wastes are important in the selection and operation of equipment and facilities, in assessing the feasibility of resource and energy recovery, and in the analysis and design of landfill disposal facilities. Published distribution data should be used cautiously because the effects of recycling activities and the use of kitchen food waste grinders are often not reflected in earlier data.

Table 4.2 Estimated distribution of all components of MSW generated in a typical community excluding industrial and agricultural wastes

Waste category	Percent by weight	
	Range	Typical
Residential and commercial, excluding special and hazardous wastes	50 ~ 75	62.0
Special (bulky items, consumer electronics, white goods, yard wastes collected separately, batteries, oil, and tires)	3 ~ 12	5.0
Hazardous	0.01 ~ 1.0	0.1
Institutional	3 ~ 5	3.4
Construction and demolition	8 ~ 20	14.0
Municipal services		

Continued Table 4.2

Street and alley cleanings	2 ~ 5	3.8
Tree and landscaping	2 ~ 5	3.0
Parks and recreational areas	1.5 ~ 3	2.0
Catch basin	0.5 ~ 1.2	0.7
Treatment plant sludges	3 ~ 8	6.0
Total		100.0

Table 4.3　Typical physical composition of residential MSW excluding recycled materials and food wastes discharged with wastewater (1990)

Component	Percent by weight			
	United States		Packaging materials	Davis, California
	Range	Typical		
Organic				
Food wastes	6 ~ 18	9.0	—	6.0
Paper	25 ~ 40	34.0	50 ~ 60	33.1
Cardboard	3 ~ 10	6.0		7.9
Plastics	4 ~ 10	7.0	12 ~ 16	10.7
Textiles	0 ~ 4	2.0	—	2.4
Rubber	0 ~ 2	0.5	—	2.5
Leather	0 ~ 2	0.5	—	0.1
Yard wastes	5 ~ 20	18.5	—	17.7
Wood	1 ~ 4	2.0	4 ~ 8	5.0
Misc. organics	—	—	—	0.4
Inorganic				
Glass	4 ~ 12	8.0	20 ~ 30	5.8
Tin cans	2 ~ 8	6.0	6 ~ 8	3.9
Aluminum	0 ~ 1	0.5	2 ~ 4	0.4
Other Metal	1 ~ 4	3.0	—	3.6
Dirt, ash, etc.	0 ~ 6	3.0	—	0.5
Total		100.0		100.0

Components that typically make up the residential portion of MSW, excluding special and hazardous wastes, and their relative distribution are reported in Table 4.3. Although any number of components could be selected, those in Table 4.3 have been selected because they are readily identifiable and consistent with component categories reported in the literature and because they have proven adequate for the characterization of solid wastes for most applications. The data in Table 4.3 are derived from both the literature and the authors' experience. For the purpose of comparison, the percentage distribution of the materials used for packaging is reported in Column 3 of Table 4.3. It is estimated that packaging wastes now account for approximately one-third of the residential and commercial MSW.

To assess the impact of waste diversions (resulting from the use of food waste grinders and waste recycling programs) on the distribution of waste components, the distribution given in Table 4.3 for as collected residential MSW must adjusted. The adjusted component distribution data are reported in Table 4.4. As shown in Table 4.4, the distribution data for the United States do not change significantly. On the other hand, the distribution data for the city of Davis, California, would change quite a bit more because of the higher percentage of recycling.

Table 4.4 Typical physical composition of residential MSW in the United States in 1990

Component	Percent by weight			
	Solid waste as collected excluding waste components now recycled and food waste that is ground up	Solid waste as collected plus ground up food waste, but excluding waste components now recycled	Solid waste as collected plus waste components now recycled excluding food waste that is ground up	Solid waste as collected plus waste components now recycled and food waste that is ground up
Organic				
Food wastes	9.0	9.4	8.0	8.4
Paper	34.0	33.8	35.8	35.6
Cardboard	6.0	6.0	6.4	6.4
Plastics	7.0	7.0	6.9	6.9
Textiles	2.0	2.0	1.8	1.8
Rubber	0.5	0.5	0.4	0.4
Leather	0.5	0.5	0.4	0.4
Yard wastes	18.5	18.4	17.3	17.2
Wood	2.0	2.0	1.8	1.8
Misc. organics	—	—	—	—
Inorganic				
Glass	8.0	7.9	9.1	9.0
Tin cans	6.0	6.0	5.8	5.8
Aluminium	0.5	0.5	0.6	0.6
Other metal	3.0	3.0	3.0	3.0
Dirt, ash, etc.	3.0	3.0	2.7	2.7
Total	100.0	100.0	100.0	100.0

4.3 Solid Waste Collection

Of the billions of dollars expended each year for municipal solid waste management, about two thirds is needed to cover the cost of waste collection. Collection includes temporary storage or containerization, transfer to a collection vehicle, and transport to a site where the waste undergoes processing and ultimate disposal. Processing and final disposal are challenging problems, but waste collection is the most expensive phase, largely because it is labor-intensive. In addition, proper collection techniques are important to protect public health, safety, and environmental quality.

Solid waste collection may be a local municipal responsibility, where by public employees and equipment are assigned to the task. Sometimes it is more economical to have private collection companies do the work under contract to the municipality. In some communities private collectors are paid for the service by the individual homeowners. Whatever the actual administrative arrangement, proper planning, operation, and regulation of the collection activity are necessary. The EPA has developed recommended procedures for the storage and collection of solid waste; these activities must be done in a way that will not cause fire, health, or safety hazards or provide food and shelter for vectors of disease (for example, rats and flies). Enforcement of these rules is left up to the states and local communities.

Proper on-site storage is of particular importance for municipal refuse that contains a significant amount of putrescible garbage. Watertight, rust-resistant containers with suitable covers reduce the incidence of rodent or insect infestation, and offensive odors and unsightly conditions may be kept to a minimum if the containers and storage areas are washed periodically. The EPA recommends that refuse be collected at least once per week, and trash should be collected at least once every 3 months. The EPA also suggests a limit of 333 N (75 lb) for the weight of manually emptied waste containers; for most residences, 115-L (30-gal) galvanized metal or plastic containers are effective. Larger containers can be used along with mechanical collection trucks. Bulk containers or dumpsters should be used where large volumes of refuse are generated, such as at shopping centers, restaurants, apartment buildings, and hotels.

The collection truck and crew make up the most important element of a collection system. Collection trucks most commonly used in the United States are of the enclosed, compacting type. The collection capacity can vary, but is typically about 24 m^3 (32 yd^3). Compaction in the collection truck significantly reduces the volume of loose refuse. Loose refuse may weigh about 0.75 to 1.5 kN/m^3 (130 to 260 lb/yd^3). After compaction in a truck, it may weigh as much as 2 to 4 kN/m^3 (350 to 700 lb/yd^3). In other words, compaction in a collection vehicle temporarily reduces the refuse volume by as much as 80 percent. (The volume will increase again to some degree when the refuse is unloaded at a processing or disposal facility.)

The majority of refuse collection truck are front-side, or rear-loading compactors, although some non-compaction closed-body trucks are also used. Typical collection trucks are shown in Figure

4.2. Crew sizes can vary from one to four people. A one-person crew generally uses a special side-loading vehicle or mechanical collection truck to minimize walking time. Four-person crews may be used where refuse is collected from the rear or side yards of houses. Two- or three person collection crews are most typical, particularly with curbside collection routes.

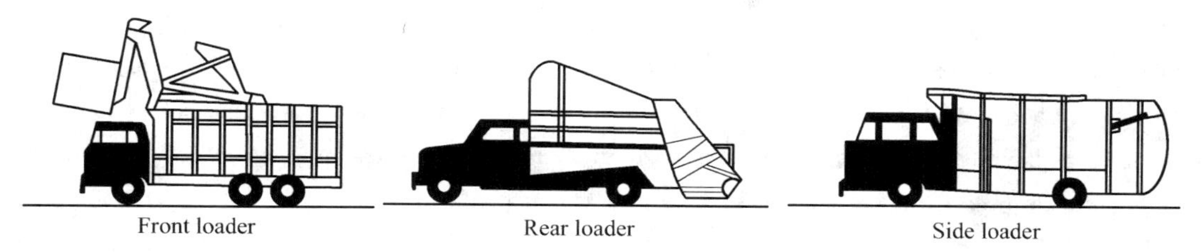

Figure 4.2 Enclosed compaction-type refuse collection vehicles reduce the volume of collected waste material by at least 50 percent. (Courtesy of the New York Department of Environmental Conservation)

Technical decisions must be made with regard to frequency of collection or pickup from each waste generation site and the point of pickup (curb, alley, backyard, or other). These decisions depend on the type of community, the type of waste (mixed or separated), the population density, and the land use in the collection area. Cost is also a major consideration; to lower costs, there is a tendency toward reduced collection frequency, an increased use of curbside (rather than walk-in) collection, and an increase in mechanical collection from standardized waste storage containers.

Mechanical collection systems are becoming popular in many communities because of improved esthetics of curbside container placement as well as lower costs. These systems consist of standardized containers and truck-mounted lifting mechanisms. In fully automatic systems, an articulated arm mechanism on the vehicle engages, lifts, empties, and replaces the container without manual assistance. Semiautomatic systems require a truck crew member to place the container in position to be automatically hoisted and emptied into the collection truck and then manually returned to its setout position Standardized MSW collection containers for single-family homes may vary from 225 to 360 L (60 to 95 gal) in size. Larger containers, up to 1 200L (315 gal) or greater, may be used at multifamily dwellings. All containers are wheel-mounted for ease of movement.

Combined collection of garbage and rubbish is generally more economical than separate collection of these types of refuse. In many communities, however, certain materials are recycled. Homeowners practice source separation, that is, they separate glass, metal, paper, and plastic from the remainder of their refuse. The recyclable materials are then picked up in a separate collection truck to prevent the refuse from contaminating the recyclable component and lowering its resale value. Recycling collection trucks often have side-loading compartments for the various recyclables.

4.3.1 Transfer Stations

It is not always feasible for individual collection trucks to haul refuse to a waste processing plant or final disposal site, especially if the ultimate destination is not in the immediate vicinity of the community in which the waste is collected. To solve this waste transport problem efficiently, one or more transfer stations may be used.

A transfer station is a facility at which solid wastes from individual collection trucks are consolidated into larger vehicles, such as tractor-trailer units. It is more economical for a few of these larger vehicles to transport the consolidated solid waste over the long-haul distance to the processing or disposal location rather than have each collection truck make the trip. A one-way haul distance of about 20 km (12 mi) may be a typical upper limit for an individual waste collection truck, but thorough engineering and cost-benefit comparison studies are generally conducted to determine the need for and advantages of a transfer station.

Individual transfer station capacities may vary from somewhat less than 100 tons to more than 500 tons of waste per day, depending on the size of the community. There are two basic modes of operation: direct discharge or storage discharge. In a storage discharge transfer station, the refuse is first emptied from the collection trucks into a storage pit or onto a large platform. Grapples or front-end loaders are then used to load or push the waste, respectively, into large trailer units.

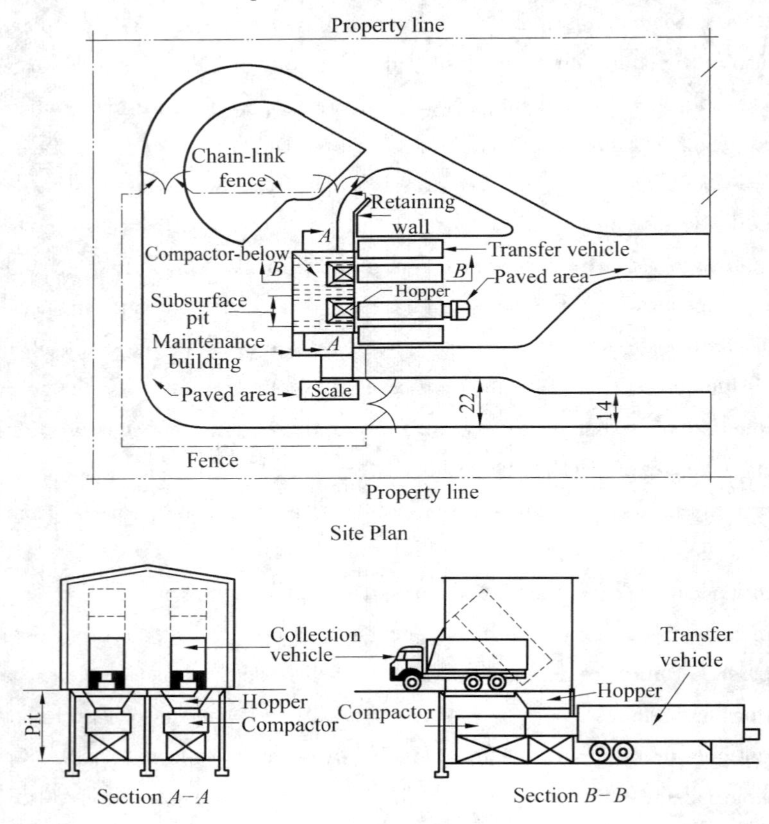

Figure 4.3　At a direct-discharge transfer station, several collection trucks deposit refuse into a large vehicle for hauling to a more distant disposal site. (From J. A. Salvato, Environmental Engineering and Sanitation, 4th ed., Wiley, New York.)

In a direct discharge station, each refuse truck empties directly into the larger transport vehicles. The trailers typically have a capacity of about 75 m^3 (100 yd^3) and hold the solid waste from four collection vehicles if it is not compacted and from up to eight collection vehicles if it is compacted. A direct discharge transfer station requires a two-level arrangement, as depicted in Figure 4.3. A backhoe equipped with a tamping device may be used to compact refuse dumped directly into an open trailer. Mechanical top-closing panels are used to produce a closed vehicle during transport.

In addition to open-top trailers, two types of closed compactor trailers are available. In one type, the compactor is built into the trailer and compacts the waste for later ejection at the process plant or disposal site. The second type of trailer is anchored to a separate compactor unit during the loading process; the trailers must be equipped with conveyors or ejection devices for unloading the compacted refuse. Another type of transfer station design, called the push pit station, includes a storage pit with a ram at one end and a hopper at the other end. After the pit is filled by the collection trucks, the ram pushes the refuse into the hopper, which loads the trailers.

4.3.2 Other Collection Methods

Before leaving the topic of municipal solid waste collection, a few other waste collection and transport methods should be mentioned. Some homes are equipped with garbage grinders, for example. These devices reduce the amount of food wastes in refuse. Since biweekly collection frequencies are usually only necessary because of the rapid decomposition of garbage, if all homes in a community had grinders, the collection frequency could be cut in half. Although the ground garbage winds up in sewage and flows to a wastewater treatment plant, most sewer systems and treatment plant can handle the extra load; an engineering study would have to confirm this.

Innovative collection systems involving pneumatic pipeline transport have been tried. In pneumatic systems, refuse is pulled by suction or vacuum through underground pipes to a central processing plant. Waste collection at the Disney World amusement park in Florida, for example, is done by a system of this type. It eliminates the need for noisy and unsightly refuse collection trucks. But complex controls, valves, and high-speed turbines are required for operation of the system, and installation costs are high. Pneumatic collection systems have also been installed in some small communities in Sweden and Japan. Sweden was the first country to use pneumatic transport of refuse in a large pipeline (300 mm), from an apartment house to an incinerator that also provided energy for space heating. Despite the high-tech appeal of pneumatic waste collection and transport systems, they are feasible only in specialized local situations and are unlikely to replace conventional methods in the foreseeable future.

4.4 Solid Waste Processing

Municipal solid waste may be treated or processed prior to final disposal. Solid waste processing provides several advantages. First, it can serve to reduce the total volume and weight of waste material that requires final disposal. Volume reduction helps to conserve land resources, since the land is the ultimate sink or repository for most waste material. It also reduces the total cost of transporting the waste to its final disposal site. In addition to volume and weight reduction, waste processing changes its form and improves its handling characteristics. Processing can also serve to recover natural resources and energy in the waste material for reuse, or recycling. The most widely used municipal waste treatment processes, including incineration, shredding, pulverizing, baling, and composting, are discussed in this section. Although incineration (burning) greatly reduces the waste volume, it is a processing rather than a disposal operation; land burial is still required for final disposal of the ashes and other unburned residue that remains behind.

4.4.1 Incineration

Incineration is the process of burning refuse in a controlled manner. The incineration of refuse was quite common in North America and Western Europe before 1940. However, many incinerators were eliminated because of aesthetic concerns, such as foul odors, noxious gases, and gritty smoke, rather than for reasons of public health. Today, about 15 percent of the municipal solid waste in the United States is incinerated; Canada incinerates about 8 percent. Most incinerators are not used just burn trash. The heat derived from the burning is converted into steam and electricity (see Figure 4.4). In 2000, about 100 combustors with energy recovery existed in the United States, with the capacity to burn up to 100 000 tonnes (110 000 U.S. tons) of MSW per day.

Most incineration facilities burn unprocessed municipal solid waste. This is not as efficient as some other technologies. About one-fourth of the incinerators use refuse-derived fuel-collected refuse that has been processed into pellets prior to combustion.

The newest means of incineration, a European concept, is called mass burn. In the mass-burn technique, municipal solid waste is fed into a furnace, where it falls onto moving grates and is burned at temperatures up to 1 300 ℃ (2 400 ℉). The burning waste heats water, and the steam drives a turbine to generate electricity, which is sold to a utility.

Incinerators drastically reduce the amount of municipal solid waste-up to 90 percent by volume and 75 percent by weight. Primary risks of incineration, however, involve air-quality problems and the toxicity and disposal of the ash.

Though mass-burn technology works efficiently in Europe, the technology is not easily transferable. North American municipal solid waste contains more plastic and toxic materials than European waste, thus creating air-pollution and ash-toxicity concerns. Modern incinerators have many pollution control devices that trap nearly all of the pollutants produced. However, tiny amounts of pollutants are released into the atmosphere, including certain metals, acid gases, and classes of chemicals known as dioxins and furans, which have been implicated in birth defects and several kinds of cancer. The long-term risks from the emissions are still a subject of debate.

Ash from incineration is also an important issue. Small concentrations of heavy metals are

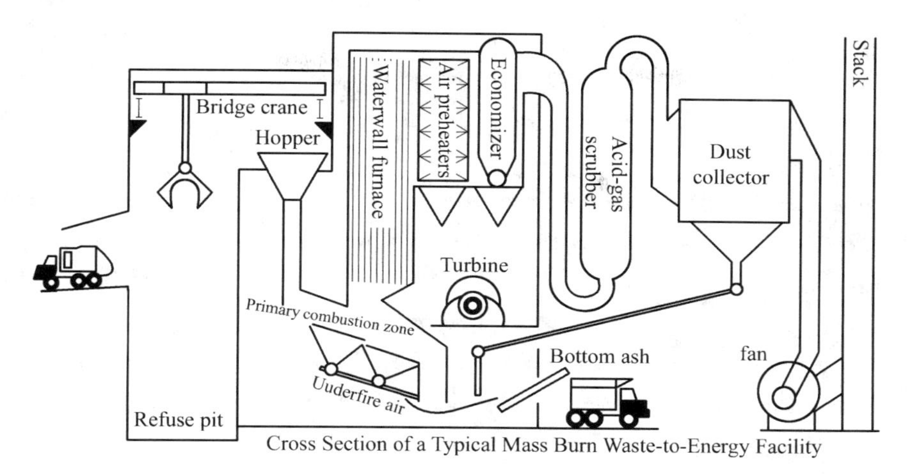

Figure 4.4 A Mass Burn Incineration System (Source: Data from the U.S. Environmental Protection Agency)

present both in the air emissions (fly ash) and residue (bottom ash) from these facilities. Because the ash contains lead, cadmium, mercury, and arsenic in varying concentrations from such items as batteries, lighting fixtures, and pigments, this ash may need to be treated as hazardous waste. The toxic substances are more concentrated in the ash than in the original garbage and can seep into groundwater from poorly sealed landfills. Many cities have had difficulty disposing of incinerator ash, and there is still considerable debate about what is the best method of disposal.

The cost of the land and construction for new incinerators are also major concerns facing many communities. Incinerator construction is often a municipality's single largest bond issue. Incinerator construction costs in North America in 2000 ranged from $45 million to $350 million, and the costs are not likely to decline.

Incineration is also more costly than landfills in most situations. As long as landfills are available, they will have a cost advantage. When cities are unable to dispose of their trash locally in a landfill and must begin to transport the trash to distant sites, incinerators become more cost effective. The U.S. Environmental Protection Agency has not looked favorably on the construction of new waste-to-energy facilities and has encouraged recycling and source reduction as more effective ways to reduce the solid waste problem. Critics have argued that cities and towns have impeded waste reduction and recycling efforts by putting a priority on incinerators and committing resources to them. Proponents of incineration have been known to oppose source reduction. They argue that incinerators need large amounts of municipal solid waste to operate and that reducing the amount of waste generated makes incineration impractical. Many communities that have opposed incineration say that they support a vigorous waste-reduction and recycling effort.

4.4.2 Composting

Composting is the process of harnessing the natural process of decomposition to transform organic materials—anything from manure and corncobs to grass and soiled paper—into compost, a humuslike material with many environmental benefits. In nature, leaves and branches that fall to the forest floor form a rich, moist layer of mulch that protects the roots of plants and provides a home for nature's most fundamental recyclers: worms, insects, and a host of bacteria, fungi, and other microorganisms.

Properly managing air and moisture provides ideal conditions for these organisms to transform large quantities of organic material into compost in a few weeks. A good small-scale example is a backyard compost pile. Green materials (grass, kitchen vegetable scraps, and flower clippings) mixed with brown materials (twigs, dry leaves, and soiled paper towels) at a ratio of 1:3 provide a balance of nitrogen and carbon that helps microbes efficiently decompose these materials.

Large-scale municipal composting uses the same principles of organic decomposition to process large volumes of organic materials. Composting facilities of various sizes and technological sophistication accept materials such as yard trimmings, food scraps, biosolids from sewage treatment plants, wood shavings, unrecyclable paper and other organic materials. These materials undergo processing—shredding, turning, and mixing—and, depending on the materials, can be turned into compost in a period ranging from 8 to 24 weeks. About 3 800 composting facilities are in use in the United States. In 2000, 57 percent of yard trimmings were composted in the United States through municipal programs. Most municipal programs involve one of three composting methods: windrows, aerated piles, or enclosed vessels.

Figure 4.5 (a) Windrow turning machine; (b) windrow turning arrangement. (L. F. Diaz, Composting and Recycling, Lewis Publishers, Boca Raton, Florida, 1993.)

Windrow systems involve placing the compostable materials into long piles or rows called windrows. Tractors with front end loaders or other kinds of specialized machinery are used to turn the piles periodically. Turning mixes the different kinds of materials and aerates the mixture (see Figure 4.5). Aerated piles are large piles of material that can be composted if they have air pumped through them (aeration), and a layer of mature compost or other material is used to insulate the pile to keep it at an optimal temperature. Air is forced through the pile, but no mechanical turning or agitation is done. Enclosed vessels can also be used to compost materials very rapidly (within days). However, these systems are much more technologically complex. In such systems, compostable material is fed into a dram, silo, or other structure where the environmental conditions

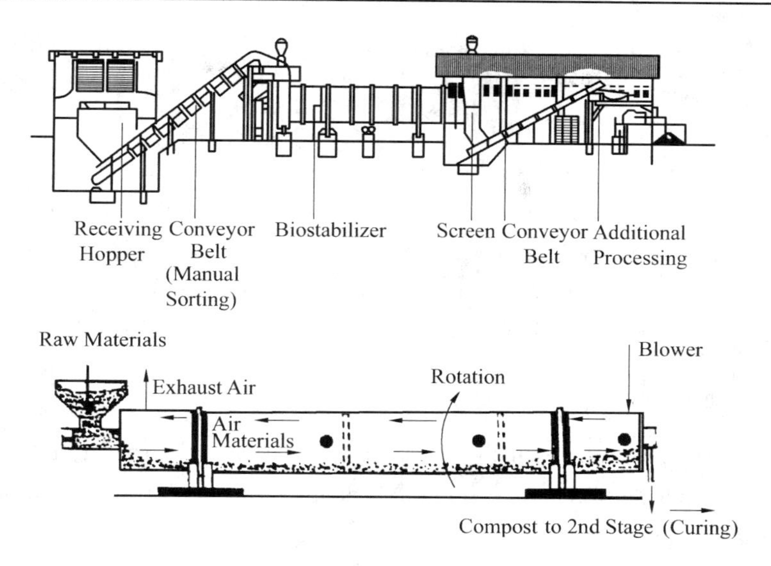

Figure 4.6 Enclosed mechanical-type composting system. (L. F. Diaz, Composting and Recycling, Lewis Publishers, Boca Raton, Florida, 1993.)

are closely controlled, and the material is mechanically aerated and mixed (see Figure 4.6).

In addition to keeping wastes from entering a landfill, composting has other significant benefits. The addition of compost to soil will improve it by making clay soils more porous or increasing the water-holding capacity of sandy soils. Nitrogen, potassium, iron, phosphorus, sulfur, and calcium are all common in compost and are beneficial to plant growth. Microorganisms are an important component of compost and play a valuable role in organic matter decomposition, which, in turn, leads to humus formation and nutrient availability.

4.4.3 Shredding and Pulverizing

Size reduction of municipal solid waste is accomplished by the physical processes of shredding or pulverizing. Shredding refers to the actions of cutting and tearing, whereas pulverizing refers to the actions of crushing and grinding. These two terms are frequently used synonymously with regard to solid waste management. Note that the size reduction obtained by shredding or pulverizing refers to the size of individual components or pieces of the solid waste material. However shredding and pulverizing also reduce the overall volume of the original or raw waste material, sometimes by as much as 40 percent.

There are many reasons for size reduction of municipal solid waste. The production of refuse-derived fuel, or RDF, requires processing of the raw solid waste; this typically includes shredding and pulverizing. Composting, which discussed in the last section, also frequently requires some type of size reduction process. Shredding and pulverizing may first be applied where the basic objective is to recover material from the waste that can be recycled and marketed. The size reduction and homogenizing processes improve the performance of the mechanical separation machinery. Finally, shredding of refuse prior to land burial can increase the capacity of the landfill; it also reduces the potential of rodent infestation, since the animal shave difficulty finding food scraps or voids for a habitat in the homogeneous material.

1. Hammer Mills

One of the most common types of equipment used for processing MSW into a uniform or homogeneous mass is the hammer mill. A hammer mill is a mechanical impact device in which the raw solid waste material is hit with a force sufficient to crush or tear individual pieces of the waste. Impact is provided by several hammers that rotate at high speeds (up to 1 500 rev/min) around a center horizontal or vertical shaft. A vertical hammer mill is shown in Figure 4.7.

In horizontal-shaft hammer mills, cutting bars or a breaker plate attached around the periphery of the mill chamber also help to reduce the size of the waste simply falls through the great at the bottom of the chamber. Not unexpectedly, repair and replacement of hammer mill components are part of a frequent maintenance routine due to the high speeds and impact action. In addition to the cost of maintenance, electric power requirements are high.

A hammer mill is a very versatile size reduction device because it will accept almost any type of waste material (except of course very bulky or dense items such as tree stumps or engine blocks). It is possible to reduce the size of solid waste material component to uniform fragments between 25 and 50 mm (1 and 2 in.) with proper operation. A typical size for a hammer mill is a 150-hp unit capable of processing about 100 kN (11 tons) of solid waste per hour. In addition to relatively high costs for operation and maintenance, the disadvantages of size reduction by hammer mills include noise and dust generation.

A modern innovation in the equipment used for MSW size reduction is the rotary shear shredder. This is a high-torque, relatively low-speed (up to 60 rev/min) machine, consisting of two or more parallel horizontal shafts that rotate in opposite directions. Each shaft has cutters that shear and tear the waste material. The high torque and shearing action allow this type of machine to shred difficult materials, such as tires. (More than 200 million used tires are discarded each year in the United States.)

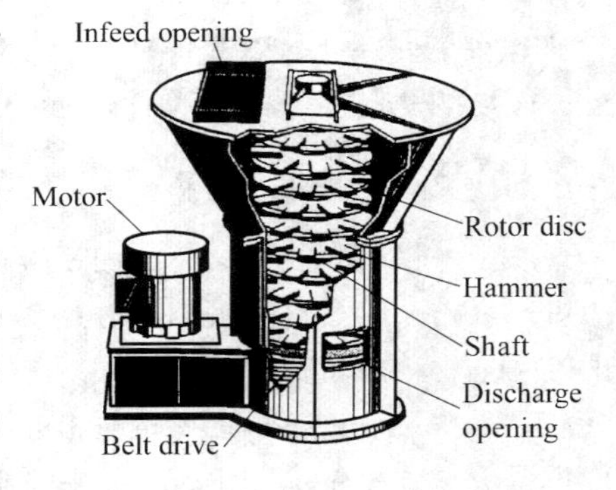

Figure 4.7 Vertical hammer mill. (L. F. Diaz, Composting and Recycling, Lewis Publishers, Boca Raton, Florida, 1993.)

2. Baling

Compacting solid waste into the form of rectangular blocks or bales is called baling. MSW bales are typically about 1.5 m^3 (2 yd^3) in size and weigh roughly 1 kN (or 1 ton). Solid waste can be compacted under high pressures (about 700 kPa or 100 psi) in either vertical or horizontal presses; the bales are frequently wrapped with steel wire to help retain their rectangular shape during handling. (They also may be enclosed in hot asphalt, plastic, or Portland cement, or tied with metal bands, depending on the intended use or disposal method. If moisture content and compaction pressures are high enough, they may retain their shape without being wire-wrapped or encased.) Semiautomatic horizontal presses can bale up to 36 kN (or 4 tons) per hour of MSW. Volume reduction can be as much as 90 percent of the original waste volume.

Solid waste volume reduction may be expressed in terms of a compaction ratio, as well as in percent. An understanding of the relationship between percent volume reduction and compaction ratio is important, particularly when reviewing and interpreting the manufacturers' data and selecting or specifying suitable compaction equipment. Appropriate formulas are

$$\text{percent volume reduction} = \frac{\text{initial volume} - \text{final volume}}{\text{initial volume}} \times 100 \tag{4.1}$$

$$\text{compaction ratio} = \frac{\text{initial volume}}{\text{final volume}} \tag{4.2}$$

The basic advantages of an MSW baling process include the significant decrease in waste volume, the ease of handling the compacted refuse, and the reduction of litter and nuisance potential. Additionally, the compacted waste can be hauled to a landfill by conventional vehicles, and the service life of the landfill can be greatly increased (by as much as 60 percent) because of the smaller volume of waste requiring burial. At the landfill, the bales can be neatly stacked in place without a problem of windblown debris, the likelihood of animal or insect infestation is decreased, soil cover requirements are reduced, and the need for onsite compaction is eliminated.

4.5　Solid Waste Recycling

As the amount of material recovered from MSW continues to increase as communities develop programs to meet waste diversion goals, materials specifications will become an important factor. In general, there is less contamination in source-separated material, but collection is more labor-intensive, and many communities are choosing to sort all materials at a central materials recovery facility (MRF). In many regions, markets for materials are not keeping pace with the volume collected, and it is expected that buyers will tighten specifications; as a result, vendors will no longer have assured markets, and will be competing to sell materials. As the specifications for recovered materials become more restrictive, recovery program managers must consider buyer specifications carefully when choosing collection and sorting systems, especially where large capital expenditures are involved.

4.5.1　Two Types of Recycling

Recycling has a number of benefits to people and the environment (see Figure 4.8). There are two types of recycling for materials: ① Primary, or closed-loop, recycling, in which wastes discarded by consumers (postconsumer wastes) are recycled to produce new products of the same type (such as newspaper into newspaper and aluminum cans into aluminum cans). This reduces pollution and use of virgin resources and saves energy. ② Secondary, or downcycling, in which waste materials are converted into different and usually lower-quality products.

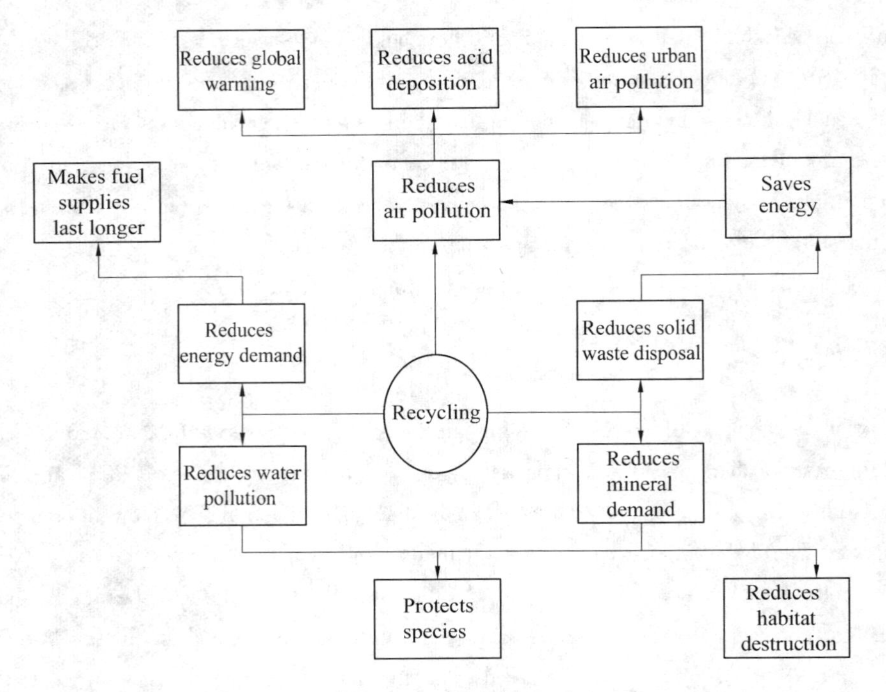

Figure 4.8 Some benefits of recycling for people and the environment.

Environmentalists urge us not to be misled by labels claiming that paper and plastic bags or other items are recyclable. Just about anything is recyclable. What counts is whether ① an item is actually recycled (ideally by primary recycling), ② it is designed for recycling by not containing mixtures of materials that are difficult and expensive to recycle, and ③ we complete the recycling loop by buying products using the maximum feasible content of postconsumer recycled materials.

Large-scale recycling can be accomplished by collecting mixed urban waste and transporting it to centralized MRFs. There, machines shred and automatically separated the mixed waste to recover valuable materials for sale to manufacturers as raw materials (see Figure 4.9). The remaining paper, plastics, and other combustible wastes are recycled or burned to produce steam or electricity to run the recovery plant or to sell to nearby industries or homes. Ash from the incinerator is buried in a landfill.

More than 225 MRFs operate in the United States. However, such plants (1) are expensive to build, operate, and maintain (which is why some have been shut down), (2) can emit toxic air pollutants if not operated properly, and (3) produce a toxic ash that must be disposed of safely. MRFs also must have a large input of garbage to make them financially successful. Thus their owners have a vested interest in increasing throughput of matter and energy resources to produce more trash, the reverse of what prominent scientists believe we should be doing.

Many solid waste experts argue that it makes more sense economically and environmentally for households and businesses to keep trash separate in recyclable and reusable categories (such as glass, paper, metals, certain types of plastics, and compostable materials). Then

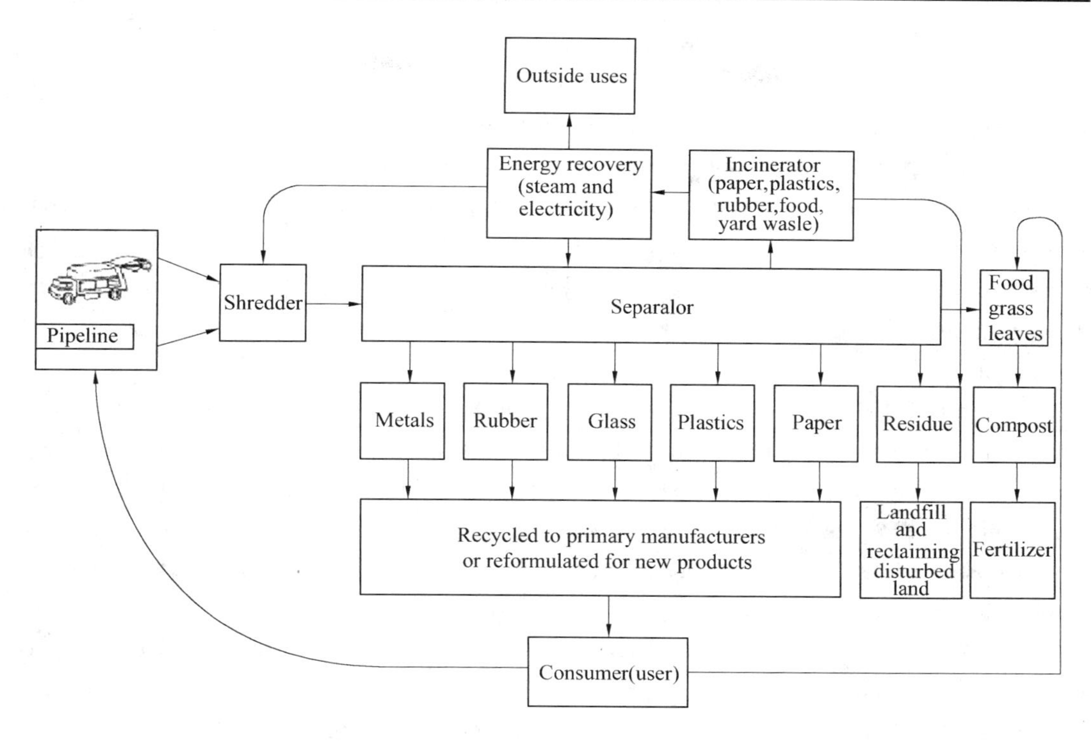

Figure 4.9 Schematic of a generalized materials-recovery facility used to sort mixed wastes for recycling and burning to produce energy

compartmentalized city collection trucks, private haulers, or volunteer recycling organizations pick up the segregated wastes and sell them to scrap dealers, compost plants, and manufacturers. Another alternative (especially in less populated areas) is to establish a net work of drop-off centers, buyback centers, and deposit-refund programs in which people deliver and sell or donate separated recyclable materials.

This source separation approach (1) produces little air and water pollution, (2) has low startup costs and moderate operating costs, (3) saves more energy and provides more jobs per unit of material than MRFs, landfills, and incinerators, (4) yields cleaner and usually more valuable recyclables, and (5) educates people about the need for waste reduction, reuse, and recycling.

Aluminum and paper separated out for recycling are worth a lot of money. As a result, in a growing number of cities people steal these materials—from curbside containers set out by residents and from unprotected recycling drop-off centers—and sell them. This undermines municipal recycling programs by lowering the income available from selling these materials.

4.5.2 Processing Recyclables

1. Materials Recycling Facility

Where communities have commingled collection, the materials must be separated by type before they can be sold. Separation and sorting at a MRF offers an efficient and reliable method for processing recyclable materials; it also allows form marketing of larger amounts of quality material. A large, centrally located MRF serving many communities can accomplish separation and sorting of recyclables on a relatively economical basis through economies of scale.

A typical MRF is an industrial-type metal or concrete building similar to a warehouse. It has high doors to allow trucks to dump inside. An average MRF building has a floor area of about 4 000 m^2 and contains separate areas for depositing paper and commingled materials. In many cases, air pollution control equipment is installed to collect and remove dust from the air. Sometimes they are heated, and special sorting rooms are constructed to provide workers with proper lighting and ventilation.

In general, the following types of materials are received at a MRF in a collection truck separated into two or three compartments: ① Paper Compartment: old newspaper (ONP), generally bundled; old corrugated cardboard (OCC); mixed paper, including envelopes, magazines, and junk mail. ② Commingled Compartment: clear, brown, and green glass; ferrous metal ("tin") food and beverage containers; aluminum food and beverage containers; HDPE and PET plastics (milk, detergent, and soft drink containers).

A MRF utilizes a combination of labor and equipment to sort the source-separated, commingled material from the collection trucks into various categories for marking. A variety of technologies are used to improve the quality of the material by carefully separating glass (by color), steel, aluminum, and plastic (by resin type) into uniform commodities. Paper is also processed. A schematic diagram of a typical 100 to 300-ton-per-day MRF is shown in Figure 4.10. MRFs may be custom-designed by consulting engineers and built by contractors or constructed as turnkey facilities by companies that design, build, and operate the systems.

2. Materials Commonly Separated from MSW

The most common ones from MSW are aluminum, paper, plastics, glass, ferrous metal, nonferrous metal, yard wastes, and construction and demolition wastes. Each is considered briefly in the following discussion.

(1) Aluminum

Aluminum recycling is made up of two sectors: aluminum cans and secondary aluminum. Secondary aluminum includes window frames, storm doors, siding, and gutters. Because secondary materials are of different grades, specifications for recycled aluminum should be checked, to recover the maximum value when selling separated materials to brokers. The demand for recycled aluminum cans is high, as it takes 95 percent less energy to produce an aluminum can from an existing can than from ore.

(2) Paper

The principal types of waste paper that are recycled are old newspaper, cardboard, high-grade paper, and mixed paper. Each of these four grades consist of indiviual grades, which are defined according to the type of fiber, source, homogeneity, extent of printing, and physical or chemical characteristics. High-grade paper includes office paper, reproduction paper, computer printout, and

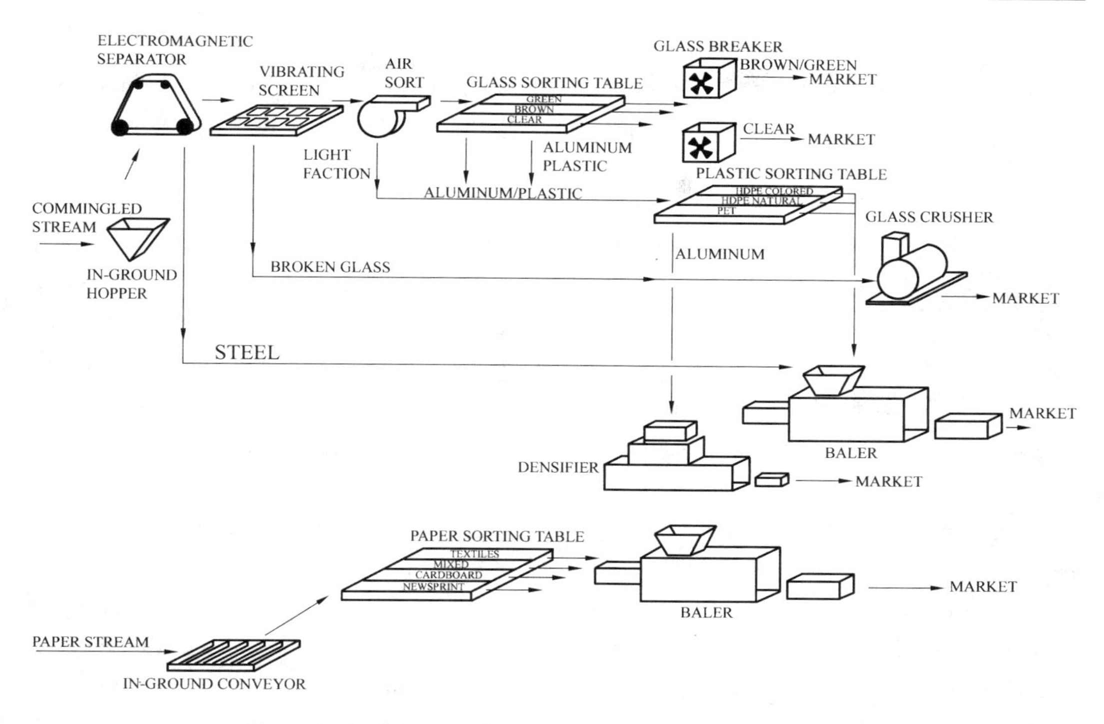

Figure 4.10　Schematic diagram of material flow at a typical MRF

other grades having a high percentage of long fibers. Mixed grades include paper with high groundwood content, such as magazines; coated paper; and individual grades containing excessive percentages of "outthrows" (papers of lower grades than the grade specified).

(3) Plastics

Plastics can be classified into two general categories: clean commercial grade scrap and post-consumer scrap. The two types of post-consumer plastics that are now most commonly recycled are polyethylene terephthalate (PETE/1), which is used for the manufacture of soft drink bottles, and high-density polyethylene (HDPE/2), used for milk and water containers and detergent bottles. In 1987, more than 150 million pounds of plastic soft drink bottles were recycled. Even so, less than five percent of the available scrap plastic is being recycled. It is anticipated that all of the other types of plastics will be recycled in greater quantities in the future, however, as processing technologies improve.

(4) Glass

Glass is also a commonly recycled material. Container glass (for food and beverage packing), flat glass (e.g., window glass), and pressed or amber and green glass are the three principal types of glass found in MSW. Glass to be reprocessed is often separated by color into categories of clear, green, and amber.

(5) Ferrous Metals (Iron and Steel)

The largest amount of recycled steel has traditionally come from large items such as cars and appliances. Many communities have large scrap metal piles at the local landfill or transfer station.

In many cases, the piles are unorganized and different metals are mixed together, making them unattractive to scrap metal buyers. Steel can recycling is also becoming more popular. Steel cans, used as juice, soft drink, and food containers, are easily separated from mixed recyclables or municipal solid waste using large magnets (which also separate other ferrous metals).

(6) Nonferrous Metals

Recyclable nonferrous metals are recovered from common household items (outdoor furniture, kitchen cookware and appliances, ladders, tools, hardware); from construction and demolition projects (copper wire, pipe and plumbing supplies, light fixtures, aluminum siding, gutters and downspouts, doors, windows); and from large consumer, commercial, and industrial products (appliances, automobiles, boats, trucks, aircraft, machinery). Virtually all nonferrous metals can be recycled if they are sorted and free of foreign materials such as plastics, fabrics, and rubber.

(7) Yard Wastes Collected Separately

In most communities yard wastes are collected separately. The composting of yard wastes has become of great interest as cities and towns seek to find ways in which to achieve mandated diversion goals. Leaves, grass clippings, bush clippings, and brush are the most commonly composted yard wastes. Stumps and wood are also compostable, but only after they have been chipped to produce a smaller more uniform size. Composting of the organic fraction of MSW is also becoming more popular.

(8) Construction and Demolition Wastes

In many locations construction and demolition (C&D) wastes are now being processed to recover marketable items such as wood chips for use as a fuel in biomass combustion facilities, aggregate for concrete in construction projects, ferrous and nonferrous metals for remanufacture, and soil for use as fill material. The reprocessing of C&D wastes is gaining in popularity as disposal fees at landfills continue to increase. When disposal fees were below 5 dollars per ton (early 1970s), reprocessing was not economically feasible. Today (1992), with average landfill disposal fees approaching 60 dollars per ton in many parts of the country, the reprocessing of C&D wastes is economically feasible.

3. Typical MRF Operation

Incoming trucks are weighed and the waste is then dumped on a concrete tipping floor. Any unsuitable materials seen on the floor are removed by hand and placed in containers for haul to the landfill. A large front-end loader then pushes the material onto inclined rubber-belt conveyors, which may pass by one or more workers who remove deleterious materials, such as large metal pots, bricks, or garbage. The conveyor then passes under an electromagnetic separator that removes the tin cans and other ferrous metals from the commingled waste stream. This metal is then conveyed to a baler for compression and baling prior to shipment to steel mills.

The remaining commingled material is then screened on a shaker table to remove the dirt and broken glass. This broken glass has little or no resale and frequently will be used in the manufacture of the material is then classified with large blowers to remove the plastic and aluminum containers from the remaining unbroken glass bottles. These glass bottles are then conveyed into a sorting room, where workers separate the bottles by color; if the glass is not separated by color, it has no resale value. In most MRFs, the glass bottles are further processed by crushing them into roughly 12 mm (1/2 in.) particles and then removing bottle caps and other such material in a rotating drum trommel screen classifier. The resultant glass product is marketable as furnace-ready cullet.

The plastic and aluminum containers are also conveyed to a sorting area. The separation of plastic is typically accomplished by hand with skilled sorters who can identify the plastic types through experience. Recently, equipment has been developed that can allow the complete automation of plastic sorting. The equipment is expensive, but it can eliminate three or four sorters who normally have to separated, it is either baled or chipped for shipment to market. Other types of plastic, such as polystyrene and styrofoam, are not generally recovered because of low tonnages and high processing costs.

The aluminum is usually separated from the plastic and other remaining nonrecyclable materials on the conveyor belt through the use of a device called an eddy current separator. This device repels aluminum up into the air and off of the belt, allowing it to be captured and baled or densified for shipment to market. While aluminum typically accounts for about 5 percent of the commingled waste stream, it produces almost 80 percent of the commingled revenue.

Quality control inspectors watch the mechanized separation processes carefully to ensure that the various materials do not mix in the plant and that the individual commodities remain clean and pure. By mechanizing the process and maintaining high quality control, a large MRF can easily process up to 300 tons of commingled material per day.

Paper is processed in a MRF on a separate conveyor line. Old newspaper is generally bundled at the curb and not allowed to mix with the other papers in the collection truck. At the MRF, the ONP is dumped in a separate area prior to being conveyed past quality control workers, who remove materials that are considered to be contaminants. It is baled and loose-loaded into tractor trailers and shipped to paper mills for use in making newspaper. Much of me remaining paper is suitable for sale to tissue mills.

4.6 Disposal of Solid Wastes and Residual Matter

The safe and reliable long-term disposal of solid waste residues is an important component of integrated waste management. Solid waste residues are waste components that are not recycled, that remain after processing at a materials recovery facility, or that remain after the recovery of conversion products and/or energy. Historically, solid waste has been placed in the soil in the earth's surface or deposited in the oceans. Although ocean dumping of municipal solid waste was officially abandoned in the United States in 1933, it is now argued that many of the wastes now placed in landfills or on land could be used as fertilizers to increase productivity of the ocean or the land. It is also argued that the placement of wastes in ocean trenches where tectonic folding is occurring is an effective method of waste disposal. Nevertheless, landfilling or land disposal is today the most commonly used method for waste disposal by far.

4.6.1 The Landfill Method of Solid Waste Disposal

Historically, landfills have been the most economical and environmentally acceptable method for the disposal of solid wastes, both in the United States and throughout the world. Even with implementation of waste reduction, recycling, and transformation technologies, disposal of residual solid waste in landfills still remains an important component of an integrated solid waste management strategy (see Figure 4.11).

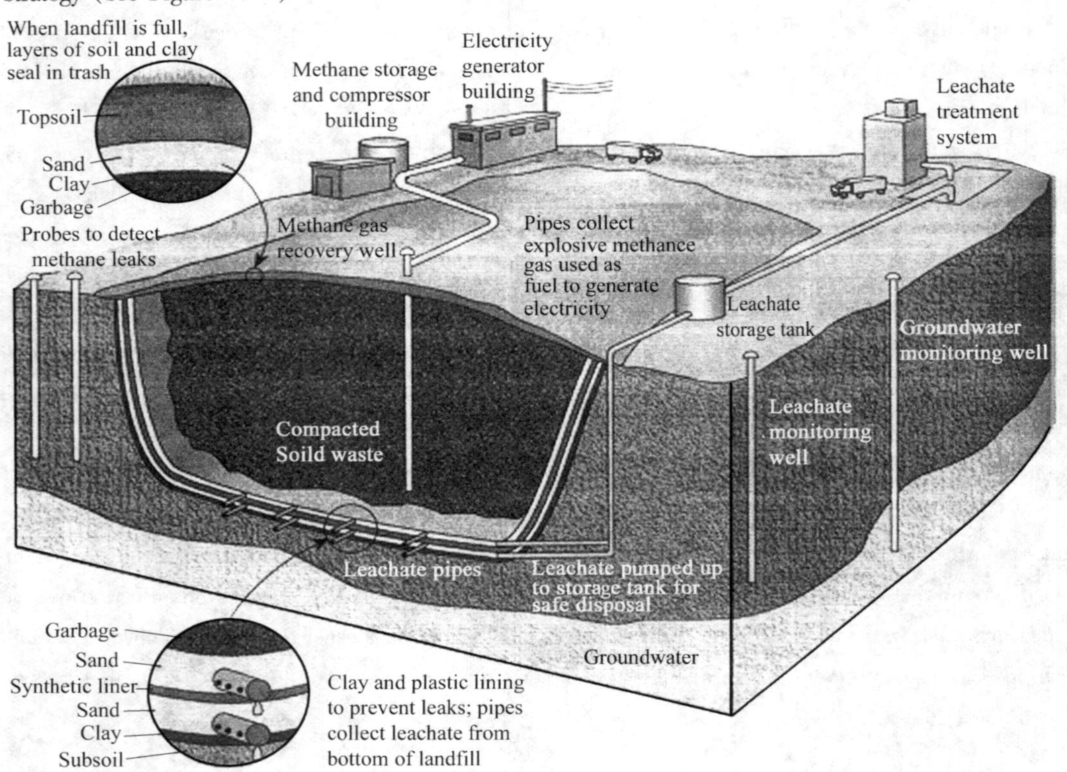

Figure 4.11 State-of the-art sanitary landfill are designed to eliminate or minimize environmental problems that plague older landfills

A modern municipal solid waste landfill is typically constructed above an impermeable clay layer that is lined with an impermeable membrane and includes mechanisms for dealing with liquid and gas materials generated by the contents of the landfill. Each day's deposit of fresh garbage is covered with a layer of soil to prevent it from blowing around and to discourage animals from scavenging for food. Selection of landfill sites is based on an understanding of local geologic conditions such as the presence of a suitable clay base, groundwater geology, and soil type. In addition, it is important to address local citizens' concerns. Once the site is selected, extensive construction activities are necessary to prepare it for use. New landfills have complex bottom layers to trap contaminant-laden water, called leachate, leaking through the buried trash. In addition, monitoring systems are necessary to detect methane gas production and groundwater contamination. In some eases, methane produced by decomposing waste is collected and used to produce heat or generate electricity. The water that leaches through the site must be collected and treated. As a result, new landfills are becoming increasingly more complex and expensive. They currently cost up to $1 million per hectare to prepare. Landfill management incorporates the planning, design, operation, closure, and postclosure control of landfills.

1. Definition of Landfilling Process Terms

Landfills are the physical facilities used for the disposal of residual solid wastes in the surface soils of the earth. In the past, the term sanitary landfill was used to denote a landfill in which the waste placed in the landfill was covered at the end of each day's operation. Today, sanitary landfill refers to an engineered facility for the disposal of MSW designed and operated to minimize public health and environmental impacts. Landfills for the disposal of hazardous wastes are called secure landfills. A sanitary landfill is also sometimes identified as a solid waste management unit. Landfilling is the process by which residual solid waste is placed in a landfill. Landfilling includes monitoring of the incoming waste stream, placement and compaction of the waste, and installation of landfill environmental monitoring and control facilities.

The term cell is used to describe the volume of material placed in a landfill during one operating period, usually one day (see Figure 4.12). A cell includes the solid waste deposited and the daily cover material surrounding it. Daily cover usually consists of 6 to 12 in of native soil or alternative materials such as compost that are applied to the working faces of the landfill at the end of each operating period. The purposes of daily cover are to control the blowing of waste materials; to prevent rats, flies, and other disease vectors from entering or exiting the landfill; and to control the entry of water into the landfill during operation.

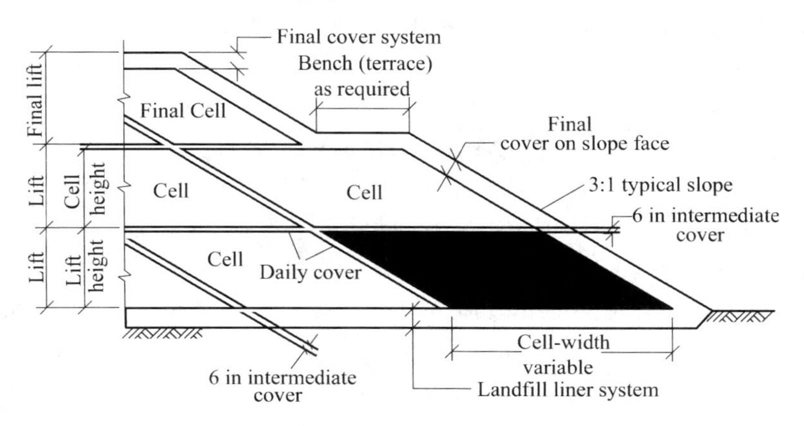

Figure 4.12 Sectional view through a sanitary landfill

A lift is a complete layer of cells over the active area of the landfill (see Figure 4.12).

Typically, landfills are comprised of a series of lifts. A bench (or terrace) is commonly used where the height of the landfill will exceed 50 to 75 ft. Benches are used to maintain the slope stability of the landfill, for the placement of surface water drainage channels, and for the location of landfill gas recovery piping. The final lift includes the cover layer. The final cover layer is applied to the entire landfill surface after all landfilling operations are complete. The final cover usually consists of multiple layers of soil and/or geomembrane materials designed to enhance surface drainage, intercept percolating water, and support surface vegetation.

The liquid that collects at the bottom of a landfill is known as leachate. In deep landfills, leachate is often collected at intermediate points. In general, leachate is a result of the percolation of precipitation, uncontrolled runoff, and irrigation water into the landfill. Leachate can also include water initially contained in the waste as well as infiltrating groundwater. Leachate contains a variety of chemical constituents derived from the solubilization of the materials deposited in the landfill and from the products of the chemical and biochemical reactions occurring within the landfill.

Landfill gas is the mixture of gases found within a landfill. The bulk of landfill gas consists of methane (CH_4) and carbon dioxide (CO_2), the principal products of the anaerobic decomposition of the biodegradable organic fraction of the MSW in the landfill. Other components of landfill gas include atmospheric nitrogen and oxygen, ammonia, and trace organic compounds.

Landfill liners are materials (both natural and manufactured) that are used to line the bottom area and below-grade sides of a landfill. Liners usually consist of layers of compacted clay and/or geomembrane material designed to prevent migration of landfill leachate and landfill gas. Landfill control facilities include liners, landfill leachate collection and extraction systems, landfill gas collection and extraction systems, and daily and final cover layers.

Environmental monitoring involves the activities, associated with collection and analysis of water and air samples that are used to monitor the movement of landfill gases and leachate at the landfill site. Landfill closure is the term used to describe the steps that must be taken to close and secure a landfill site once the filling operation has been completed. Postclosure care refers to the activities associated with the long-term monitoring and maintenance of the completed landfill (typically 30 to 50 years).

2. Overview of Landfill Planning, Design, and Operation

Concerns with the landfilling of solid waste are related to ① the uncontrolled release of landfill gases that might migrate off-site and cause odor and other potentially dangerous conditions, ② the impact of the uncontrolled discharge of landfill gases on the greenhouse effect in the atmosphere, ③ the uncontrolled release of leachate that might migrate down to underlying groundwater or to surface waters, ④ the breeding and harboring of disease vectors in improperly managed landfills, and ⑤ the health and environmental impacts associated with the release of the trace gases arising from the hazardous materials that were often placed in landfills in the past. The goal for the design and operation of a modern landfill is to eliminate or minimize the impacts associated with these concerns.

The principal elements that must be considered in the planning, design, and operation of landfills include: ① landfill layout and design; ② landfill operations and management; ③ the reactions occurring in landfills; ④ the management of landfill gases; ⑤ the management of leachate; ⑥ environmental monitoring; and ⑦ landfill closure and postclosure care.

(1) Preparation of the site for landfilling

The first step in the process involves the preparation of the site for landfill construction. Existing site drainage must be modified to route any runoff away from the intended landfill area. Rerouting of drainage is particularly important for ravine landfills where a significant watershed may drain through the site. In addition, drainage of the landfill area itself must be modified to route water away from the initial fill area. Other site preparation tasks include construction of access roads and weighing facilities, and installation of fences.

The next step in the development of a landfill is the excavation and preparation of the landfill bottom and subsurface sides. Modern landfills typically are constructed in sections. Working by sections allows only a small part of the unprotected landfill surface to be exposed to precipitation at any time. In addition, excavations are carried out over time, rather than preparing the entire landfill bottom at once. Excavated material can be stockpiled on unexcavated soil near the active area and the problem of precipitation collecting in the excavation is minimized. Where the entire bottom of the landfill is lined at once, provision must be made to remove stormwater runoff from the portion of the landfill that is not being used.

To minimize costs, it is desirable to obtain cover materials from the landfill site whenever possible. The initial working area of the landfill is excavated to the design depth, and the excavated material stockpiled for later use. Vadose zone (zone between ground surface and permanent groundwater) and groundwater monitoring equipment is installed before the landfill liner is laid down. The landfill bottom is shaped to provide drainage of leachate, and a low-permeability liner is installed. Leachate collection and extraction facilities are placed within or on top of the liner. Typically, the liner extends up the excavated walls of the landfill.

Horizontal gas recovery trenches may be installed at the bottom of the landfill, particularly if emissions of volatile organic compounds (VOCs) from the newly placed waste is expected to be a problem. To minimize the release of VOCs, a vacuum is applied and air is drawn through the completed portions of the landfill. The gas that is removed must be burned under controlled conditions to destroy the VOCs. Before the fill operation begins, a soil berm is constructed at the downwind side of the planned fill area. The berm serves as a windbreak to control blowing materials and as a face against which the waste can be compacted. For excavated landfills, the wall of the excavation usually serves as the initial compaction face.

(2) The placement of wastes

Once the landfill site has been prepared, the next step in the process involves the actual placement of waste material. The waste is placed in cells beginning along the compaction face, continuing outward and upward from the face. The waste deposited in each operating period, usually one day, forms an individual cell. Wastes deposited by the collection and transfer vehicles are spread out in 18- to 215-in layers and compacted. Typical cell heights vary from 8 to 12 ft. The length of the working face varies with the site conditions and the size of the operation. The working face is the area of a landfill where solid waste is being unloaded, placed and compacted during a given operating period. The width of a cell varies from 10 to 30 ft, again depending on the design and capacity of the landfill. All exposed faces of the cell are covered with a thin layer of soil (6 to 12 in) or other suitable material at the end of each operating period.

After one or more lifts have been placed, horizontal gas recovery trenches can be excavated in the completed surface. The excavated trenches are filled with gravel, and perforated plastic pipes are installed in the trenches. Landfill gas is extracted through the pipes as the gas is produced. Successive lifts are placed on top of one another until the final design grade is reached. Depending on the depth of the landfill, additional leachate collection facilities may be placed in successive lifts. A cover layer is applied to the completed landfill section. The final cover is designed to minimize infiltration of precipitation and to route drainage away from the active section of the landfill. The cover is landscaped to control erosion. Vertical gas extraction wells may be installed at this time through the completed landfill surface. The gas extraction system is tied together and the extracted gas may be flared or routed to energy recovery facilities as appropriate.

Additional sections of the landfill are constructed outward from the completed sections, repeating the construction steps outlined above. As organic materials deposited within the landfill decompose, completed sections may settle. Landfill construction activities must include refilling and repairing of settled landfill surfaces to maintain the desired final grade and drainage. The gas and leachate control systems also must be extended and maintained. Upon completion of all fill activities, the landfill surface is repaired and upgraded with the installation of a final cover. The site is landscaped appropriately and prepared for other uses.

(3) Postclosure management

Monitoring and maintenance of the completed landfill must continue by law for some time after closure (30 to 50 years). It is particularly important that the landfill surface be maintained and repaired to enhance drainage, that gas and leachate control systems be maintained and operated, and that the pollution detection system be monitored.

3. Reactions Occurring in Landfills

Solid wastes placed in a sanitary landfill undergo a number of simultaneous and interrelated biological, chemical, and physical changes. The most important biological reactions occurring in landfills are those involving the organic material in MSW that lead to the evolution of landfill gases and, eventually, liquids. The biological decomposition process usually proceeds aerobically for some short period immediately after deposition of the waste until the oxygen initially present is depleted. During aerobic decomposition CO_2 is the principal gas produced. Once the available oxygen has been consumed, the decomposition becomes anaerobic and the organic matter is converted to CO_2, CH_4, and trace amounts of ammonia and hydrogen sulfide. Many other chemical reactions are biologically mediated as well. Because of the number of interrelated influences, it is difficult to define the conditions that will exist in any landfill or portion of a landfill at any stated time.

Important chemical reactions that occur within the landfill include dissolution and suspension of landfill materials and biological conversion products in the liquid percolating through the waste, evaporation and vaporization of chemical compounds and water into the evolving landfill gas, sorption of volatile and semivolatile organic compounds into the landfilled material, dehalogenation and decomposition of organic compounds, and oxidation-reduction reactions affecting metals and the solubility of metal salts. The dissolution of biological conversion products and other compounds, particularly of organic compounds, into the leachate is of special importance because these materials can be transported out of the landfill with the leachate. These organic compounds can subsequently be released into the atmosphere either through the soil (where leachate has move away from an unlined landfill) or from uncovered leachate treatment facilities. Other important chemical reactions include those between certain organic compounds and clay liners, which may alter the structure and permeability of the liner material. The interrelationships of these chemical reactions within a landfill are not well understood.

Among the more important physical changes in landfills are the lateral diffusion of gases in the landfill and emission of landfill gases to the surrounding environment, movement of leachate within the landfill and into underlying soils, and settlement caused by consolidation and decomposition of landfilled material. Landfill gas movement and emissions are particularly important considerations in landfill management. As gas is evolved within a landfill, internal pressure may build, causing the landfill cover to crack and leak. Water entering the landfill through the leaking cover may enhance the gas production rate, causing still more cracking. Escaping landfill gas may carry trace carcinogenic and teratogenic compounds into the surrounding environment. Because landfill gas usually has a high methane content, there may be a combustion and/or explosion hazard. Leachate migration is another concern. As leachate migrates downward in the landfill, it may transfer compounds and materials to new locations where they may react more readily. Leachate occupies pore spaces in the landfill and in doing so may interfere with the migration of landfill gas.

4.6.2 Composition and Control of Landfill Gases

A solid waste landfill can be conceptualized as a biochemical reactor, with solid waste and water as the major inputs, and with landfill gas and leachate as the principal outputs. Material stored in the landfill includes partially biodegraded organic material and the other inorganic waste materials originally placed in the landfill. Landfill gas control systems are employed to prevent unwanted movement of landfill gas into the atmosphere or the lateral and vertical movement through the surrounding soil. Recovered landfill gas can be used to produce energy or can be flared under controlled conditions to eliminate the discharge of harmful constituents to the atmosphere.

1. Composition and Characteristics of Landfill Gas

Landfill gas is composed of a number of gases that are present in large amounts (the principal gases) and a number of gases that are present in very small amounts (the trace gases). The principal gases are produced from the decomposition of the organic fraction of MSW. Some of the trace gases, although present in small quantities, can be toxic and could present risks to public health.

Gases found in landfills include ammonia (NH_3), carbon dioxide (CO_2), carbon monoxide (CO), hydrogen (H_2), hydrogen sulfide (H_2S), methane (CH_4), nitrogen (N_3), and oxygen (O_2). Methane and carbon dioxide are the principal gases produced from the anaerobic decomposition of the biodegradable organic waste components in MSW. When methane is present in the air in concentrations between 5 and 15 percent, it is explosive. Because only limited amounts of oxygen are present in a landfill when methane concentrations reach this critical level, there is little danger that the landfill will explode. However, methane mixtures in the explosive range can form if landfill gas migrates off-site and mixes with air. The concentration of these gases that may be expected in the leachate will depend on their concentration in the gas phase in contact with the leachate. Because carbon dioxide will affect the pH of the leachate, carbonate equilibrium data can be used to estimate the pH of the leachate.

2. Management of Landfill Gas

Typically, landfill gases that have been recovered from an active landfill are either flared or used for the recovery of energy in the form of electricity, or both. More recently, the separation of the carbon dioxide from the methane in landfill gas has been suggested as an alternative to the production of heat and electricity.

A common method of treatment for landfill gases is thermal destruction; that is, methane and any other trace gases (including VOCs) are combusted in the presence of oxygen (contained in air) to carbon dioxide (CO_2), sulfur dioxide (SO_2), oxides of nitrogen, and other related gases. The thermal destruction of landfill gases is usually accomplished in a specially designed flaring facility. Because of concerns over air pollution, modem flaring facilities are designed to meet rigorous operating specifications to ensure effective destruction of VOCs and other similar compounds that may be present in the landfill gas. For example, a typical requirement might be a minimum combustion temperature of 1 500 °F and a residence time of 0.3 to 0.5 s, along with a variety of controls and instrumentation in the flaring station.

Landfill gas is usually converted to electricity. In smaller installations (up to 5 MW), it is common to use dual fuel internal combustion piston engines or gas turbines. In larger installations, the use of steam turbines is common. Where piston-type engines are used, the landfill gas must be processed to remove as much moisture as possible so as to limit damage to the cylinder heads. If the gas contains H_2S, the combustion temperature must be controlled carefully to avoid corrosion problems. Alternatively, the landfill gas can be passed through a scrubber containing iron shavings, or through other proprietary scrubbing devices, to remove the H_2S before the gas is combusted.

Combustion temperatures will also be critical where the landfill gas contains VOCs released from wastes placed in the landfill before the disposal of hazardous waste in municipal landfills was banned. The typical service cycle for dual fuel engines running on landfill gas varies from 3 000 to 10 000 hours before the engine must be overhauled. The typical service cycle for gas turbines running on landfill gas is approximately 10 000 hours.

Where there is a potential use for the CO_2 contained in the landfill gas, the CH_4 and CO_2 in landfill gas can be separated. The separation of the CO_2 from the CH_4 can be accomplished by physical adsorption, chemical adsorption, and by membrane separation. In physical and chemical adsorption, one component is adsorbed preferentially using a suitable solvent. Membrane separation involves the use of a semipermeable membrane to remove the CO_2 from the CH_4. Semipermeable membranes have been developed that allow CO_2, H_2S, and H_2O to pass while CH_4 is retained. Membranes are available as flat sheets or as hollow fibers. To increase efficiency of separation, the flat sheets are spiral wound on a support medium while the hollow fibers are grouped together in bundles.

4.6.3 Composition and Control of Leachate in Landfills

Leachate may be defined as liquid that has percolated through solid waste and has extracted dissolved or suspended materials. In most landfills leachate is composed of the liquid that has entered the landfill from external sources, such as surface drainage, rainfall, groundwater, and water from underground springs and the liquid produced from the decomposition of the wastes, if any.

1. Composition of Leachate

When water percolates through solid wastes that are undergoing decomposition, both biological materials and chemical constituents are leached into solution. Representative data on the characteristics of leachate are reported in Table 4.5 for both new and mature landfills. Because the range of the observed concentration values for the various constituents reported in Table 4.5 is rather large, especially for new landfills, great care should be exercised in using the typical values that are given.

Note that the chemical composition of leachate will vary greatly depending on the age of landfill and the events preceding the time of sampling. For example, if a leachate sample is collected during the acid phase of decomposition, the pH value will be low and the concentrations of BOD_5, TOC, COD, nutrients, and heavy metals will be high. If, on the other hand, a leachate sample is collected during the methane fermentation phase, the pH will be in the range from 6.5 to 7.5, and the BOD_5, TOC, COD, and nutrient concentration values will be significantly lower. Similarly the concentrations of heavy metals will be lower because most metals are less soluble at neutral pH values. The pH of the leachate will depend not only on the concentration of the acids that are present but also on the partial pressure of the CO_2 in the landfill gas that is in contact with the leachate.

Table 4.5　Typical data on the composition of leachate from new and mature landfills

Constituent	Value, mg/L[a]		
	New landfill (less than 2 years)		Mature landfill (greater than 10 years)
	Range[b]	Typical[c]	
BOD_5 (5-day biochemical oxygen demand)	2 000 ~ 30 000	10 000	100 ~ 200
TOC (total organic carbon)	15 000 ~ 20 000	6 000	80 ~ 160
COD (chemical oxygen demand)	3 000 ~ 60 000	18 000	100 ~ 500
Total suspended solids	200 ~ 2 000	500	100 ~ 400
Organic nitrogen	10 ~ 800	200	80 ~ 120
Ammonia nitrogen	10 ~ 800	200	20 ~ 40
Nitrate	5 ~ 40	25	5 ~ 10
Total phosphorous	5 ~ 100	30	5 ~ 10
Ortho phosphorous	4 ~ 80	20	4 ~ 8
Alkalinity as $CaCO_3$	1 000 ~ 10 000	3 000	200 ~ 1 000
pH	4.5 ~ 7.5	6	6.6 ~ 7.5
Total hardness as $CaCO_3$	300 ~ 10 000	3 500	200 ~ 500
Calcium	200 ~ 3 000	1 000	100 ~ 400
Magnesium	50 ~ 1 500	250	50 ~ 200
Potassium	200 ~ 1 000	300	50 ~ 400
Sodium	200 ~ 2 500	500	100 ~ 200
Chloride	200 ~ 3 000	500	100 ~ 400
Sulfate	50 ~ 1 000	300	20 ~ 50
Total iron	50 ~ 1 200	60	20 ~ 200

[a]Except pH, which has no units.

[b]Representative range of values. Higher maximum values has been reported in the literature for some of the constituents.

[c]Typical values for new landfills will vary with the metabolic state of the landfill.

2. Control of Leachate in Landfills

As leachate percolates through the underlying strata, many of the chemical and biological constituents originally contained in it will be removed by the filtering and adsorptive action of the material composing the strata. In general, the extent of this action depends on the characteristics of the soil, especially the clay content. Because of the potential risk involved in allowing leachate to percolate to the groundwater, best practice calls for its elimination or containment. Landfill liners are now commonly used to limit or eliminate the movement of leachate and landfill gases from the landfill site. To date (1992), the use of clay as a liner material has been the favored method of reducing or eliminating the seepage (percolation) of leachate from landfills. Clay is favored for its ability to adsorb and retain many of the chemical constituents found in leachate and for its resistance to the flow of leachate. However, the use of combination composite geomembrane and clay liners is gaining in popularity, especially because of the resistance afforded by geomembranes to the movement of both leachate and landfill gases. The characteristics, advantages, and disadvantages of the geomembrane liners (also known as flexible membrane liners, FMLs) that have been used for MSW landfills are summarized in Table 4.6.

Table 4.6 Guidelines for leachate control facilities

Item	Comments
Synthetic flexible membrane liners (FMLs)	Liners must be designed and constructed to contain fluids, which include wastes and leachates. For MSW waste management units, synthetic liners are not required. However, if this alternative is selected, synthetic liners must have a minimum thickness of 40 miles. These liners must be installed to cover all natural geologic materials that are likely to be in contact with waste or leachate at a waste management unit.
Bottom seals	No specific regulations exist governing the application of bottom seals at MSW waste management units. Design, construction, and installation of bottom seals are subject to the approval of the local enforcement agencies.
Artificial earthen liners	Clay liners are optional for MSW landfills. If required by site conditions, clay liners for MSW waste management units must be a minimum of 1 ft thick and must be installed at a relative compaction of at least 90 percent. A clay liner must exhibit a maximum permeability of 1×10^{-6} cm/s. Clay liners, if installed, must cover all natural geologic materials that are likely to be in contact with waste or leachate at a waste management unit.
Subsurface barriers	A subsurface barrier is intended to be used in conjunction with natural geological materials to assure that lateral permeability standers are satisfied. Barriers may be required by regional agencies at MSW waste management units where there is potential for lateral movement of fluid, including waste and leachate, and the permeability of natural geologic materials is used for waste containment in lieu of a liner. Barriers must be a minimum of 2 ft thick for clay material or a minimum of 40 mils for synthetic materials. These structures are required to be keyed a minimum of 5 ft into natural geologic materials that satisfy permeability requirements of 1×10^{-6} to 1×10^{-7}cm/s. If cutoff walls are used, excavations for waste management units must also be keyed into natural geologic materials exhibiting permeabilities of no greater than 1×10^{-6}cm/s. Barriers are required to have fluid collection systems upgradient of the structure. The systems must be designed, constructed, operated, and maintained to prevent the buildup of hydraulic head against the structure. The collection system must be inspection regularly and accumulated fluid removed.

3. Leachate Collection Systems

The design of a leachate collection system involves (1) the selection of the type of liner system to be used, (2) the development of the grading plan including the placement of the leachate collection and drainage channels and pipelines for the removal of leachate, and (3) the layout and design of the leachate removal, collection, and holding facilities.

The type of liner system selected will depend to a large extent on the local geology and environmental requirement of the landfill site. For example, in locations where there is no groundwater, a single compacted lay liner may be sufficient in locations where both leachate and gas migration must be controlled, a combined liner comprising a clay liner and a geomembrane liner with an appropriate drainage and soil protection layer will be necessary.

Two methods have been used for the removal of leachate that accumulates within a landfill. The leachate collection pipe is passed through the side of the landfill. Where this method is used, great care must be taken to ensure that the seal where the pipe penetrates the landfill liner is sound. An alternative method used for the removal of leachate from landfills involves the use of an inclined collection pipe located within the landfill. Leachate collection facilities are used where the leachate is to be recycled from or treated at a central location. The capacity of the holding tank will depend on the type of treatment facilities that are available and the maximum allowable discharge rate to the treatment facility. Typically, leachate holding tanks are designed to hold from 1 to 3 days of leachate production during the peak leachate production period. Both double-and single-walled tanks have been used, but the double walled tanks are preferred over single-walled tanks because of the added safety afforded. Although both plastic and metallic tanks have been used, plastic tanks are more corrosion resistant.

4. Leachate Management Options

The management of leachate, when and if it forms, is key to the elimination of the potential for a landfill to pollute underground aquifers. A number of alternative have been used to manage the leachate collected from landfills including: (1) leachate recycling, (2) leachate evaporation, (3) treatment followed by disposal.

An effective method for the treatment of leachate is to collect, and recirculate the leachate through the landfill. During the early stages of landfill operation the leachate will contain significant amounts of TDS, BOD_5, COD, nutrients, and heavy metals. When the leachate is recirculated, the constituents are, attenuated by the biological activity and by other chemical and physical reactions occurring within the landfill. An additional benefit of leachate recycling is the recovery of landfill gas that contains CH_4. Ultimately, it will be necessary to collect, treat, and dispose of the residual leachate. In large landfills it may be necessary to provide leachate storage facilities.

One of the simplest leachate management systems involves the use of lined leachate evaporation ponds. Leachate that is not evaporated is sprayed on the completed portions of the landfill. In locations with high rainfall, the lined leachate storage facility is covered with a geomembrane during the winter season to exclude rainfall. The accumulated leachate is disposed of by evaporation during the warm summer months, by uncovering the storage facility, and by spraying the leachate on the surface of the operating and completed landfill. Odorous gases that may accumulate under the surface cover are vented to a compost or soil filter.

The principal biological and physical/chemical treatment operations and processes used for the treatment of leachate are summarized in Table 4.7. The treatment process or processes selected will depend to a large extent on the contaminant(s) to be removed. In those locations where a landfill is located near a wastewater collection system or where a pressure sewer can be used to connect the landfill leachate collection system to a wastewater collection system, leachate is often discharged to the wastewater collection system. In many cases pretreatment, using one or more of the methods reported in Table 4.7, may be required to reduce the organic content before the leachate can be discharged to the sewer.

Table 4.7 Representative biological, chemical, and physical processes and operations used for the treatment of leachate

Treatment process	Application	Comments
Biological processes		
Activated sludge	Removal of organics	Defoaming additives may be necessary; separate clarifier needed
Sequencing batch reactors		Similar to activated sludge, but no separate clarifier needed; only applicable to relatively low flow rates
Aerated stabilization basins		Requires large land area
Fixed film processes (trickling filters, rotating biological contactors)		Commonly used on industrial effluents similar to leachates, but untreated on actual landfill leachates
Anaerobic lagoons and contactors		Lower power requirements and sludge production than aerobic systems; requires heating; greater potential for process instability; slower than aerobic systems
Nitrification/ denitrification	Removal of nitrogen	Nitrification/ denitrification can be accomplished simultaneously with the removal of organics
Chemical processes		
Neutralization	pH control	Of limited applicability to most leachates
Precipitation	Removalof metals and some anions	Produces a sludge, possibly requiring disposal as a hazardous waste
Oxidation	Removal of organics; detoxification of some inorganic species	Works best on dilute waste streams, use of chlorine can result in formation of chlorinated hydrocarbons
Physical operations		
Sedimentation/flotation	Removal of suspended matter	Of limited applicability alone; may be used in conjunction with other treatment processes
Filtration		Useful only as a polishing step
Air stripping	Removal of ammonia or volatile organics	May require air pollution control equipment
Adsorption	Removal of organics	Proven technology; variable costs depending on leachate
Evaporation	Where leachate discharge is not permissible	Resulting sludge may be hazardous; can be costly except in arid regions

4.6.4 Environmental Quality Monitoring at Landfills

Environmental monitoring is conducted at sanitary landfills to ensure that no contaminants that may affect public health and the surrounding environment are released from the landfill. The monitoring required may be divided into three general categories: (1) vadose zone monitoring for gases and liquids, (2) groundwater monitoring, and (3) air quality monitoring.

1. Vadose Zone Monitoring

The vadose zone is defined as that zone from the ground surface to where the permanent groundwater is found. An important characteristic of the vadose zone is that the pore spaces are not filled with water, and that the small amounts of water that are present coexist with air. Vadose zone monitoring at landfills involves both liquids and gases.

Monitoring for liquids in the vadose zone is necessary to detect any leakage of leachate from the bottom of a landfill. In the vadose zone, moisture held in the interstices of the soil particles or within porous rock is always held at pressures below atmospheric pressure. To remove the moisture it is necessary to develop a negative pressure or vacuum to pull the moisture away from the soil particles. Because suction must be applied to draw moisture out of the soil in the vadose zone, conventional wells or other open cavities cannot be used to collect samples in this zone. Monitoring for gases in the vadose zone is necessary to detect the lateral movement of any landfill gases. In many monitoring systems, gas samples are collected from multiple depths in the vadose zone.

2. Groundwater Monitoring

Monitoring of the groundwater is necessary to detect changes in water quality that may be caused by the escape of leachate and landfill gases. Both down-and up-gradient wells are required to detect any contamination of the underground aquifer by leachate from the landfill. To obtain a representative sample, the liquid in permanent sample collection tubing, where used, must be purged before the sample is collected.

3. Landfill Air Quality Monitoring

Ambient air quality is monitored at landfill sites to detect the possible movement of gaseous contaminants from the boundaries of the landfill site. Gas sampling devices can be divided into three categories: (1) passive, (2) grab, and (3) active. Passive sampling involves the collection of a gas sample by passing a stream of gas through a collection device in which the contaminants contained in the gas stream are removed for subsequent analysis. Commonly used in the past, passive sampling is seldom used today. Grab samples are collected using an evacuated flask, gas syringe, or an air collection bag made of a synthetic material. An active sampler involves the collection and analysis of a continuous stream of gas. Landfill gas is monitored to assess the composition of the gas, and to determine the presence of trace constituents that may pose a health or environmental risk. Monitoring off-gases from treatment and energy recovery facilities is done to determine compliance with local air pollution control requirements. Both grab and continuous sampling have been used for this purpose.

4.7 Hazardous Waste

In a modern society, large quantities of dangerous wastes are generated by chemical manufacturing companies, petroleum refineries, paper mills, smelters, and other industries. Even commercial establishments, such as dry cleaners, machine shops, and automobile repair shops, generate some dangerous waste. These hazardous wastes, as they are called, can result in serious illness, injury, or even death of the individuals and populations exposed to them; they can also pose an immediate and significant threat to environmental quality when improperly stored, transported, or disposed of.

4.7.1 Definition of Hazardous Waste

Published definitions and classifications of hazardous waste vary. In the United States, wastes are defined under the Resource Conservation and Recovery Act (RCRA) as being hazardous if they "cause or significantly contribute to an increase in mortality or an increase in serious irreversible, or incapacitating reversible illness; or pose a substantial present or potential hazard to human health or the environment when improperly treated, stored, transported, or disposed of, or otherwise managed." This may give a broad view of hazardous waste, but it is definitely not a workable definition; more specific criteria are needed to facilitate an accurate identification of a particular waste material as being hazardous or not.

The necessary criteria for identifying hazardous wastes have been set up by the EPA. First, a material may be defined as hazardous if it is specifically named or listed as such in the federal regulations. This is a very direct and unambiguous method of definition and makes it easy for waste generators to know if they must manage their waste as hazardous; if it is on the list, it is hazardous, period. The list adopted by the EPA includes nonspecific source wastes, specific source wastes, and certain commercial chemical products.

Nonspecific source wastes commonly generated by industrial processes include materials such as degreasing solvents, dioxin wastes, and other very dangerous materials that come from a wide variety of manufacturing plants. Source-specific wastes, on the other hand, come from identifiable industries, such as petroleum-refining or wood-preserving facilities, and include wastewaters, sludges, and other residues. Commercial chemical products include discarded acids, chloroform, creosote, and pesticides such as DDT. Also, any waste material that contains a listed waste, regardless of the percentage, is considered a hazardous waste.

If a material is not on the EPA list of hazardous substances that does not imply that it is nonhazardous. It still may be defined as a hazardous waste if it exhibits any of the measurable characteristics of a hazardous waste. The four primary characteristics, under the RCRA, are based on the physical or chemical properties of toxicity, reactivity, ignitability, and corrosivity. Two additional types of hazardous materials include waste products that are either infectious or radioactive.

1. Toxicity

Toxic wastes are poisons, even in very small or trace amounts. Some may have an acute or immediate effect on humans or animals, causing death or violent illness. Other may have a chronic or long-term effect, slowly causing irreparable harm to exposed persons. Certain toxic wastes are known to be carcinogenic, causing cancer (sometimes many years after initial exposure). Other may be mutagenic, causing biological change in the children or offspring of exposed people and animals.

Most toxic wastes are generated by industrial activities, including the manufacture of chemicals, pesticides paints, petroleum products, metals, textiles, and many other products. The toxicity of any particular waste is determined by an EPA-specified test called the toxicity characteristics leaching procedure (TCLP). The TCLP is used to determine the mobility of organic and inorganic compounds present in the waste. This procedure attempts to mimic conditions a waste may be exposed to in a landfill, thus projecting the potential mobility of these compounds. In general, the extract from a representative sample of the waste is analyzed to see if it contains more than the allowable concentrations of one or more of the specific toxic substances listed by the EPA. Currently 39 specific organic and inorganic chemicals are listed. An abbreviated list showing some selected toxic chemicals is provided in Table 4.8.

Table 4.8　Maximum concentration of contaminants for the TCLP[a]

Contaminant	Maximum level(mg/L)
Arsenic	5.0
Benzene	0.5
Carbon tetrachloride	0.5
Chlordane	0.03
Chloroform	6.0
Chromium	5.0
Endrin	0.02
Lead	5.0
Mercury	0.2
Pentachlorophenol	100.0
Silver	5.0
Trichloroethylene	0.5
Vinyl chloride	0.2

[a]This is an abbreviated list, for illustration purposes only.

2. Other Characteristics of Hazardous Wastes

Reactive wastes are unstable and tend to react vigorously with air, water, or other substances. The reactions cause explosions or form very harmful vapors and fumes. Ignitable wastes are those that burn at relatively low temperatures (less than 60 ℃) and are capable of spontaneous combustion (that is, they present an immediate fire hazard) during storage, transport, or disposal. Many waste oils and solvents are ignitable. Corrosive wastes, including strong alkaline or acidic substances, destroy material and living tissue by chemical reaction. The pH value is used as an indicator of this characteristic; typically, liquids with pH less than 2 or greater than 12.5 are considered to be corrosive. Such wastes can rust or corrode unprotected steel at a rate more than 6 mm per year at temperature of about 55 ℃.

Infectious or medical waste includes human tissue from surgery, used bandages and hypodermic needles, microbiological material, and other substances generated by hospitals and biological research centers. This type of material must be handled and disposed of properly, following EPA guidelines, to avoid infection and the spread of communicable disease.

Radioactive waste, particularly high-level radioactive waste from nuclear power plants, is also of special concern as a hazardous waste. Excessive exposure to ionizing radiation can harm living organisms. Radioactive material may persist in the environment for thousands of years before it decays appreciably. Because of the scope and technical complexity of this problem, radioactive waste disposal is always considered separately from other forms of hazardous waste.

The EPA estimates that, until the mid-1980s, only about 10 percent of hazardous waste was actually disposed of in an environmentally sound manner. Much of it was disposed of in unlined landfills, waste piles, or lagoons and poses a potential threat to public health and environmental quality. The cost of cleanup is expected to reach billions of dollars and is currently being paid for through industrial contributions and federal funds allocated through the Superfund program.

4.7.2 Waste Sources and Amounts

Under RCRA, the generator or creator of hazardous waste is responsible for identifying it as such. The generator is any person or company that produces material that is listed by the EPA as hazardous or has any of the defined characteristics of a hazardous waste. It is the generator's task to analyze all solid wastes to determine if they meet the RCRA definitions or are on the list. Once a waste is identified as being hazardous, it becomes subject to RCRA rules, and the generator assumes legal responsibilities for its proper management and disposal.

Three different categories of generators are recognized by RCRA, including large-quantity, small-quantity, and conditionally exempt small-quantity generators. Large-quantity generators create more than 1 000 kg of hazardous waste per month or more than 1 kg of acutely hazardous waste per month. Acutely hazardous wastes are those considered to be so dangerous that even small amounts are regulated in the same way as are larger amounts of other hazardous wastes; they are specifically identified as being acutely hazardous on the EPA list.

Small-quantity generators (SQGs) create between 100 and 1 000 kg of hazardous waste per month or less than 1 kg of acutely hazardous waste per month. Typical SQGs include dry cleaning facilities and automobile service stations. Conditionally exempt small-quantity generators create less than 100 kg of hazardous waste per month.

Large-quantity waste generators and SQGs must comply with the RCRA regulations, including obtaining an EPA identification (EPA ID) number, properly handling the waste before transport, manifesting the waste, and properly keeping records and reporting. Conditionally exempt SQGs do not require EPA ID numbers. Appropriate pretransport handling requires suitable packaging to prevent leakage and labeling of the packaged waste to identify its characteristics and dangers. Large-quantity generators may accumulate waste on site for a maximum of 90d if it is properly stored and labeled and if a written emergency plan is prepared. SQGs may accumulate waste on site for up to 180 d. The complete details of these regulations are extensive and complex and require full and careful attention by the responsible parties.

1. Hazardous Waste Quantities

Databases regarding the total amount of hazardous waste generated each year in the United States differ with respect to scope, content, and time frame. It can be said, however, that of the several hundred millions of tons of hazardous waste generated each year, most of it is corrosive liquid industrial waste that is discharged into sewerage systems after appropriate pretreatment. Roughly 35 million tons per year of additional types of hazardous waste, though, must be stored, incinerated, treated, or disposed of on or in the land. Much of the total production of hazardous waste in the United States comes from the northeastern and midwestern regions of the country.

Hazardous wastes are generated by a wide variety of manufacturing and nonmanufacturing facilities and processes, much too numerous to give a comprehensive list here. About half of the hazardous waste comes from the chemical products industry; the rest of it is generated by the petroleum, electronics, and metal-related industries, as well as numerous others.

There are about 750 listed hazardous wastes and countless more of the characteristic hazardous wastes. Examples of the types of hazardous wastes produced by the chemical industry include spent solvents and still bottoms (acetone, benzene, toluene, trichloroethylene, and others), strong acids (nitric, sulfuric, and hydrochloric acid, and others), strong alkaline wastes (ammonium and potassium hydroxide, and others), and reactive wastes (sodium permanganate, potassium sulfide, and others).

Metal manufacturing produces spent plating wastes, heavy metal sludges, cyanide wastes, chromic acid, and many others. The paper industry creates carbon tetrachloride, acids, ammonium hydroxide, and petroleum distillates. The list of examples can go on and on, and the few given here only serve to indicate the scope and wide variety of hazardous wastes that must be managed. Adding to the vast scope of the problem is the introduction of new products by industry and therefore new wastes; the tremendous pace and volume of hazardous waste generation point to the need to intensify waste minimization and recycling efforts.

2. Household Hazardous Wastes

About 1.6 million tons of household hazardous wastes is generated each year in the United States. This is a small fraction (less than 1 percent) of the total amount of MSW generated, but when it is improperly disposed of it can create potential risks to people and the environment. Household hazardous wastes should not be intermingled with municipal refuse.

As much as 45 kg of hazardous substances can accumulate in the basements, garages and storage closets of the average American home. These substances include leftover paint, stains, and varnishes, batteries, motor oil, and pesticides. They are sometimes disposed of improperly in sewers, on the ground, or by intermingling them with ordinary household refuse. Many of these wastes can injure sanitation workers, contaminate wastewater treatment systems, and pollute the air, surface water, and groundwater. Individuals can minimize these risks by reducing the use of hazardous products and by taking their hazardous wastes to local waste collection facilities, if available.

4.8 Transportation of Hazardous Waste

Hazardous waste generated at a particular source requires movement to a licensed facility for proper treatment, storage, or disposal. Because these three activities are so closely related, the RCRA legislation links them all together and refers to their locations as treatment, storage, and disposal facilities (or TSDFs). Moving or transporting the waste from its source to a TSDF requires special care and attention due to the potential threats to environmental quality and public health and safety. Not only is there a possibility of an accidental spill, there has been (before RCRA) a tendency for unscrupulous waste transporters to abandon the wastes at random locations or to open a valve and dump them "on the run." This illegal practice (often called "midnight dumping") has been significantly curtailed by the enactment of RCRA and other legislation. In the United States, the transport of hazardous materials and hazardous waste is regulated by the EPA, as well as by the Department of Transportation (DOT). A law that specifically focuses on transport activities is the Hazardous Materials Transportation Act (HMTA), passed by Congress in 1975. The HMTA requires proper labeling and transport of all hazardous materials, including hazardous waste.

4.8.1 Common Vehicles for Transportation

Hazardous waste that is moved from its source is transported mostly by truck over public highway routes that are less than 160 km long. Only a very small amount of hazardous waste is transported by rail and almost none by air or inland waterways. Highway shipment is used the most because the trucks can access most industrial sites and TSDFs; trains require expensive siding facilities and are suitable only for very large waste shipment.

Hazardous wastes can be shipped in cargo tank trucks made of steel or aluminum alloy, with capacities between 7 600 and 34 000 L. They also can be containerized and shipped in 210 L drums. Specifications and standards for cargo tank trucks and shipping containers are included in RCRA and HMTA and are enforced by the EPA and the DOT. The DOT regulations deal mostly with the "hardware," that is, with the containers, trucks, and shipping descriptions; the EPA regulations pertain primarily to a comprehensive tracking or manifest system, which follow the waste from "cradle to grave." All transporters of hazardous waste are required to have an EPA ID number.

4.8.2 Manifest System

One of the key features of RCRA with regard to the transport of hazardous waste is the "cradle-to-grave" manifest system for monitoring the journey of the waste from its point of origin to point of final disposal. The manifest is a record-keeping document that must be prepared by the generator of the hazardous waste, such as a chemical manufacturing plant or other industry. The generator has primary responsibility for managing the waste and gives the manifest to a licensed waste transporter, along with the waste itself. The transporter must comply with the regulations of the DOT and the EPA with regard to transport vehicle standards, operation, and response to accidental spills.

The manifest must be delivered by the waste transporter or hauler to the recipient of the waste at an authorized off-site storage, processing, or disposal facility. Each time the waste changes hands, the manifest form must be signed. Copies of the manifest are kept by each party involved, and copies are also sent to the appropriate state environmental agency. Figure 4.13 illustrates a typical manifest cycle followed by the state of New Jersey.

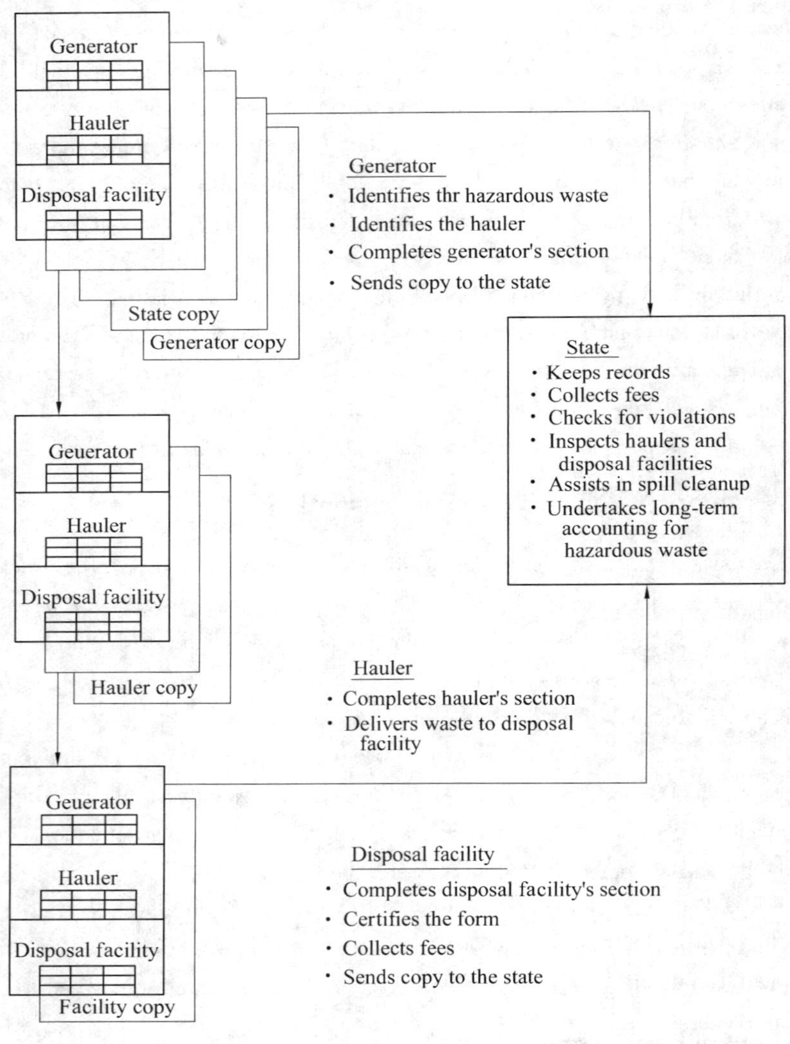

Figure 4.13　Hazardous waste "cradle-to-grave" manifest system

(From G. F. Bennett, et al., Hazardous Materials Spills Handbook, McGraw-Hill, New York, 1983.)

In addition to curtailing the practice of improper disposal (for example, midnight dumping), the manifest system serves several other purposes. It provides data with regard to sources, types, and quantifies of hazardous wastes. These data are valuable for future planning and design of hazardous waste management systems. The manifest ensures that the nature of the waste is described to haulers and operators of processing or disposal facilities; this prevents accidents due to improper handling or disposal of the waste. It also provides information regarding recommended emergency response procedures in the event of accidental spills or leaks.

In the event of a leak or accidental spillage of hazardous waste during its transport, the EPA and the DOT require the truck driver to take immediate and appropriate actions, including notifying local authorities of the discharge. An area may have to be diked to contain the wastes and cleanups may have to be started to remove it and reduce environmental or public health hazards.

4.9 Treatments, Storage, and Disposal of Hazardous Waste

The RCRA links together the activities of hazardous waste treatment, storage, and disposal. A site to which hazardous wastes are transported and at which any one or a combination of these three activities occur is called a TSDF (treatment, storage, and disposal facility). TSDFs are subject to the RCRA rules and regulations and must obtain an EPA ID number and operating permit. Treatment, storage, and disposal include a wide variety of methods and technologies, and, as a result, TSD regulations are even more extensive than those for hazardous waste generators and transporters. Personnel at TSDFs must be well trained to ensure that the wastes are correctly identified and handled, and the facilities must have controlled entry systems of 24 h security surveillance to prevent the possibility of unauthorized entry. In addition, TSDFs must have contingency plans and emergency procedures, equipment and facility inspections, closure and postclosure plans, and other requirements.

4.9.1 Hazardous Waste Treatment Methods

Some types of hazardous waste can be detoxified or made less dangerous by chemical, biological, or physical treatment methods. Treatment of hazardous waste may be costly, but it can serve to prepare the material for recycling or ultimate disposal in a manner that is safer than disposal without treatment. It also can reduce the volume needing final disposal. Many effective treatment methods are available. The hazardous waste treatment industry is in a phase of rapid development and innovation because of the need for even more economical treatment techniques.

1. Chemical Treatment Processes

Chemical processes often used for treatment of hazardous waste include incineration, ion exchange, neutralization, precipitation, and oxidation-reduction.

Incineration is a thermal-chemical process that not only can detoxify certain organic wastes, but can essentially destroy them as well. It is preferred by some people in the waste management industry over most other hazardous waste treatment processes, particularly because of the economic and public pressures to reduce or eliminate land disposal. The burning of organic wastes at very high temperatures converts them to an ash residue and gaseous emissions. Combustion detoxifies hazardous waste material by altering its molecular structure and breaking it down into simpler chemical substances. Although the ash itself may have to be treated as a hazardous waste (if so indicated after a TCLP test), a much small volume of waste is left for ultimate disposal. Stack emissions from a properly designed and operated incinerator burning organics, such as chlorinated hydrocarbons, for example, include CO_2, H_2O, N_2, and HCl (hydrochloric acid). Only the HCl is hazardous, but it is readily reacted with lime to produce nonhazardous salts, which can be landfilled.

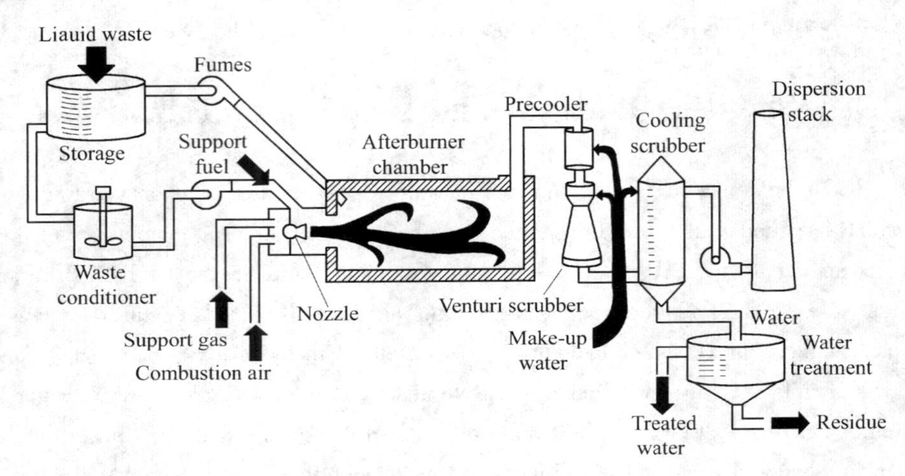

Figure 4.14　Schematic of a liquid injection incinerator

Not all hazardous waste can be incinerated. Heavy metals, for example, are not destroyed, but enter the atmosphere in vapor form. However, incineration been successfully applied to potent hazardous wastes such as chlorinated hydrocarbon pesticides, polychlorinated biphenyl(PCBs), and many other organic substances. Special types of thermal processing equipment, such as the rotary kiln, the fluidized bed incinerator, the multiple-hearth furnace, and the liquid injection incinerator, are available for burning hazardous waste in either solid, sludge, or liquid form. Figure 4.14 is a diagram of a liquid injection incinerator. It is of interest to note that incineration of hazardous wastes on ocean-going ships (to reduce air pollution concerns) has been tried, but the future of this practice is doubtful due to technical difficulties.

In the ion-exchange process, industrial waste water is passed through a bed of resin that selectively adsorbs charged metal ions. An example of its use in the metal finishing industry is for removal of waste chromic acid from production rinse water. Neutralization refers to pH adjustment for reducing the strength and reactivity of acidic or alkaline wastes. Limestone, for example, can be used to neutralize acids, and compressed carbon dioxide can be used to neutralize strong bases. Precipitation refers to a type of reaction in which certain chemicals are made to settle out of solution

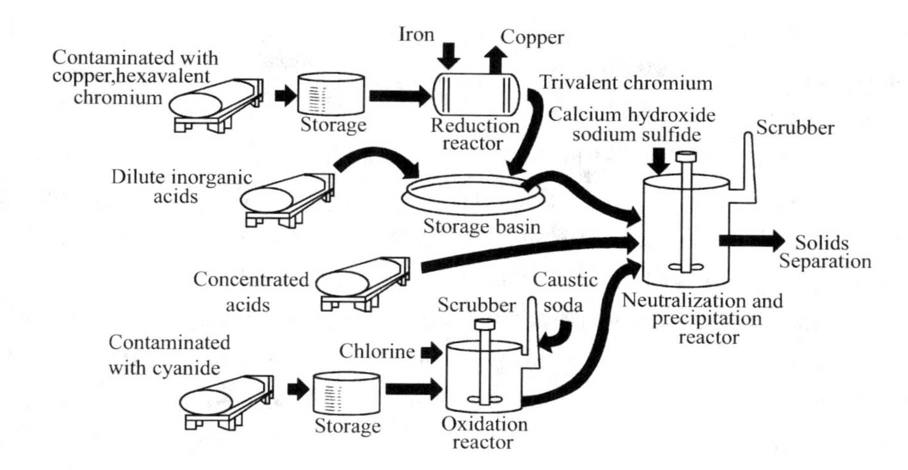

Figure 4.15 Schematic chemical treatment: neutralization, precipitation, and chemical oxidation-reduction

as a solid material. An example of its application is the battery industry, where the addition of lime and sodium hydroxide to acidic battery waste causes lead and nickel (both are toxic heavy metals) to precipitate out of solution. Oxidation and reduction are complementary chemical reactions involving the transfer of electrons among ions. Oxidation of waste cyanide by chlorine, for example, renders it less hazardous. A schematic diagram showing a chemical treatment facility is shown in Figure 4.15.

2. Biological treatment Processes

Biological treatment involves the action of living microorganisms. The microbes utilize the waste material as food and convert it, by natural metabolic processes, into simpler substances. It is most commonly used for stabilizing the organic waste in municipal sewage, but certain types of hazardous industrial waste can also be treated by this method, Organic waste from the petroleum industry, for example, can be treated biologically. It is necessary, though, to inoculate the waste with bacteria that are readily acclimated to it and can use it for food. In some cases, genetically engineered species of bacteria may be used.

In addition to the traditional biological treatment systems, including the activated sludge and trickling filter processes, a treatment method called landfarming or land treatment (which is not the same as landfilling) may be used. The waste is carefully applied to and mixed with surface soil; microorganisms and nutrients may also be added to the mixture, as needed. The toxic organic material is degraded biologically, whereas inorganics are adsorbed and retained in the soil.

Landfarming is a relatively inexpensive method for treatment, as well as being a way to ultimately dispose of certain types of hazardous waste. But food or forage crops must not be grown on the same site because they could take up toxic material. In this regard, a disadvantage of land treatment for hazardous waste is that relatively large tracts of land may have to be withdrawn from potentially productive agricultural use. The surface topography and subsurface geological conditions of the site must be suitable so that surface or groundwater contamination will not occur.

Certain organic hazardous wastes can be treated in slurry form in an open lagoon or in a closed vessel called a bioreactor. A bioreactor may have fine bubble diffusers to provide oxygen and a mixing device to keep the slurry solids in suspension.

3. Physical Treatment Processes

Physical treatment can be used to concentrate, solidify, or minimize the volume of hazardous waste material. Solidification can be accomplished by encapsulating the waste in concrete, asphalt, or plastic. This produces a solid mass of material that is resistant to leaching. Hazardous wastes can also be solidified by mixing them with lime, fly ash, and water to form a solid, cement-like product. Another process that is used to solidify and stabilize contaminated soil is vitrification. This involves the melting and fusion of the materials at high temperatures (about 1 600℃, or 2 900℉), thereby reducing the potential for leaching of contaminants. It is possible to remediate old waste dumps using in situ (on-site) vitrification; the required high temperatures can be achieved by inserting electrodes in the ground and allowing heat to build up as current flows through the soil.

The simplest physical processes that can concentrate and reduce wastewater volume is evaporation, which may be facilitated by using mechanical sprayers. Other physical processes utilized to separate hazardous waste from a liquid include sedimentation, flotation, and filtration.

Two examples of physical processes used to remove specific hazardous components from a liquid waste include activated carbon adsorption and air/gas stripping. Hazardous substances can be adsorbed onto a porous granular or powdered carbon matrix. Used or spent carbon is regenerated or activated for reuse. Air/gas stripping in cascade or countercurrent towers has been used to remove volatile organics from wastewater and contaminated groundwater.

4.9.2 Storage Tanks and Impoundments

Proper storage of hazardous waste is imperative because of the potential for serious harm to public health and environmental damage in the event of an accidental discharge. Many generators of hazardous waste store the material on site for varying periods of time. Relatively large quantities may be stored in above ground basins or lagoons. Aboveground basins may be constructed of steel or concrete, but they are subject to corrosion or cracking and are not suitable for storing reactive or ignitable waste.

Relatively small amounts of hazardous waste that are generated on an intermittent basis may be placed in 210 L fiberglass, plastic, or steel drums for ease of handling, temporary storage, and transportation. Corrosive material is stored in fiberglass or glasslined containers to reduce deterioration and leakage. Toxic chemical liquids may be stored in metal drums.

Containers or drums of hazardous waste must be labeled properly before transport to a processing or disposal facility. The label must identify the contents as an explosive, flammable, corrosive, or toxic material. Appropriate signs or placards must be placed on the transport vehicle to warn the public of potential danger and to assist emergency response workers if there is an accidental spill along the transport route.

1. Underground Storage Tanks

The design and construction of tanks used to store hazardous materials as well as hazardous waste is a major environmental concern. Thousands of cases of environmental damage (particularly groundwater contamination) caused by leaking tanks are known to have occurred in the past. The bulk of these cases involved leaks from old underground gasoline or oil storage tanks, but many old underground tanks holding hazardous wastes have also been known to corrode and leak. Since more than haft of the population in the United States depends on groundwater for public and household use, leaking waste storage tanks are a threat to public health as well as to environmental quality. Underground storage tanks (USTs) pose even more of a threat than aboveground tanks, USTs are unseen, and their existence is often unknown to people living in their vicinity until a problem occurs, damage is done, expensive cleanup or remediation actions are required.

Under RCRA and its 1984 amendments, standards have been set by the EPA to control the use of UST system. These standards apply to tanks with more than 10 percent of their volume below ground, containing hazardous wastes, petroleum products, or other hazardous substances; underground piping and pumping systems associated with the tanks are also regulated. UST regulations are intended to prevent leaks and spills as well as to detect them if they do occur.

Regulations governing the design, installation, and operation of aboveground as well as underground storage tanks are numerous and complex. New UST systems must be made of fiberglass or cathodically protected steel for corrosion control. A secondary containment system, including either double-wall tanks or cutoff walls with impervious underlayment, must be provided.

Automatic leak-detection systems and alarms must be installed as well as devices at prevent over-fill (see Figure 4.16). Owners or operators of the tanks are responsible for monthly inspections and periodic removal of sludge from the tanks. Existing storage tanks that are not protected from corrosion or leakage must be taken permanently out of service; they must be emptied of all liquids, vapors, and sludge and then either removed or filled with sand or concrete.

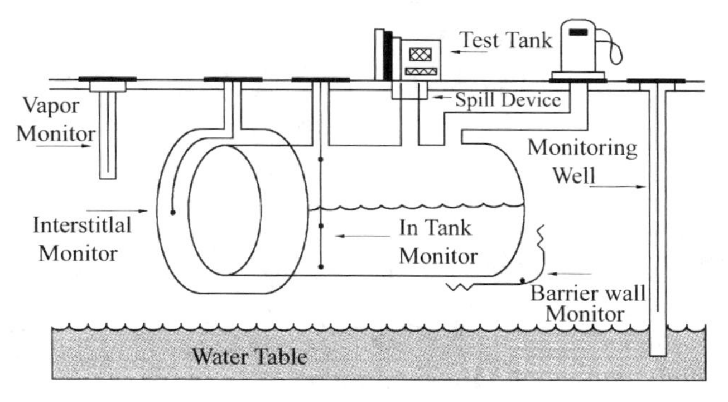

Figure 4.16 Leak-detection alternatives include automatic monitors to detect loss of volume, vapors from leaked petroleum products, and leaked liquids. (Courtesy of the EPA)

2. Surface Impoundments

Before implementation of the RCRA regulations, large volumes of liquid hazardous wastes were deposited in surface excavations such as pits, ponds, and lagoons (PPLs, in waste management jargon). Most PPLs were unlined impoundments or holding facilities, which provided no protection against leakage and groundwater contamination. Except for sedimentation, evaporation of volatile organics, and possibly some surface aeration, they provided no treatment of the waste. A large number of these old PPLs are today's hazardous waste remediation sites.

Surface impoundments of liquid hazardous waste are allowed under the RCRA if they meet stringent design criteria, and they are still widely used by generators of liquid waste. All existing surface impoundments must have at least one liner and be located over an impermeable base if they are to continue in use. New impoundments must have at least two liners and a leachate collection system; groundwater monitoring is also required for all surface impoundments. A schematic diagram of a surface impoundment for liquid hazardous waste is shown in Figure 4.17. Accumulated sludge must be periodically removed and provided further handing as a hazardous waste.

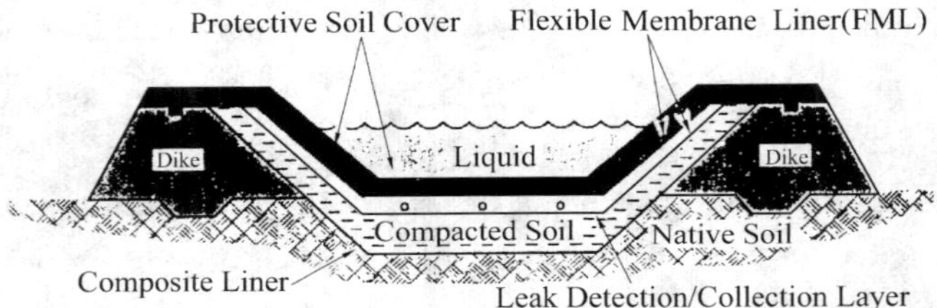

Figure 4.17　Schematic of a cross section of a liquid waste impoundment double liner system

3. Waste Piles

As with unlined lagoons used for liquid wastes, many old piles of solid hazardous waste are now in need of cleanup or remediation. But with proper precautions, generators of certain hazardous wastes are allowed to use a waste pile for temporary accumulation of hazardous waste; the material must be landfilled when the pile size becomes unmanageable. Only noncontainerized solid, nonflowing material can be stored in a waste pile; examples of such materials include the nonmagnetic materials from automobile shredding operations, which contain fabric, rubber, plastic, insulation, lead, and cadmium. Waste from aluminum salvage operations may also be stored temporarily in waste piles. Waste piles must be carefully constructed over an impervious base and must comply with requirement for landfills. The pile must be protected from wind dispersion and erosion; if hazardous leachate or runoff is generated, monitoring and control systems must be provided.

4.9.3　Land Disposal of Hazardous Waste

Certain hazardous wastes may be disposed of in the ground. Nonliquid or containerized hazardous waste can be buried in a secure landfill; liquid waste can be disposed of using deep-well injection systems. Disposal in underground mines, caves, or concrete vaults may sometimes be allowed, but only after proper treatment and containment.

Land disposal of hazardous waste is not an attractive option because of the inherent environmental dangers involved in its practice and because of future liability. In the past, it did not provide sufficient environmental protection, and many uncontrolled and abandoned waste sites now need to be cleaned up. But with proper site selection, engineering design, and operational safeguards, land disposal is the least expensive alternative form any types of hazardous waste. It will continue to be necessary in the future. Even with waste minimization and incineration, it is not possible to completely eliminate the generation of hazardous waste.

To discourage land disposal activities, particularly when other treatment or destruction methods exist, the EPA and many states have placed strict limitations on their use. Certain hazardous wastes, such as dioxins, PCBs, cyanides, halogenated organic compounds and acids with pH less than 2, are banned from land disposal. These wastes must be treated or stabilized before land disposal or meet certain concentration limits on the hazardous constituents.

1. Secure Landfills

A secure landfill must have a minimum of 3 m (10 ft) of height separating the base of the landfill from underlying bedrock or a groundwater aquifer, this is twice the minimum separation needed for a municipal solid waste landfill. All secure landfills also must have a double-liner and leachate collection system for increased safety, a network of monitoring wells to sample groundwater, and an impermeable cap or cover when completed. A cross-sectional view of a typical secure landfill is shown in Figure 4.18.

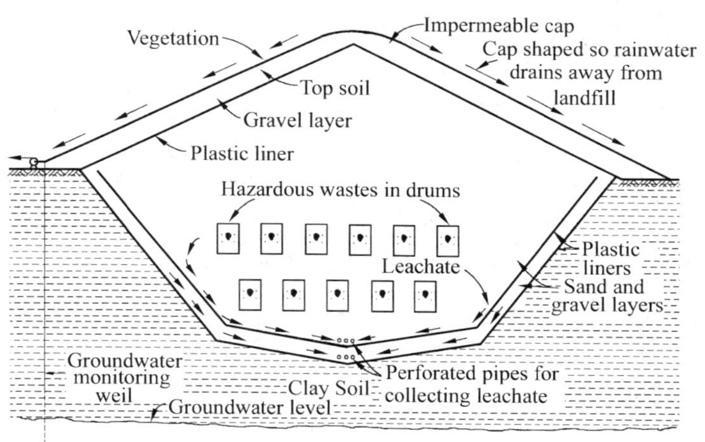

Figure 4.18 Cross section of a typical secure landfill

An impermeable liner on the bottom and sides of a secure hazardous waste landfill serves as a barrier preventing groundwater from entering or any leachate from leaving the fill. But no liner can be considered 100 percent effective; some leakage of liquid can always be expected. The double-liner and leachate collection system at a secure landfill provide redundancy to ensure protection of environmental quality and public health.

A synthetic flexible filter called a geotextile may be used to separate soil and waste material from the upper or primary leachate piping system. Under that is the primary liner, which must be a synthetic material with a permeability coefficient equivalent to 10^{-6} mm/s or less. It is called a geomembrane or flexible membrane liner (FML). Geotextiles and FMLs are generally made from plastic polymer resins. The secondary or lowermost leachate collection piping network serves as a leak detection and backup system. The lower or secondary liner is a composite layer of compacted clay and an FML. Collected leachate can be pumped to a treatment facility for processing.

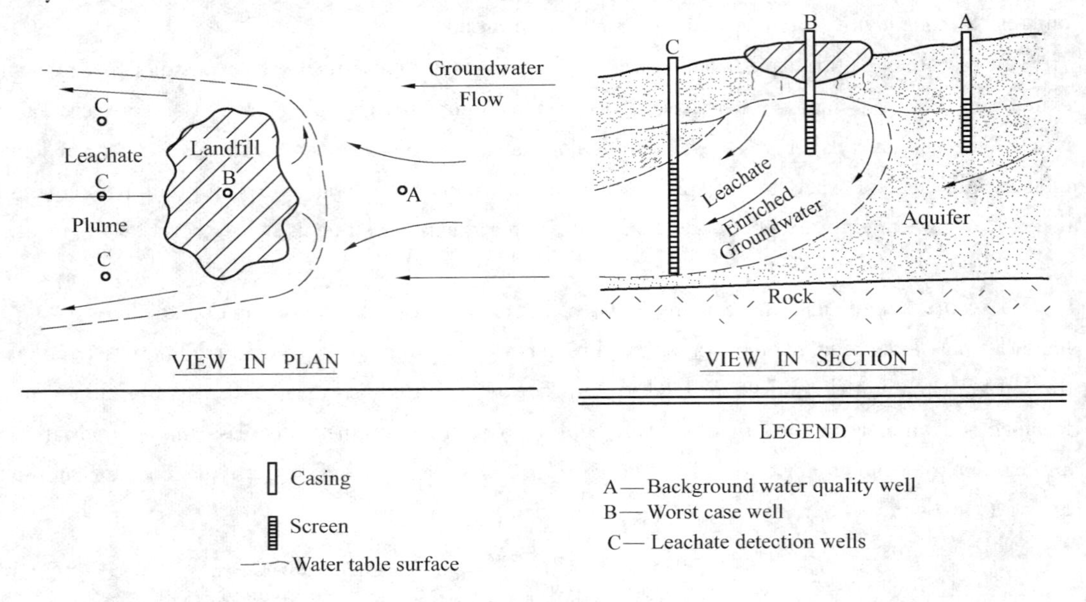

Figure 4.19 A groundwater monitoring system consists of a series of wells strategically located around the waste disposal site

A secure landfill is considered to have four phases in its total operation. During the first or active phase, the hazardous wastes are deposited in the prepared fill area. Incompatible chemical wastes are placed at separate locations in the fill to avoid explosions or other dangerous reactions. The waste material should first be solidified or containerized in drums, and care must be taken to avoid rupturing the individual containers as they are placed in the fill.

During the second or closure phase, the impermeable cap or cover is constructed over the landfill site. The composite bottom liner, the impermeable cap, and the double leachate collection systems serve as a first line of defense against leakage during both the active and the closure phases of the landfill operation. A third or postclosure phase, defined as the 30-year period after closure of the site, involves continuous operation of the monitoring well system (see Figure 4.19). This is a second line of defense. A routine program of sampling and testing must be implemented to detect any plumes of chemical leakage or contaminated groundwater.

It is believed that no landfill can be completely secure forever; the natural degradation of the protective liner or natural geological forces, including possibility of earthquake, will eventually destroy the structural integrity of the landfill. A fourth and last phase for the landfill, called the eternity phase, is expected to involve some leakage of the waste material into the environment. When the leakage is detected, pumps can be installed in the monitoring wells to intercept the polluted water and bring it to the surface for treatment. This last line of defense is intended to protect the underlying aquifer from any significant damage from the landfill operation.

2. Underground Injection

An option for land disposal of liquid hazardous waste is called deep-well or underground injection. This involves pumping the liquid down through a drilled well into a porous layer of rock. The liquid is then injected under high pressure into the pores and fissures of the rock. The layer or stratum of rock in which the waste is stored, usually limestone or sandstone, must lie between impervious layer of clay or rock; this injection zone can be from a few hundred meters to a few kilometers below the surface. The capacity of the geological strata to accept an injected waste depends on its porosity (void space), permeability (ability to transmit a liquid), and other factors.

The injection well must be at least 0.4 km from an underground source of drinking water, and the waste must be injected into a separate geological formation free of faults or fractures. The well must be cased and cemented for added protection against contamination of any drinking water supplies. The casing usually includes three concentric pipes. The outermost or surface casing should extend below the deepest usable water aquifer, as illustrated in Figure 4.20. A long string casing pipe and inner injection tubing extend into the injection layer or strata. Well injection pressures and flow rates must be carefully monitored; injection pressure must not be so great as to fracture the geological formation receiving the waste.

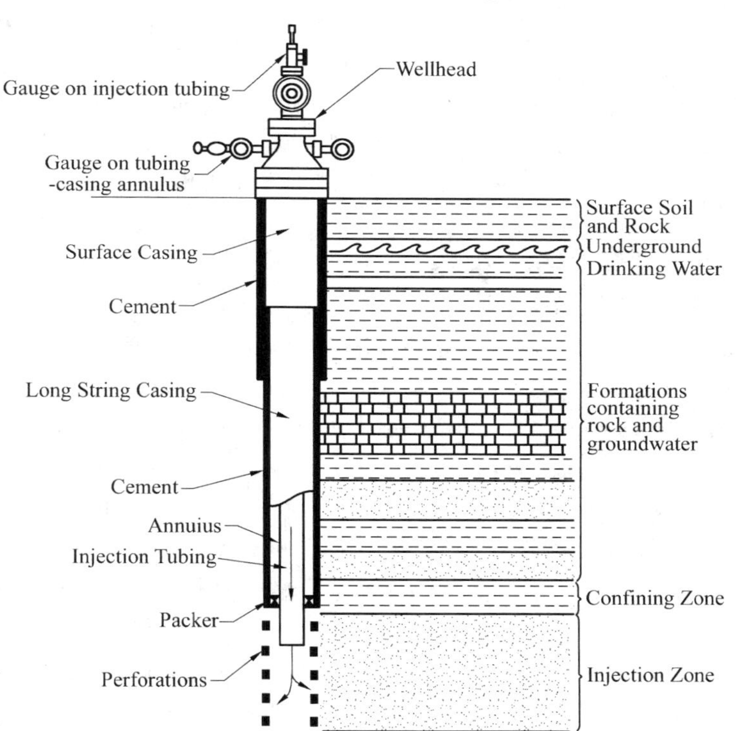

Figure 4.20 Typical hazardous waste injection well

The chemical and petroleum refining industries are the largest users of injection wells, together accounting for about 90 percent of their use as a disposal option. Most active waste injection wells are located in heavily industrialized states or in oil-and gas-producing states. Sitting criteria for new wells focus on the geological requirements. There must be a water-bearing stratum of nonbeneficial use that has sufficient volume, porosity, and permeability to accept the injected waste. Confining strata below and above the injection zone must be sufficiently thick and impermeable so as to confine the waste.

Comprehensive engineering and geological studies must be conducted to show that underground drinking water supplies will not be degraded by operation of the injection well. In addition to the requirements of the Underground Injection Control program (under the SDWA), each state may have regulatory requirements for injection wells. Underground injection takes up little land area and requires little or no waste pretreatment, but it is unlikely to become a predominant method for disposal of hazardous waste because of the inherent risk of eventual leakage into drinking water aquifers, despite the most though engineering and subsurface geological investigations.

4.10 Site Remediation Techniques

Before the enactment of legislation prohibiting uncontrolled disposal of hazardous wastes, many industrial and manufacturing facilities had long been storing or disposing of their hazardous waste materials in unlined waste piles, lagoons, and landfills. Many thousands of these old abandoned and unlined waste sites exist in the United States, posing a serious threat to public health and environmental quality. Efforts to remediate or clean them up will continue for years to come. The cradle-to-grave hazardous waste tracking or manifest provision of the RCRA was implemented primarily to eliminate this problem in the future.

The basic objective of site remediation is to eliminate any immediate danger caused by the spread or migration of waste material as well as to reduce any long-term threat to public health and environmental quality (especially to groundwater). The required course of action depends on the type of waste material, the extent of existing contamination, the location of the site, and other factors. Since each site is unique, particularly with regard to hydrogeological conditions, no two hazardous waste remediation projects are identical; sound engineering judgment must be used in each case.

A site-specific goal regarding the appropriate degree of remediation is necessary, but there is no general method for determining "how clean is clean." Some states simply require that site cleanups achieve background levels of the waste contaminants. For known carcinogens, however, where risk assessment data are available, the EPA may set specific exposure risk levels.

4.10.1 Removal of the Waste

One possible course of action is to physically remove the waste material from the site by excavation or dredging and transport it to some other location for treatment, incineration, or final disposal in a secure landfill. This off-site solution may be the most desirable for people living in the vicinity of the site, but it can be one of the most expensive options. Also, moving the waste from one location to another still involves some risk of environmental pollution. Contaminated soil may be removed using standard earth-moving equipment, but special equipment is often needed to remove buried drums or other containers. Extreme care must be taken to prevent releases of contaminants during the excavation or removal operation.

Dredging contaminated sediments from hazardous waste ponds or lagoons also requires special precautions to prevent further pollution. Clamshell, dragline, or backhoe machines are used to dredge consolidated sediments; in some cases, stream diversion or diking may first be necessary. When sediments are unconsolidated or have a high water content, the material may be pumped in slurry form; this is called hydraulic dredging. Dredged material can be dewatered, and both the solid and liquid fractions can be subjected to appropriate treatment and disposal operations.

4.10.2 On-Site Remediation

On-site remediation, in which the waste is not removed to another location, generally focuses on the need to minimize the production of leachate and to eliminate groundwater pollution. Another primary goal is to contain or prevent the further migration of any groundwater pollution that may have already occurred. This could involve the temporary removal of the waste, construction of a lined, secure landfill on the same site, and replacement of the waste in the new landfill. It could involve the extraction of soil or groundwater, treatment or destruction of the pollutants, and replacement or reinjection of the cleaned soil or groundwater. Finally, on-site actions could involve the isolation and containment of the waste, without moving it, by the construction of impermeable barriers to block the flow of water or other liquids.

1. Extraction, Treatment, and Replacement

Groundwater at an old waste site may be pumped or extracted to lower the water table below the waste material. In many cases, the groundwater is extracted from a contaminated area or plume and treated to remove the pollutants. The treated water may be discharged on the ground surface or reinjected around the perimeter the plume. This pump-and-treat method is illustrated in Figure 4.21.

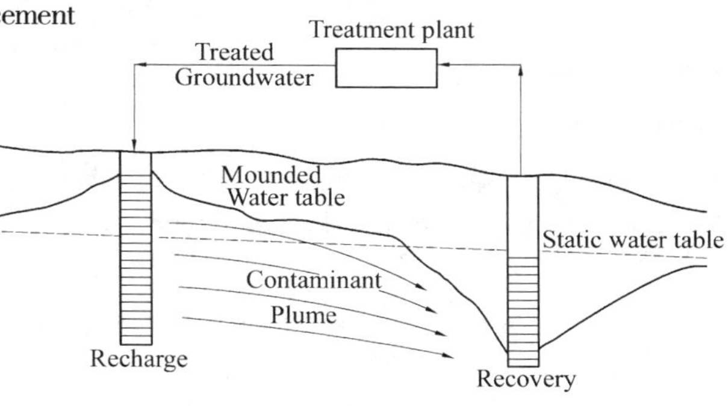

Figure 4.21 The "pump-and-treat" method for remediating contaminated groundwater

Toxic or flammable soil gas generated from anaerobic decomposition or volatilization of the buried waste can be removed using an induced draft extraction fan and treated using granular activated carbon adsorption; in some cases, the gases can be destroyed by flares or combustion devices. Contaminated soils can also be excavated from a hazardous waste site, treated, and then replaced on site, as illustrated in Figure 4.22.

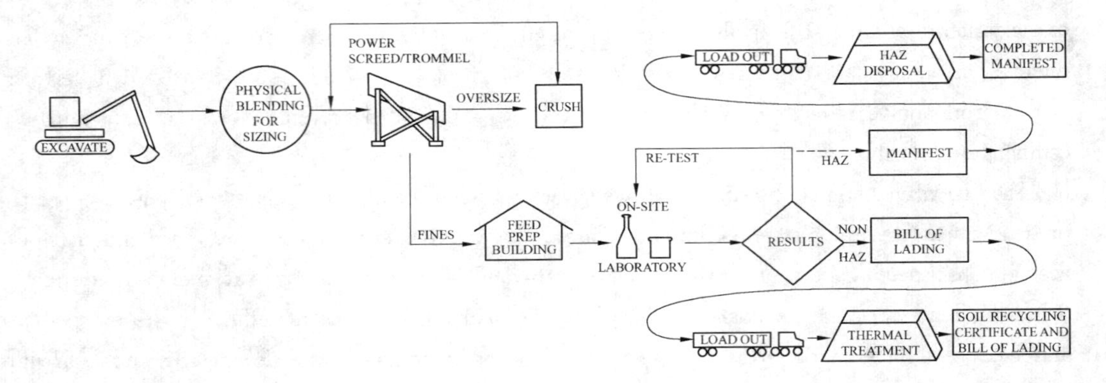

Figure 4.22　Soil management flow chart for remediation of a former "town gas" plant

2. Permanent Containment Methods

Vertical walls can be constructed of steel sheet piling or slurry trenches to block the horizontal movement of liquid. But these subsurface cutoff walls are feasible only if there is an aquiclude, or naturally occurring impermeable layer, below the waste site. The cutoff wall must penetrate or be keyed into the aquiclude, which serves as a natural bottom liner. A wall made of steel sheet piling can be driven to depths of about 30 m if the soil is not too rocky, but steel piling is not suitable if the waste contains corrosive liquids. A slurry trench cutoff wall can be built to contain most hazardous wastes, as shown schematically in Figure 4.23.

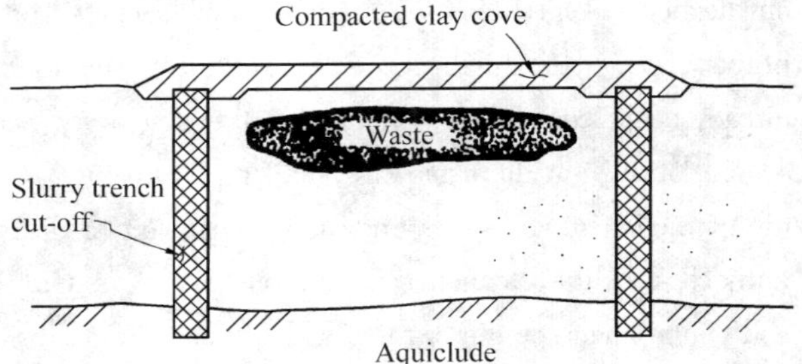

Figure 4.23　Slurry trench cutoff walls can be used to prevent the spread of polluted groundwater at an abandoned hazardous waste dump site

A slurry trench cutoff wall can be excavated from the ground surface with a clamshell or backhoe excavator without moving or otherwise disturbing the waste material. The trenches, roughly 1 m wide and up to 30 m deep, can be dug without collapse by filling them temporarily with a bentonite-clay slurry. The dense slurry (a mixture of clay and water) maintains the stability of the trench during excavation until it is backfilled with a material that forms the vertical barrier. The backfill material may be a mixture of soil and cement or soil and clay; it must be blended properly to have a permeability of 10^{-6} mm/s or less to block the flow of water. In some cases, a double trench wall, or a trench wall with an installed synthetic liner sheet, may be built for extra insurances that there will be no liquid or water movement into or out of the site.

3. In Situ Bioremediation

Bioremediation is a technique used for cleaning up contaminated soil or groundwater at hazardous waste sites. It relies on the biological action of microorganisms to convert the contaminants into harmless substances. The use of microorganisms for environmental purposes is not new; it has existed for decades at secondary wastewater treatment plants. What is unique about bioremediation is the way in which biological processes are applied in situ, that is, on the site of old hazardous waste dumps.

In soil above the water table, contamination may exist in four phases. It may be in the vapor phase, within the pore spaces; the adsorbed phase, attached to soil particles; the aqueous phase, dissolved in water, and the liquid phase, in the form of nonaqueous-phase liquids (NAPLs). Below the water table, contaminants may exist in all but the vapor phase. Certain contaminants, such as petroleum hydrocarbons (for example, benzene, gasoline, and oil), are less dense than water and are called light nonaqueous-phase liquids (LNAPLs). These tend to float on the surface of the water table and spread laterally, as illustrated in Figure 4.24.

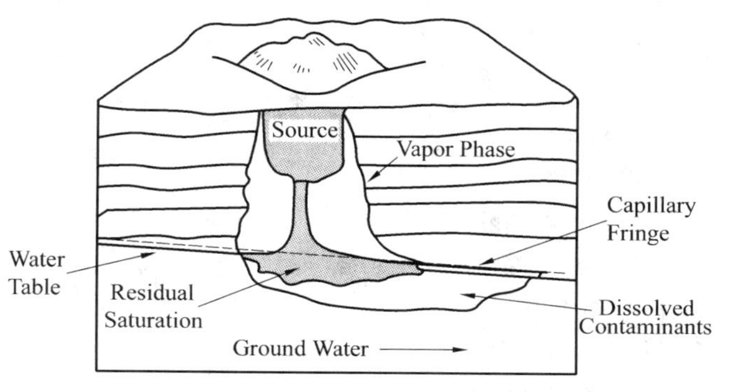

Figure 4.24 Underground distribution of LNAPLs

Other contaminants, including chlorinated hydrocarbons (for example, carbon tetrachloride and trichloroethane), are denser than water and are called dense nonaqueous-phase liquids (DNAPLs). These substances tend to migrate downward, sometimes penetrating deep into the groundwater. Chlorinated hydrocarbons are used widely as industrial degreasing solvents. They have often been disposed of improperly in refuse sites, lagoons, and storage tanks; as a group, chlorinated solvents are the most prevalent type of groundwater contaminants found in the United States. They are not as readily biodegradable as petroleum products.

Bioremediation of sites contaminated with petroleum hydrocarbons often relies on indigenous (native) bacteria found in soil. The addition of suitable bacterial mixtures to "jump start" the process can significantly reduce the treatment time. This is called bioaugmentation. If oxygen and nutrients (that is, nitrogen and phosphorus) are not present in sufficient quantities one or more of these elements may also be added. Systems for treating ground water typically comprise recovery wells, monitoring wells, and injection wells. Recovered groundwater is often treated prior to the addition of nutrients or oxygen and reinjection. Treatment may involve the use of an air stripper tower or activated carbon unit an oil-water separator, a biological unit or a combination of these. Wells are located so that nutrient or oxygen-enriched groundwater flows to ward the recovery wells, and only a portion of the recovered groundwater may be reinjected; to control flow rate and flow path, a portion may be discharged into surface waters.

4. Bioventing and Air Sparging

To stimulate aerobic biodegradation of petroleum hydrocarbon contaminants air may be injected below the ground surface. When air is injected through boreholes above the water table, the process is known as bioventing. When it is injected below the water table, the technique is termed air sparging, as illustrated in Figure 4.25.

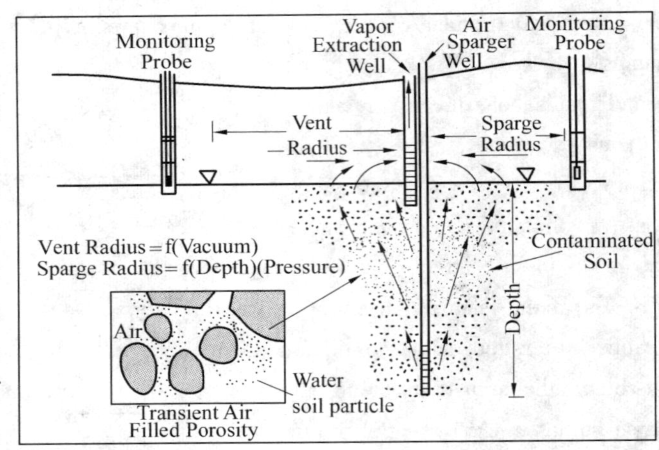

Figure 4.25　Air sparging

Air sparging is an effective means of treating petroleum hydrocarbons because it also enhances physical removal by direct extraction of VOCs. Airflow must be carefully controlled during a sparging operation to avoid the possible increase in the spread of contaminants. Too low a flow can significantly reduce treatment efficiency, and too high a flow can result in a loss of control. Because of this, installation of an air sparging system must always be preceded by a pilot test.

5. Remediation Methods for Chlorinated Solvents

Chlorinated solvents can be transformed biologically by microorganisms, thereby remediating contaminated soils. In addition, soils contaminated with chlorinated solvents may be treated effectively by soil vacuum extraction to remove volatile vapors; groundwater can be treated by air stripping. These are physical treatment processes; the vapors are discharged into the atmosphere or treated with activated carbon, catalytic combustion, or incineration.

6. Natural Bioremediation

Naturally occurring microorganisms can degrade contaminants in the subsurface environment. Studies have shown that plumes of dissolved hydrocarbons will eventually degrade without human intervention. The natural assimilative capacity of an aquifer depends on the local hydrogeology, the geochemistry, and the metabolic characteristics of indigenous microorganisms. In some cases, removal of leaking tanks or contaminated soil maybe all that is necessary for natural bioremediation to complete the cleanup. In most cases, though, natural bioremediation is used only to supplement the methods discussed previously.

Hydrocarbons that do not completely degrade in the unsaturated zone will be transposed within the water table aquifer, where the extent of further biodegradation is likely to be limited by the available oxygen supply; oxygen is only slightly soluble in water. As the plume migrates, though, contaminated water will disperse and mix with clean, oxygenated water.

Depending on site conditions, problems related to risk analysis, property rights, third-party liability, and the possible need for variances from existing regulations will have to be considered if natural bioremediation is used. A comprehensive monitoring system is needed, including interior wells to monitor the plume and guardian wells at the outside edge of the contaminated area; guardian wells are needed to monitor potential off-site migration of contaminants and to determine if additional remedial steps are necessary. Computer models may also be used to predict the extent of plume migration and rate of biodegradation in an aquifer. In some cases, particularly for chlorinated contaminants, the inoculation or addition of nonindigenous microorganisms into the ground can speed up the rate of biological action and remediation.

4.11 Hazardous Waste Minimization

The most desirable solution to the hazardous waste problem is to reduce or minimize the quantity of waste at its source. Waste minimization is, in fact, a primary waste management goal in the United States. In the 1984 Hazardous and Solid Waste Amendments, it is stated that "The Congress hereby declares it to be the national policy of the United States that, wherever feasible, the generation of hazardous waste is to be reduced or eliminated as expeditiously as possible." Waste minimization policies are also followed by Japan and many European countries.

Waste minimization techniques focus on source reduction and recycling activities that reduce either waste volume or hazards. Recycling is an increasingly attractive option, particularly for waste that includes high concentrations of metals, oils, acids, and other substances that may have economic value. A diagram outlining waste minimization techniques is shown in Figure 4.26.

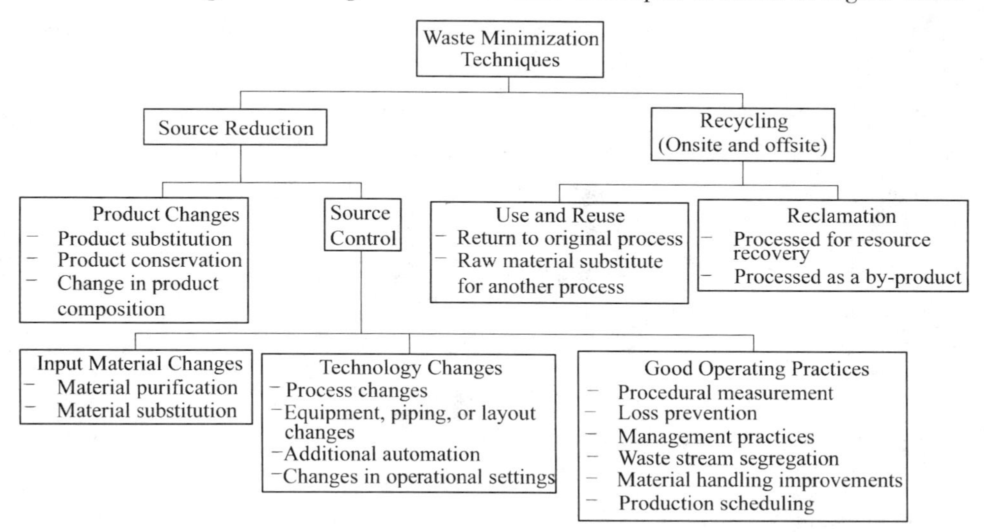

Figure 4.26 Methods to reduce waste

Waste minimization is now a top priority or goal for industry, not because of mandatory requirements (there are none), but because of the mandatory requirements of the other options, including disposal on or in the ground. Disposal sites are scarce and prices keep rising. It has become clear that simply storing or burying hazardous waste in the land is not the ultimate solution or management option for industry. Even treatment and stabilization of the waste prior to its land disposal will not remedy the problem. Because of restriction on land disposal, many untreated wastes that were sent to landfills in the past must now be incinerated, at costs much higher than those for landfilling. Waste minimization also can reduce a generator's financial and legal liability; the less waste generated, the lower are the costs and environmental risks.

Waste minimization is not regulated as a mandatory program; the United States and other countries that promote minimization and recycling as a management option rely on cooperative, voluntary efforts by the industrial sector. Mandatory performance standards and other regulatory approaches have been rejected because they would be counterproductive. They would second-guess industry's production decisions and would be very difficult to administer. It is considered best to rely on existing strong economic incentives for industry to reduce their wastes. The most constructive role for government in this effort is to provide technical information and other assistance to waste generators. Since the states deal first-hand with the generators, they also should play a central role in assisting industry with waste minimization programs. A few states provide incentives and encourage waste reduction through tax preferences.

4.11.1 Waste Minimization Audits

The first step in establishing a waste minimization program is to conduct a waste minimization assessment or audit. This is a careful review of an industry's potential opportunities to reduce or recycle its waste; it may be done by in-house staff or an independent consultant. An effective way to begin such an audit is to select a few waste streams or processes for intensive review, rather than to attempt to cover all waste streams at once. Waste minimization assessments identify and characterize waste streams, the production processes that are generating the waste, and the quantity of waste generated. The EPA suggests the following steps for a waste minimization assessment:

(1) Prepare background material for the assessment.

(2) Conduct a preassessment visit to identify candidate waste streams.

(3) Select waste streams for detailed analysis.

(4) Conduct a detailed site visit to collect data on selected waste streams and control sand related process data.

(5) Develop a series of potential waste minimization options.

(6) Undertake preliminary option evaluations (including development of preliminary cost estimates).

(7) Rank options according to waste reduction effectiveness, extent of current use in the industry, and potential for future application at the facility.

(8) Present preliminary results to plant personnel along with a ranking of options.

(9) Prepare a final report, including recommendations to plant management.

(10) Develop an implementation plan and schedule.

(11) Conduct periodic reviews and updates of assessments.

4.11.2 Waste Reduction Methods

Methods of reducing hazardous waste quantities can be grouped into several categories, including product changes, input material changes, technology changes, good operating practices, and recycling (see Figure 4.26). These methods can be applied in a wide range of industries and manufacturing operations. Most involve source reduction, the EPA's preferred option. Others involve on-and off-site recycling or reuse. The most suitable method for any particular industry can be determined after a waste minimization audit.

Improved operations may include the purchase of fewer toxic production materials, improved material storage and handling practices, employee training, and better housekeeping practices. Hazardous waste can be separated from nonhazardous waste to save money for disposal and find new opportunities for reuse. Production equipment can be modified to produce less waste and to enhance material recovery operation; an improved preventive maintenance program can also be helpful. End products call be redesigned or reformulated to be less hazardous, and the sources of leaks and spills can be eliminated. Closed-loop production systems can be designed and implemented to increase on site recycling.

Metal parts cleaning and paint application processes are essential for many industries and businesses. Hazards from metal parts cleaning can be minimized by reducing the volume or the toxicity of the cleaning agents used. Low-toxicity paints (for example, water-based) that do not contain heavy metals can be used to reduce paint-related hazardous wastes.

Sometimes, one company's waste can be used as the feedstock or raw material for another company's production process. Organizations that serve as hazardous waste clearinghouses or material exchanges are able to facilitate exchange and recycling efforts. A clearinghouse can help to make arrangements between waste generators and potential users of the waste material. It can serve as a matchmaker or directly as a transfer agent, purchasing the waste from the generator, reprocessing it if needed, and selling it for reuse by some other industry. Waste clearinghouses and exchanges have found some success in several European and Scandinavian countries; private or government-funded waste exchange organizations are likely to play an increasingly important role in managing hazardous waste in the United States in the coming years.

Review Questions

1. Definition the MSW, including the meaning of the terms refuse, garbage, rubbish, and trash.

2. What is a transfer station and when is it needed?

3. What is composting? Briefly describe two methods by which it is accomplished, and discuss any advantages or disadvantages.

4. What materials in the MSW stream are recyclable? Briefly discuss some of the factors related to the recovery and reuse of each type of material.

5. What are key characteristics of an MSW sanitary landfill that distinguish it from an open dump? What are some potential disadvantages of landfilling as a waste disposal technique?

6. Give a workable definition of hazardous waste, including the basic properties that are characteristics of such waste. Briefly describe these properties.

7. Briefly discuss hazardous waste storage methods.

8. Briefly discuss biological treatment methods for hazardous waste.

9. Briefly discuss two basic alternatives for on-site remediation.

10. Define the term of bioremediation.

Chapter 5 Noise Pollution and Control

5.1 Introduction of Sound

Noise, commonly defined as unwanted sound, is an environmental phenomenon to which we are exposed before birth and throughout life. Noise is an environmental pollutant, a waste product generated in conjunction with various anthropogenic activities. Under the latter definition, noise is any sound-independent of loudness-that can produce an undesired physiological or psychological effect in an individual, and that may interfere with the social ends of an individual or group. These social ends include all of our activities-communication, work, rest, recreation, and sleep.

It has long been known that noise of sufficient intensity and duration can induce temporary or permanent hearing loss, ranging from slight impairment to nearly total deafness. In general, a pattern of exposure to any source of sound that produces high enough levels can result in temporary hearing loss. If the exposure persists over a period of time, this can lead to permanent hearing impairment. Short-term, but frequently serious, effects include interference with speech communication and the perception of other auditory signals, disturbance of sleep and relaxation, annoyance, interference with an individual's ability to perform complicated tasks, and general diminution of the life.

The engineering and scientific community has already accumulated considerable knowledge concerning noise, its effects, and its abatement and control. In that regard, noise differs from most other environmental pollutants. Generally, the technology exists to control most indoor and outdoor noise. As a matter of fact, this is one instance in which knowledge of control techniques exceeds the knowledge of biological and physical effects of the pollutant.

5.1.1 Properties of Sound Waves

Sound waves result from the vibration of solid objects or the separation of fluids as they pass over, around, or through holes in solid objects. The vibration and/or separation causes the surrounding air to undergo alternating compression and rarefaction, much in the same manner as a piston vibrating in a tube (see Figure 5.1). The compression of the air molecules causes a local increase in air density and pressure. Conversely, the rarefaction causes a local decrease in density and pressure. These alternating pressure changes are the sound detected by the human ear.

Let us assume that you could stand at Point A in Figure 5.1. Also let us assume that you have an instrument that will measure the air pressure every 0.000 010 seconds and plot the value on a graph. If the piston vibrates at a constant rate, the condensations and rarefactions will move down the tube at a constant speed. That speed is the speed of sound (c). The rise and fall of pressure at point A will follow a cyclic or wave pattern over a period of time (see Figure 5.2). The wave pattern is called sinusoidal. The time between successive peaks or between successive troughs of the oscillation is called the period (P). The inverse of this, that is, the number of times a peak arrives in one second of oscillations, is called the frequency (f). Period and frequency are then related as follows:

$$P = 1/f \tag{5.1}$$

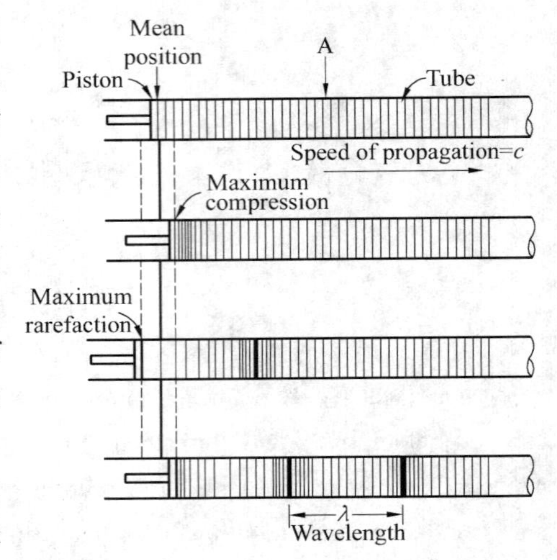

Figure 5.1 Alternating compression and rarefaction of air molecules resulting from a vibrating piston

Since the pressure wave moves down the tube at a constant speed, you would find that the distance between equal pressure readings would remain constant The distance between adjacent crests or troughs of pressure is called the wavelength (λ). Wavelength and frequency are then related as follows:

$$\lambda = c/f \tag{5.2}$$

The amplitude (A) of the wave is the height of the peak or depth of the trough measured from the zero pressure line (see Figure 5.2). From Figure 5.2, we can also note that the average pressure could be zero if all averaging time was selected that correspond to the period of the wave. This would result regardless of the amplitude. This ,of course, is not an acceptable state of affairs. The root mean square (rms) sound pressure (p_{rms}) is used to overcome this difficulty. [Sound pressure = (total atmospheric pressure) − (barometric pressure)] The rms sound pressure is obtained squaring the value of the amplitude at each instant in time; summing the squared values; dividing the total by the averaging time; and taking the square root of the total. The equation for rms is

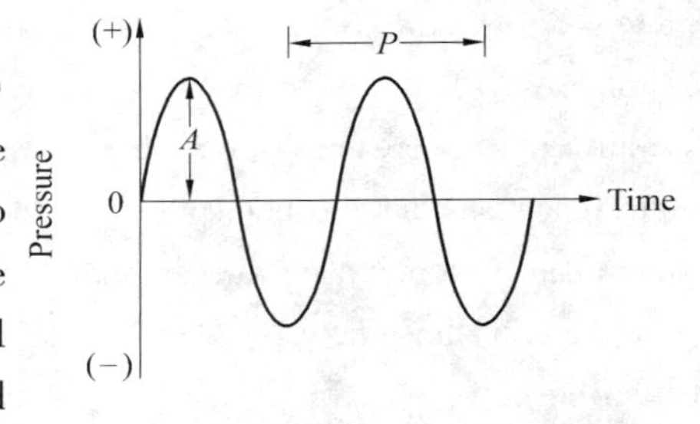

Figure 5.2 Sinusoidal wave that results from alternating compression and rarefaction of air molecules. The amplitude is shown as *A* and the period is *P*

$$p_{rms} = (\overline{p^2})^{1/2} = [\frac{1}{T}\int_0^T P^2(t)\,dt]^{1/2} \tag{5.3}$$

where the overbar refers to the time-weighted average and T is the time period of the measurement.

5.1.2 Sound Power and Intensity

Work is defined as the product of the magnitude of the displacement of a body and the component of force in the direction of the displacement. Thus, traveling waves sound pressure transmit energy in the direction of propagation of the wave. The rate at which this work is done is defined as the sound power (W).

Sound intensity (I) is defined as the time-weighted average sound per unit area normal to the direction of propagation of the sound wave. Intensity and power are related as follows:

$$I = \frac{W}{A} \tag{5.4}$$

where A is a unit area perpendicular to the direction of wave motion. Intensity, and hence, sound power, is related to sound pressure in the following manner:

$$I = \frac{(p_{rms})^2}{\rho c} \tag{5.5}$$

where I = intensity, W/m^2

p_{rms} = root mean square sound pressure, Pa

ρ = density of medium, kg/m^3

c = speed of sound in medium, m/s

Both the density of air and speed of sound are a function of temperature. Given the temperature and pressure, the density of air (1.185 kg/m^3 at 101.325 kPa and 298K) may be determined using the gas laws. The speed of sound in air at 101.325 kPa may be determined from the following equation:

$$c = 20.05\sqrt{T} \tag{5.6}$$

where T is the absolute temperature in kelvins (K) and c is in m/s.

5.1.3 Level and the Decibel

The sound pressure of the faintest sound that a normal healthy individual can hear is about 0.000 02 pascal. The sound pressure produced by a Saturn rocket at liftoff is greater than 200 pascal. Even in scientific notation this is an "astronomical" range of numbers.

In order to cope with this problem, a scale based on the logarithm of the ratios of the measured quantities is used. Measurements on this scale are called levels. The unit or these types of measurement scale is the bel, which was named after Alexander Graham Bell:

$$L' = \lg \frac{Q}{Q_0} \tag{5.7}$$

where L' = level, bels

Q = measured quantity

Q_0 = reference quantity

lg = logarithm in base 10

A bel turns out to be a rather large unit, so for convenience it is divided into 10 subunits called decibels (dB). Levels in dB are computed as follows:

$$L = 10\lg \frac{Q}{Q_0} \tag{5.8}$$

The dB does not represent any physical unit. It merely indicates that a logarithmic transformation has been performed.

If the reference quantity (Q_0) is specified, then the dB takes on physical significance. For noise measurements the reference power level has been established as 10^{-12} watts. Thus, sound power level may be expressed as

$$L_w = 10 \lg \frac{W}{10^{-12}} \tag{5.9}$$

For noise measurements, the reference sound intensity (see Equation 5.4) is 10^{-12} W/m^2. Thus, the sound intensity level is given as

$$L_I = 10 \lg \frac{I}{10^{-12}} \tag{5.10}$$

Because sound-measuring instruments measure the root mean square pressure, the sound pressure level is computed as follows:

$$L_P = 10 \lg \frac{(p_{rms})^2}{(p_{rms})_0^2} \tag{5.11}$$

which, after extraction of the squaring term, is given as

$$L_P = 20 \lg \frac{p_{rms}}{(p_{rms})_0} \tag{5.12}$$

The reference pressure has been established as 20 micropascals (μPa).

Because of their logarithmic heritage, decibels don't add and subtract the way apples and oranges do. Remember: adding the logarithms of members is the same as multiplying them. If you take a 60-decibel noise (re: 20μPa) and add another 60-decibel noise (re: 20μPa). If you're strictly an apple-and-orange mathematician, you may take this on faith. For skeptics, this can be demonstrated by converting the dB to sound power level, adding them, and converting back to dB. Figure5.3 provides a graphical solution of this type of problem. For noise pollution work, results should be reported to the nearest whole number. When there are several levels to be combined, they should be combined two at a time, starting with lower-valued levels and continuing two at a time with each successive pair until one number remains.

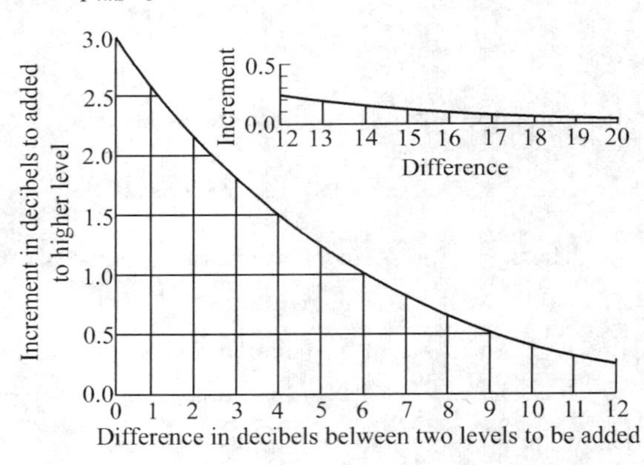

Figure 5.3 Graph for solving decibel addition problems

5.2 Characterization of Noise

5.2.1 Weighting Networks

Weighting networks are used to account for the frequency of a sound. They are electronic filtering circuits built into the meter to attenuate certain frequencies. They permit the sound level meter to respond more to some frequencies than to others with a prejudice something like that of the human ear. Writers of the acoustical standards have established three weighting characteristics: A, B, and C. The chief difference among them is that very low frequencies are filtered quite severely by the A network, moderately by the B network, and hard at all by the C network. Therefore, if the measured sound level of a noise is much higher on C weighting than on A weighting, much of the noise is probably of low frequency. If you really want to know the frequency distribution of a noise (and most serious noise measurers do), it is necessary to use a sound analyzer. But if you are unable to justify the expense of an analyzer, you can still find out something about the frequency of a noise by shrewd use of the weighting networks of a sound level meter.

Figure 5.4 shows the response characteristics of the three basic networks as prescribed by the American National Standards Institute (ANSI) specification number S1. 4-1971. When a weighting network is used, the sound level meter electronically subtracts or adds the number of dB shown at each frequency shown in Table 5.1 from or to the actual sound pressure level at that frequency. It then sums all the resultant numbers by logarithmic addition to give a single reading. Readings taken when a network is in use are said to be "sound levels" rather than "sound pressure levels." The reading taken are designated in decibels in one of the following forms: dB(A); dBa; dBA; dB(B); dBb; dBB; and so on. Tabular notations may refer to L_A, L_B, L_C.

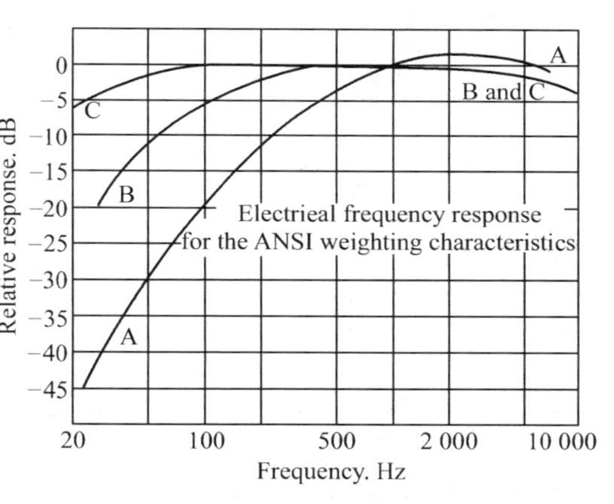

Figure 5.4 Response characteristics of the three basic weighting networks

Table 5.1 Sound level meter network weighting values

Frequency (Hz)	Curve A (dB)	Curve B (dB)	Curve C (dB)
10	-70.4	-38.2	-14.3
12.5	-63.4	-33.2	-11.2
16	-56.7	-28.5	-8.5
20	-50.5	-24.2	-6.2
25	-44.7	-20.4	-4.4

31.5	-39.4	-17.1	-3.0
40	-34.6	-14.2	-2.0
50	-30.2	-11.6	-1.3
63	-26.2	-9.3	-0.8
80	-22.5	-7.4	-0.5
100	-19.1	-5.6	-0.3
125	-16.1	-4.2	-0.2
160	-13.4	-3.0	-0.1
200	-10.9	-2.0	0
250	-8.6	-1.3	0
315	-6.6	-0.8	0
400	-4.8	-0.5	0
500	-3.2	-0.3	0
630	-1.9	-0.1	0
800	-0.8	0	0
1 000	0	0	0
1 250	0.6	0	0
1 600	1.0	0	-0.1
2 000	1.2	-0.1	-0.2
2 500	1.3	-0.2	-0.3
3 150	1.2	-0.4	-0.5
4 000	1.0	-0.7	-0.8
5 000	0.5	-1.2	-1.3
6 300	-0.1	-1.9	-2.0
8 000	-1.1	-2.9	-3.0
10 000	-2.5	-4.3	-4.4
12 500	-4.3	-6.1	-6.2
16 000	-6.6	-8.4	-8.5
20 000	-9.3	-11.1	-11.2

5.2.2 Octave Bands

To completely characterize a noise, it is necessary to break it down into its frequency components or spectra. Normal practice is to consider 8 to 11 octave bands. An octave is the frequency interval between a given frequency and twice that frequency. For example, given the frequency 22 Hz, the octave band is from 22 to 44 Hz. A second octave band would then be from 44 to 88 Hz.

The standard octave bands and their geometric mean frequencies (center band frequencies) are given in Table 5.2. Octave analysis is performed with a combination precision sound level meter and an octave filter set.

While octave band analysis is frequently satisfactory for community noise control (that is, identifying violators), more refined analysis is required or corrective action and design. One-third octave band analysis provides a slightly more refined picture of the noise source than the full octave band analysis (see Figure 5.5(a)). This improved resolution is usually sufficient for determining corrective action for community noise problems. Narrow band analysis is highly refined and may imply band widths down to 2 Hz (see Figure 5.5(b)). This degree of refinement is only justified in product design and testing or in troubleshooting industrial machine noise and vibration).

Table 5.2 Octave bands

Octave frequency range (Hz)	Geometric mean frequency (Hz)
22 ~ 44	31.5
44 ~ 88	63
88 ~ 175	125
175 ~ 350	250
350 ~ 700	500
700 ~ 1 400	1 000
1 400 ~ 2 800	2 000
2 800 ~ 5 600	4 000
5 600 ~ 11 200	8 000
11 200 ~ 22 400	16 000
22 400 ~ 44 800	31 500

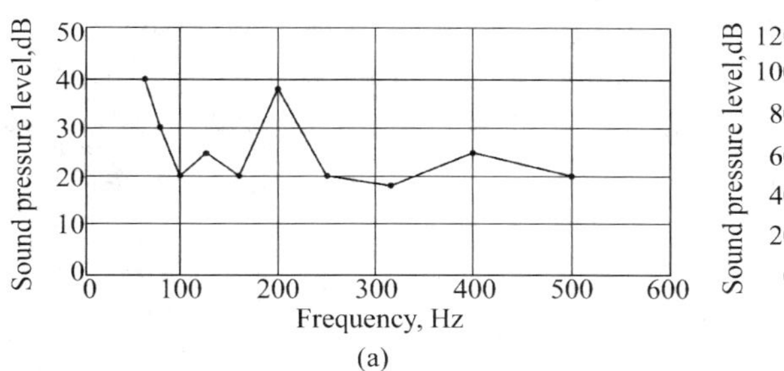

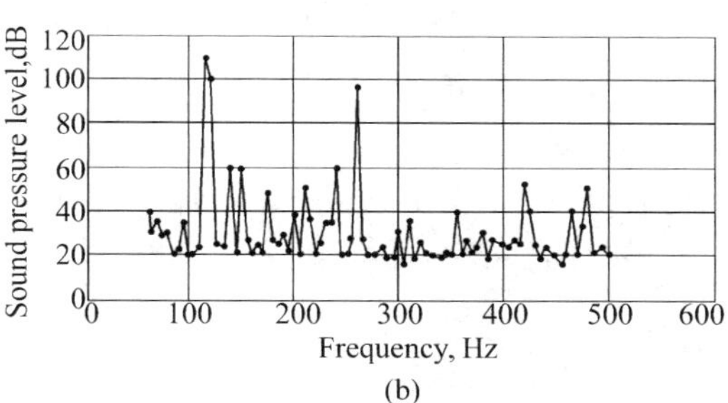

Figure 5.5 (a) One-third octave band analysis of a small electric motor. (b) Narrowband analysis of a small electric motor

5.2.3 Averaging Sound Pressure Levels

Because of the logarithmic nature of the dB, the average value of a collection of sound pressure level measurements cannot be computed in the normal fashion. Instead, the following equation must be used:

$$\overline{L_P} = 20 \log \frac{1}{N} \sum_{j=1}^{N} 10^{(L_j/20)} \qquad (5.13)$$

where L_p = average sound pressure level, dB re: 20 μPa

N = number of measurements

L_j = the jth sound pressure level, dB re: 20 μPa

$j = 1, 2, 3, \ldots, N$

This equation is equally applicable to sound levels in dBA. It may also be used to compute average sound power levels if the factors of 20 are replaced with 10s.

5.2.4 Types of Sounds

Patterns of noise may be qualitatively described by one of the following terms: steady-state or continuous; intermittent; and impulse or impact. Continuous noise is all uninterrupted sound level that varies less than 5 dB during the period of observation. An example is the noise from a household fan. Intermittent noise is a continuous noise that persists for more than one second that is interrupted for more than one second. A dentist's drilling would be an example of an intermittent noise. Impulse noise is characterized by a change of sound pressure of 40 dB or more with 0.5 second with a duration of less than one second. The Occupational Safety and Health Administration (OSHA) classifies repetitive events, including impulses, as steady noise if the interval between events is less than 0.5 seconds. The noise from firing a weapon would be an example of an impulsive noise.

Two types of impulse noise generally are recognized. The type A impulse is characterized by a rapid rise to a peak sound pressure level followed by small negative pressure wave or by decay to background level (see Figure 5.6). The type B impulse is characterized by a damped (oscillatory) decay (see Figure 5.7). Where the duration of the type A impulse is simply the duration of the initial peak, the duration of the type B impulse is the time required for the envelope to decay to 20 dB below the peak. Because of the short duration of the impulse, a special sound-level meter must be employed to measure impulse noise. You should note that the peak sound pressure level is different than the impulse sound level because of the time-averaging used in the latter.

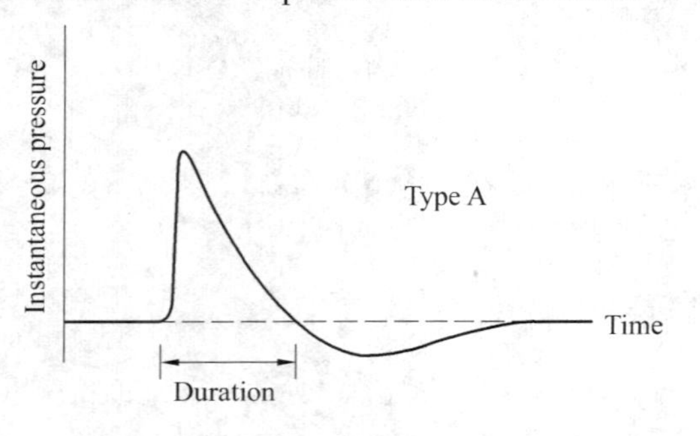

Figure 5.6 Type A impulse noise

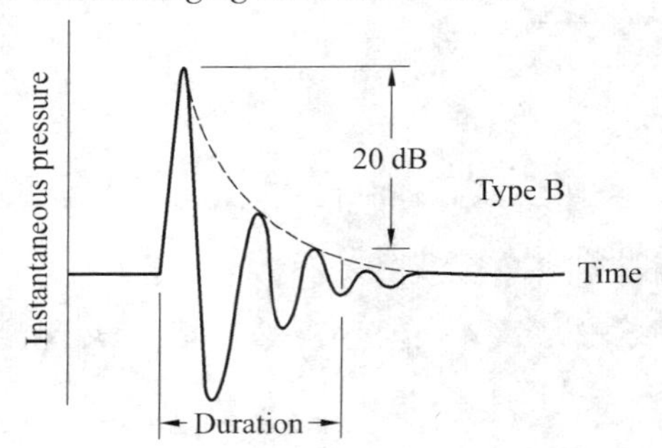

Figure 5.7 Type B impulse noise

5.3 Effects of Noise on People

For the purpose of our discussion, we have classified the effects of noise on people into the following two categories: auditory effects and psychological/sociological effects. Auditory effects include both hearing loss and speech interference. Psychological/sociological effects include annoyance, sleep interference, effects on performance, and acoustical privacy.

5.3.1 Hearing Impairment

With the exception of eardrum rupture from intense explosive noise, the outer and middle ear rarely are damaged by noise. More commonly, hearing loss is a result of neural damage involving injury to the hair cells (see Figure 5.8). Two theories are offered to explain noise-induced injury. The first is that excessive shearing forces mechanically damage the hair cells. The second is that intense noise stimulation forces the hair cells into high metabolic activity, which overdrives them to the point of metabolic failure and consequent cell death. Once destroyed, hair cells are not capable of regeneration.

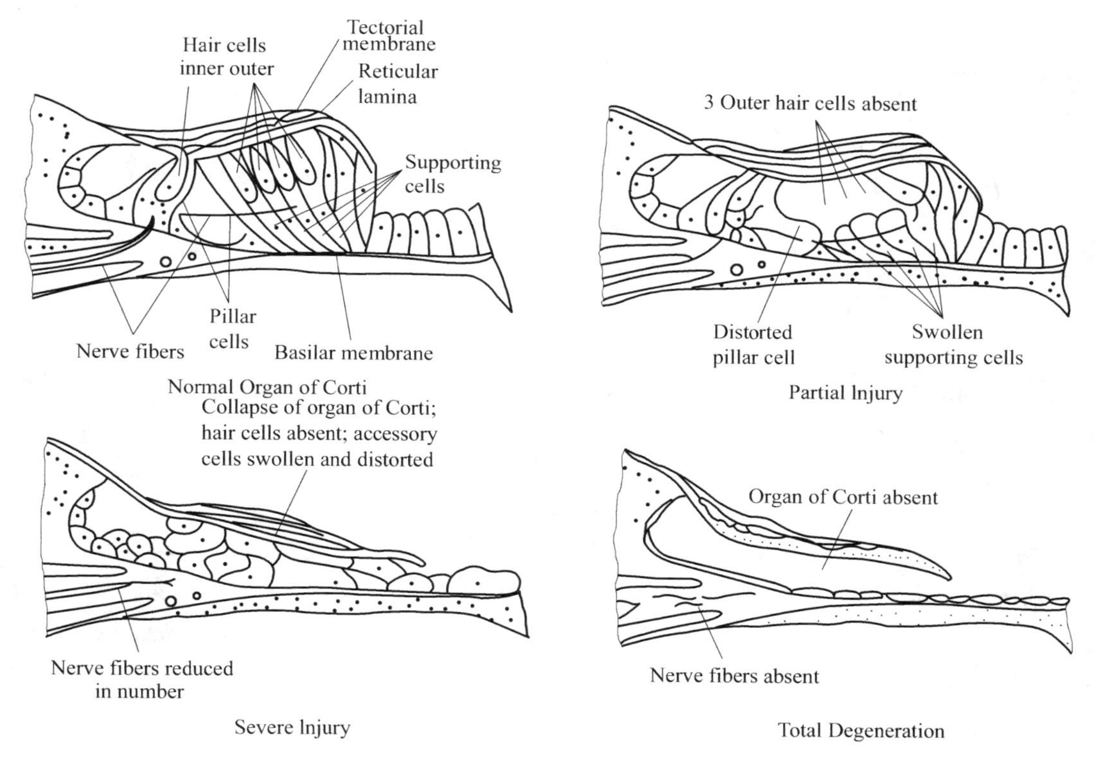

Figure 5.8 Various degrees of injury to the hair cells

Since direct observation of the organ of Corti in persons having potential hearing loss is impossible, injury is inferred from losses in their hearing threshold level (HTL). The increase sound pressure level required to achieve a new HTL is called threshold shift. Obviously, any measurement of threshold shift is dependent upon having a baseline audiogram taken before the noise exposure. Heating losses may be either temporary or permanent. Noise-induced losses must be separated from other causes of hearing loss such as age (presbycusis), drugs, disease, and blows on the head. Temporary threshold shift (TTS) is distinguished from permanent threshold shift (PTS) by the fact that in TTS removal of the noise over stimulation will result in a gradual return to baseline heating thresholds.

The outer and middle ear rarely are damaged by intense noise. However, explosive sounds can rupture the tympanic membrane or dislocate the ossicular chain. The permanent hearing loss that results from very brief exposure to a very brief exposure to a very loud noise is termed acoustic trauma. Damage to the outer and middle ear may or may not accompany acoustic trauma.

5.3.2 Damage-Risk Criteria

A damage-risk criterion specifies the maximum allowable exposure to which a person may be exposed if risk of hearing impairment is to be avoided. The American Academy of Ophthalmology and Otolaryngology has defined hearing impairment as an average HTL in excess of 25 dB (ANSI-1969) at 500, 1 000, and 2 000 Hz. This is called the low fence. Total impairment is said to occur when the average HTL exceeds 92 dB. Presbycusis is included in setting the 25 dB ANSI low fence. Two criteria have been set to provide conditions under which nearly all workers may be repeatedly exposed without adverse effect on their ability to hear and understand normal speech.

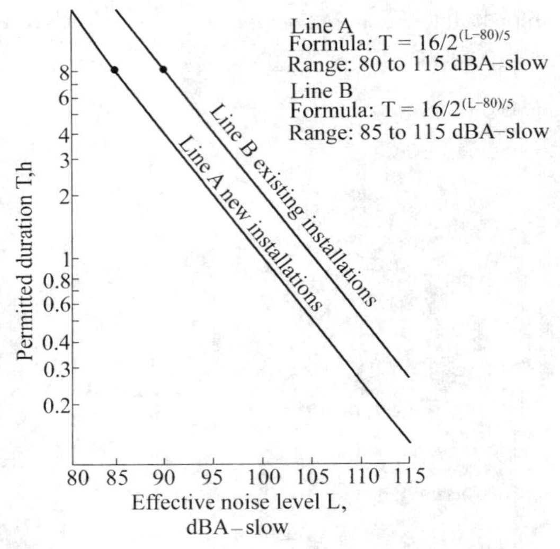

Figure 5.9 NIOSH occupational noise exposure limits for continuous or intermittent noise exposure

The National Institute for Occupational Safety and Health (NIOSH) has recommended that occupational noise exposure be controlled so that no worker is exposed in excess of the limits defined by line B in Figure 5.9. In addition, NIOSH recommends that new installations be designed to hold noise exposure below the limits defined by line A in Figure 5.9. The Walsh-Healey Act, which was enacted by Congress in 1969 to protect workers, used a damage-risk criterion equivalent to the line A criterion.

5.3.3 Speech Interference

As we all know, noise can interfere with our ability to communicate. Many noises that are not intense enough to cause hearing impairment can interfere with speech communication. The interference, or masking,

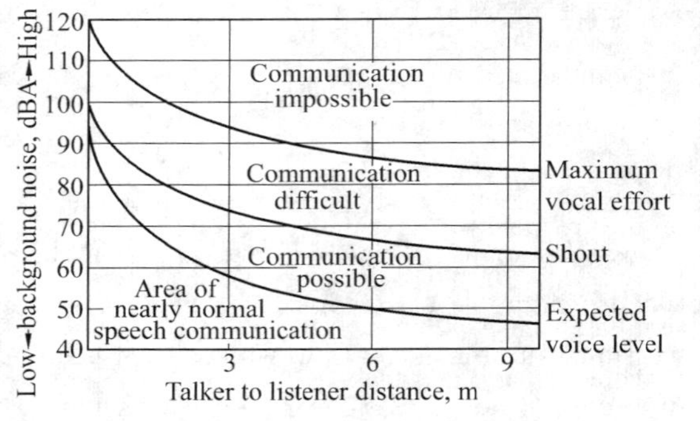

Figure 5. 10 Quality of speech communication as a function of sound level and distance

effect is a complicated function of the distance between the speaker and listener and the frequency components of the spoken words. The Speech Interference Level (SIL) was developed as a measure of the difficulty in communication that could he expected with different background noise levels. It is now more convenient to talk in terms of A-weighted background noise levels and the quality of speech communication (see Figure 5.10).

5.3.4 Annoyance

Annoyance by noise is a response to auditory experience. Annoyance has its base in the unpleasant nature of some sounds, in the activities that are disturbed or disrupted by noise, in the physiological reactions to noise, and in the responses to the meaning of "messages" carried by the noise. For example, a sound heard at night may be more annoying than one heard by day, just as one that fluctuates may be more annoying than one that does not. A sound that resembles another sound that we already dislike and that perhaps threatens us may be especially annoying. A sound that we know is mindlessly inflicted and will not be removed soon may be more annoying than one that is temporarily and regretfully inflicted. A sound, the source of which is visible, May be more annoying than one with an invisible source. A sound that is new may be less annoying. A sound that is locally a political issue may have a particularly high or low annoyance.

The degree of annoyance and whether that annoyance leads to complaints, product rejection, or action against an existing or anticipated noise source depend upon many factors. Some of these factors have been identified, and their relative importance has been accessed. Responses to aircraft noise have received the greatest attention. There is less information available concerning responses to other noises, such as those of surface transportation and industry, and those from recreational activities. Many of the noise rating or forecasting systems that are now in existence were developed in an effort to predict annoyance reactions.

5.3.5 Sleep Interference

Sleep interference is a special category of annoyance that has received a great deal of attention and study. Almost, all of us have been wakened or kept from falling asleep by loud, strange, frightening, or annoying sounds. It is commonplace to be wakened by an alarm clock or clock radio. But it also appears that one can get used to sounds and sleep through them. Possibly, environmental sounds only disturb sleep when they are unfamiliar. If so, disturbance of sleep would depend only on the frequency of unusual or novel sounds. Everyday experience also suggests that sound can help to induce sleep and, perhaps, to maintain it. The soothing lullaby, the steady hum of a fan, or the rhythmic sound of the surf can serve to induce relaxation. Certain steady sounds can serve as an acoustical shade and mask disturbing transient sounds.

The effects of relatively brief noises (about three minutes of less) on a person sleeping in a quiet environment have been studied the most thoroughly. Typically, presentations of the sounds are widely spaced throughout a sleep period of 5 to 7 hours. A summary of some of these observations is presented in Figure 5.11. The dashed lines are hypothetical curves that represent the percent of awakenings under conditions in which the subject is a normally rested young adult male who has been adapted for several nights to the procedures of a quiet sleep laboratory. He has been instructed to press an easily reached button to indicate that he has awakened, and had been moderately motivated to awake and respond to the noise.

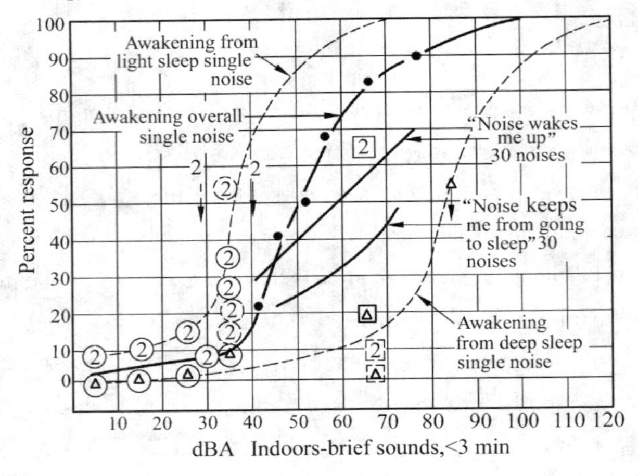

Figure 5.11 Effects of brief noise on sleep

While in light sleep, subject can awake to sounds that are about 30 ~ 40 decibels above the level at which they can be detected when subjects are conscious, alert, and attentive. While in deep sleep, the stimulus may have to be 50 ~ 80 decibels above the level at which they can be detected by conscious, alert, attentive subjects before they will awaken the sleeping subject.

5.3.6 Effects on Performance

When a task requires the use of auditory signals, speech or nonspeech, then noise at any intensity level sufficient to mask or interfere with the perception of these signals will interfere with the performance of the task.

Where mental or motor tasks do not involve auditory signals, the effects of noise on their performance have been difficult to assess. Human behavior is complicated, and it has been difficult to discover exactly how different kinds of noises might influence different kinds of people doing different kinds of tasks. Nonetheless, the following general conclusions have emerged. Steady noises without special meaning do not seem to interfere with human performance unless the A-weighted noise level exceeds about 90 decibels. Irregular bursts of noise (intrusive noise) are more disruptive than steady noise. Even when the A-weighted sound levels of irregular bursts are below 90 decibels, they may sometimes interfere with performance of a task. High-frequency components of noise, above about 1 000 ~ 2 000 hertz, may produce more interference with performance than low-frequency components of noise. Noise does not seem to influence the overall rate of work, but high-levels of noise may increase the variability of the rate of work. There may be noise pauses followed by compensating increases in work rate. Noise is more likely to reduce the accuracy of work than to reduce the total quantity of work. Complex tasks arc more likely to be adversely influenced by noise than are simple tasks.

5.3.7 Acoustic Privacy

Without opportunity for privacy, either everyone must conform strictly to an elaborate social code or everyone must adopt highly permissive attitudes. Opportunity for privacy avoids the necessity for either extreme. In particular, without opportunity for acoustical privacy, one may experience all of the effects of noise previously described and, in addition, one is constrained because one's own activities may disturb others. Without acoustical privacy, sound, like a faulty telephone exchange, reaches the "wrong number." The result disturbs both the sender and the receiver.

5.4 Rating Systems

5.4.1 Goals of a Noise-Rating System

An ideal noise-rating system is one that allows measurements by sound level meters or analyzers to be summarized succinctly and yet represent noise exposure in a meaningful way. In our previous discussions on loudness and annoyance, we noted that our response to sound is strongly dependent on the frequency of the sound. Furthermore, we noted that the type of noise (continuous, intermittent, or impulsive) and the time of day that it occurred (night being worse than day) were significant factors in annoyance. Thus, the ideal system must take frequency into account. It should differentiate between daytime and nighttime noise. And, finally, it must be capable of describing the cumulative noise exposure. A statistical system can satisfy these requirements.

The practical difficulty with a statistical rating system is that it would yield a large set of parameters for each measuring location. A much larger array of numbers would be required to characterize a neighborhood. It is literally impossible for such an array of numbers to be used effectively in enforcement. Thus, there has been a considerable effort to define a single number measure of noise exposure. The following paragraphs describe two of the systems now being used.

5.4.2 The L_N Concept

The parameter L_N is a statistical measure that indicates how frequently a particular sound level is exceeded. If, for example, we write L_{30} = 67dBA, then we know that 67dB(A) was exceeded for 30 percent of the measuring time. A plot of L_N against N where N = 1 percent, 2 percent, 3 percent, and so forth, would look like the cumulative distribution curve shown in Figure 5.12.

Allied to the cumulative distribution curve is the probability distribution curve. A plot of this will show how often the noise levels fall into certain class intervals. In Figure 5.13 we can see that 35 percent of the time the measured noise levels ranged between 65 and 67dBA; for 15 percent of the time they ranged between 67 and 69dBA; and so on. The relationship between this picture and the one for L_N is really quite simple. By adding the percentages given in successive class intervals from right to left, we can arrive at a corresponding L_N where N is the sum of the percentages and L is the lower limit of the left—most class interval added, thus, L_{30}

$$L\ (1 + 2 + 12 + 15)\ =\ 67\text{dBA}$$

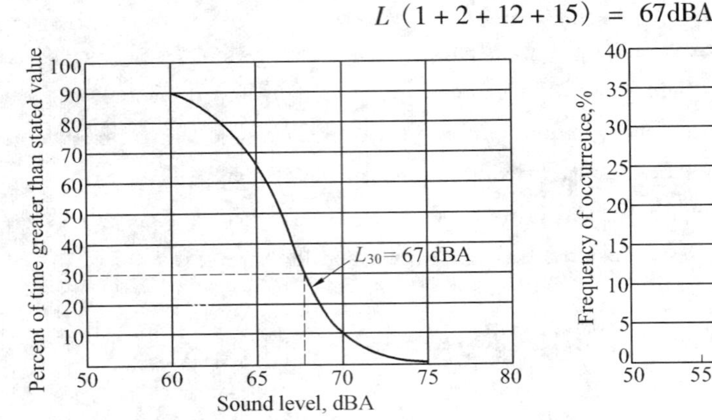

Figure 5.12 Cumulative distribution curve

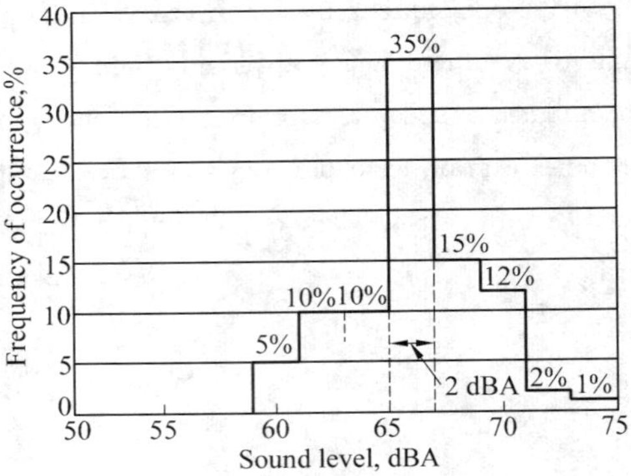

Figure 5.13 Probability distribution plot

5.4.3 The L_{eq} Concept

The equivalent continuous equal energy level (L_{eq}) can be applied to any fluctuating noise level. It is that constant noise level that, over a given time, expends the same amount of energy as the fluctuating level over the same time period. It is expressed as follows:

$$L_{eq} = 10\log \frac{1}{t}\int_0^t 10^{L(t)/10}\mathrm{d}t \tag{5.14}$$

Where t = the time over which L_{eq} is determined

$L(t)$ = the time varying noise level in dBA

Generally speaking, there is no well-defined relationship between $L(t)$ and time, so a series of discrete samples of $L(t)$ have to be taken. This modifies the expression to:

$$L_{eq} = 10\log \sum_{i=t}^{i=n} 10^{L(t)/10}(t_i) \tag{5.15}$$

where n = the total number of samples taken

L_i = the noise level in dBA of the ith sample

t_i = fraction of total sample time

The equivalent noise level was introduced in 1965 in Germany as a rating specifically to evaluate the impact of aircraft noise upon the neighbors of airports (Burck et al, 1965). It was almost immediately recognized in Austria as appropriate for evaluating the impact of street traffic noise in dwellings and schoolrooms. It has been embodied in the National Test Standards of Germany for rating the subjective effects of fluctuating noises of all kinds, such as from street and road traffic, rail traffic, canal and river ship traffic, aircraft, industrial operations (including the noise from individual machines), sports stadiums, playgrounds, and the like.

5.4.4 The L_{dn} Concept

The L_{dn} is the L_{eq} computed over a 24-hour period with a "penalty" of 10 dBA for a designated nighttime period. Thus, it is a day-night average and the subscript "dn" is assigned instead of "eq." In applications to airport noise, the L_{dn} may be referred to as DNL. The nighttime period is from 10 PM. to 7 AM. The L_{dn} equation is derived from the L_{eq} equation with the time increment specified as 1 second. Because the time over which the L_{dn} is computed is a day, the total time period is 86 400 seconds. Equation 5.15 is then written as

$$L_{dn} = 10\log[\frac{1}{86\ 400}\sum 10^{L_i/10}t_i + \sum 10^{(L_j+10)/10}t_i] \tag{5.16}$$

Because 10(log86 400)≈49.4, the day-night average sound level may be written as

$$L_{dn} = 10\log[\sum 10^{L_i/10}t_i + \sum 10^{(L_j+10)/10}t_i] - 49.4 \tag{5.17}$$

5.5 Noise Control

5.5.1 Source-Path-Receiver Concept

If you have a noise problem and want to solve it, you have to find out something about what the noise is doing, where it comes from, how it travels, and what can be done about it. A straightforward approach is to examine the problem in terms of its three basic elements: that is, sound arises from a source, travels over a path, and affects a receiver or listener.

The source may be one or any number of mechanical devices that radiate noise or vibratory energy. Such a situation occurs when several appliances or machines are in operation at a given time in a home or office.

The most obvious transmission path by which noise travels is simply a direct line-of-sight air path between the source and the listener. For example, aircraft fly over noise reaches an observer on the ground by the direct line-of-sight air path. Noise also travels along structural paths. Noise can travel from one point to another via any one path or a combination of several paths. Noise from a washing machine operating in one apartment may be transmitted to another apartment along air passages such as open windows, doorways, corridors, or duct work. Direct physical contact of the washing machine with the floor or walls sets these building components into vibration. This vibration is transmitted structurally throughout the building, causing walls in other areas to vibrate and to radiate noise.

The receiver may be, for example, a single person, a classroom of students, or a suburban community.

Solution of a given noise problem might require alteration or modification of any or all of these three basic elements:

(1) Modifying the source to reduce its noise output

(2) Altering or controlling the transmission path and the environment to reduce the noise level reaching the listener

(3) Providing the receiver with personal protective equipment

5.5.2 Control of Noise Source by Design

1. Reduce Impact Forces

Many machines and items of equipment are designed with parts that strike forcefully against other parts, producing noise. Often, this striking action or impact is essential to the machine's function. A familiar example is the typewriter—its keys must strike the ribbon and paper in order to leave an inked impression. But the force of the key also produces noise as the impact falls on the ribbon, paper, and platen.

Several steps can be taken to reduce noise from impact forces. The particular remedy to be applied will be determined by the nature of the machine in question. Not all of the steps listed below are practical for every machine and for every impact-produced noise. But application of even one suggested measure can often reduce the noise appreciably.

Some of the more obvious design modifications are as follows:

(1) Reduce the weight, size, or height of fall of the impacting mass.

(2) Cushion the impact by inserting a layer of shock-absorbing material between the impacting surfaces. (For example, insert several sheets of paper in the typewriter behind the top sheet to absorb some of the noise-producing impact of the keys.) In some situations, you could insert a layer of shock-absorbing material behind each of the impacting heads or objects to reduce the transmission of impact energy to other parts of the machine.

(3) Whenever practical, one of the impact heads or surfaces should be made of nonmetallic material to reduce resonance (ringing) of the heads.

(4) Substitute the application of a small impact force over a long time period for a large force over a short period to achieve the same result.

(5) Smooth out acceleration of moving parts by applying accelerating forces gradually. Avoid high, jerky acceleration or jerky motion.

(6) Minimize overshoot, backlash, and loose play in cams, followers, gears, link-ages, and other parts. This can be achieved by reducing the operational speed of the machine, better adjustment, or by using spring-loaded restraints or guides. Machines that are well made, with parts machined to close tolerances, generally produce a minimum of such impact noise.

2. Reduce Speeds and Pressures

Reducing the speed of rotating and moving parts in machines and mechanical systems results in smoother operation and lower noise output reducing pressure and flow velocities in air, gas, and liquid circulation systems lessens turbulence, resulting in decreased noise radiation. Some specific suggestions that may be incorporated in design are the following:

(1) Fans, impellers, rotors, turbines, and blowers should be operated at the lowest bladetip speeds that will still meet job needs. Use large-diameter, low-speed fans rather than small-diameter, high-speed units for quiet operation. In short, maximize diameter and minimize tip speed.

(2) All other factors being equal, centrifugal squirrel-cage type fans are less noisy than vane axial or propeller type fans.

(3) In air ventilation systems, a 50 percent reduction in the speed of the air flow may lower the noise output by 10 to 20 dB, or roughly one-quarter to one-half of the original loudness. Air speeds less than 3 m/s measured at a supply or return grille produce a level of noise that usually is unnoticeable in residential or office areas. In a given system, reduction of air speed can be achieved by operating at lower motor or blower speeds, installing a greater number of ventilating grilles, or increasing the cross-sectional area of the existing grilles.

3. Reduce Frictional Resistance

Reducing friction between rotating, sliding, or moving parts in mechanical systems frequently results in smoother operation and lower noise output. Similarly, reducing flow resistance in fluid distribution systems results in less noise radiation.

Four of the more important factors that should be checked to reduce frictional resistance in moving parts are the following:

(1) Alignment: Proper alignment of all rotating, moving, or contacting parts results in less noise output. Good axial and directional alignment in pulley systems, gear trains, shaft couplings, power transmission systems, and bearing and axle alignment are fundamental requirements for low noise output.

(2) Polish: Highly polished and smooth surfaces between sliding, meshing, or contacting parts are required for quiet operation, particularly where bearings, gears, cams, rails, and guides are concerned.

(3) Balance: Static and dynamic balancing of rotating parts reduces frictional resistance and vibration, resulting in lower noise output.

(4) Eccentricity (out-of-roundness): Off-centering of rotating parts such as pulleys, gears, rotors, and shaft/bearing alignment causes vibration and noise. Likewise, out-of-roundness of wheels, rollers, and gears causes uneven wear, resulting in flat spots that generate vibration and noise.

The key to effective noise control in fluid systems is streamline flow. This holds true regardless of whether one is concerned with air flow in ducts or vacuum cleaners, or with water flow in plumbing systems. Streamline flow is simply smooth, non-turbulent, low-friction flow.

The two most important factors that determine whether flow will be streamline or turbulent are the speed of the fluid and the cross-sectional area of the flow path, that is, the pipe or duct diameter. The rule of thumb for quiet operation is to use a low-speed, large-diameter system to meet a specified flow capacity requirement. However, even such a system can inadvertently generate noise if certain aerodynamic design features are overlooked or ignored. A system designed for quiet operation will employ the following features:

(1) Low fluid speed: Low fluid speeds avoid turbulence, which is one of the causes of noise.

(2) Smooth boundary surfaces: Duct or pipe systems with smooth interior edges, and joints generate less turbulence and noise than systems with or jagged walls or joints.

(3) Simple layout: A well-designed duct or pipe system with a minimum of branches, turns, fittings, and connectors is substantially less noisy than a complicated layout.

(4) Long-radius turns: Changes in flow direction should be made gradually in a duct smoothly. It has been suggested that turns should be made with a curve radius equal to about five times the pipe diameter or major cross-sectional dimension of the duct.

(5) Flared sections: Flaring of intake and exhaust openings, particularly in a duct system, tends to reduce flow speeds at these locations, often with substantial reductions in noise output.

(6) Streamline transition in flow path: Changes in flow path dimensions or cross-sectional areas should be made gradually and smoothly with tapered or flared transition sections to avoid turbulence. A good rule of thumb is to keep the cross-sectional area of the flow path as large and as uniform as possible throughout the system.

(7) Remove unnecessary obstacles: The greater the number of obstacles in the flow path, the more tortuous, turbulent, and hence noisier, the flow. All other required and functional devices in the path, such as structural supports, deflectors, and control dampers, should be made as small and as possible to smooth out the flow patterns.

4. Reduce Radiating Area

Generally speaking, the larger the vibrating part or surface, the greater the noise output. The rule of thumb for quiet machine design is to minimize the effective radiating surface areas of the parts without impairing their operation or structure strength. This can be done by making parts smaller, removing excess material, or by cutting openings, slots, or perforations in the parts. For example, replacing a large, vibrating sheet-metal safety guard on a machine with a guard made of wire mesh or metal webbing might result in a substantial reduction in noise because of the drastic reduction in surface area of the part.

5. Reduce Noise Leakage

In many cases, machine cabinets can be made into rather effective soundproof enclosures through simple design changes and the application of some sound-absorbing treatment. Substantial reductions in noise output may be achieved by adopting some of the following recommendations:

(1) All unnecessary holes or cracks, particularly at joints, should be caulked.

(2) All electrical or plumbing penetrations of the housing or cabinet should be sealed with rubber gaskets or a suitable non-setting caulk.

(3) If practical, all other functional or required openings or ports that radiate noise should be covered with lids or shields edged with soft rubber gaskets to effect all airtight seal.

(4) Other openings required for exhaust, cooling, or ventilation purposes should be equipped

with mufflers or acoustically lined ducts

(5) Openings should be directed away from the operator and other people.

6. Isolate and Dampen Vibrating Elements

In all but the simplest machines, the vibrational energy from a specific moving part is transmitted through the machine structure, forcing other component parts and surfaces to vibrate and radiate sound-often with greater intensity than that generated by the originating source itself.

Generally, vibration problems can be considered in two parts. First, we must prevent energy transmission between the source and surfaces that radiate the energy. Second, we must dissipate or attenuate the energy somewhere in the structure. The first part of the problem is solved by isolation. The second part is solved by damping.

The most effective method of vibration isolation involves the resilient mounting of the vibrating component on the most massive and structurally rigid part of the machine. All attachments or connections to the vibrating part, in the form of pipes, conduits, and shaft couplers, must be made with flexible or resilient connectors or couplers. For example, pipe connections to a pump that is resiliently mounted on the structure frame of a machine should be made of resilient tubing and be mounted as close to the pump as possible. Resilient pipe supports or hangers may also be required to avoid bypassing the isolated system (see Figure 5.14).

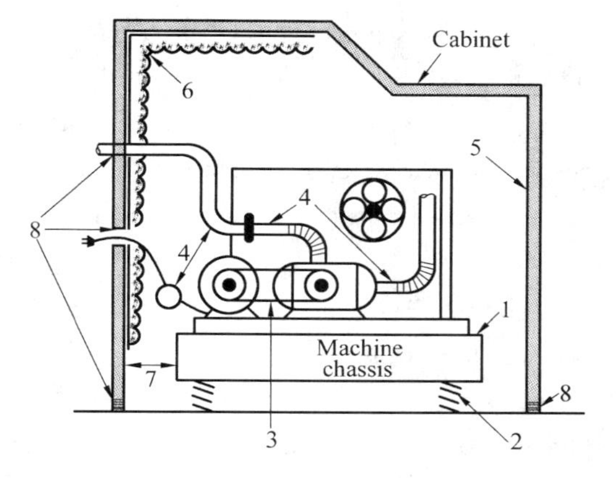

Figure 5.14 Examples of vibration isolation

Code:

1. Motors. pumps and fans installed on most massive part of the machine.

2. Resilient mounts or vibration isolators used for the installation

3. Belt-drive or roller-drive systems used in place of gear trains.

4. Flexible hoses and wiring used instead of rigid piping and stiff wiring.

5. Vibration-damping materials applied to surfaces undergoing most vibration.

6. Acoustical lining installed to reduce noise buildup inside machine.

7. Mechanical contact minimized between the cabinet and the machine chassis.

8. Openings at the base and other parts of the cabinet scaled to prevent noise leakage.

Damping material or structures are those that have some viscous properties. They tend to bend or distort slightly, thus consuming part of the noise energy in molecular motion. The use of spring mounts on motors and laminated galvanized steel and plastic in air-conditioning ducts are two examples.

When the vibrating noise source is not amenable to isolation, as, for example, in ventilation ducts, cabinet panels, and covers, then damping materials can be used to reduce the noise.

The type of material best suited for a particular vibration problem depends on factors such as size, mass, vibrational frequency, and operational function of the vibrating structure. Generally speaking, the following guidelines should be observed in the selection and use of such materials to maximize vibration damping efficiency:

(1) Damping materials should be applied to those sections of a vibrating surface where the most flexing, bending, or motion occurs. These usually are the thinnest sections.

(2) For a single layer of damping material, the stiffness and mass of the material should be comparable to that of the vibrating surface to which it is applied. This means that single-layer damping materials should be about two or three times as thick as the vibrating surface to which they are applied.

(3) Sandwich materials (laminates) made up of metal sheets bonded to mastic (sheet metal viscoelastic composites) are much more effective vibration dampers than single-layer materials; the thickness of the sheet-metal constraining layer and the viscoelastic layer should each be about thickness of the vibrating surface to which they are applied. Ducts and panels can be purchased already fabricated as laminates.

7. Provide Mufflers/silencers

There is no real distinction between mufflers and silencers. They are often used interchangeably. They are, in effect, acoustical filters and are used when fluid flow noise is to be reduced. The devices can be classified into two fundamental groups: absorptive mufflers and reactive mufflers. An absorptive muffler is one whose noise reduction is determined mainly by the presence of fibrous or porous materials, which absorb the sound. A reactive muffler is one whose noise reduction is determined mainly by geometry. It is shaped to reflect or expand the sound waves with resultant self-destruction.

Although there are several terms used to describe the performance of mufflers, the most frequently used appears to be insertion loss (1 L). Insertion loss is the difference between two sound pressure levels that are measured at the same point in space before and after a muffler has been inserted. Since each muffler's 1 L is highly dependent on the selection of materials and configuration, we will not present general 1 L prediction equations.

5.5.3 Noise Control in The Transmission Path

After you have tried all possible ways of controlling the noise at the source, your next line of defense is to set up devices in the transmission path to block or reduce the flow of sound energy before it reaches your ears. This can be done in several ways: (a) absorb the sound along the path, (b) deflect the sound in some other direction by placing a reflecting barrier in its path, or (c) contain the sound by placing the source inside a sound-insulating box or enclosure. Selection of the most effective technique will depend upon various factors, such as the size and type of source, intensity and frequency range of the noise, and the nature and type of environment.

1. Separation

We can make use of the absorptive capacity of the atmosphere, as well as divergence, as a simple, economical method of reducing the noise level. Air absorbs high-frequency sounds more effectively than it absorbs low-frequency sounds. However, if enough distance is available, even low-frequency sounds will be absorbed appreciably.

If you can double your distance from a point source, you will have succeeded in lowering the sound pressure level by 6 dB. It takes about a 10 dB drop to halve the loudness. If you have to contend with a line source such as a railroad train, the noise level drops by only 3 dB for each doubling of distance from the source. The main reason for this lower rate of attenuation is that line sources radiate sound waves that are cylindrical in shape. The surface area of such waves only increases two-fold for each doubling of distance from the source. However, when the distance from the train becomes comparable to its length, the noise level will begin to drop at a rate of 6 dB for each subsequent doubling of distance.

Indoors, the noise level generally drops only from 3 to 5 dB for each doubling of distance in the near vicinity of the source. However, further from the source, reductions of only 1 or 2 dB occur for each doubling of distance due to the reflections of sound off hard walls and ceiling surfaces.

2. Absorbing Materials

Noise, like light, will bounce from one hard surface to another. In noise control work, this is called reverberation. If a soft, spongy material is placed on the walls, floors, and ceiling, the reflected sound will be diffused and soaked up (absorbed). Sound-absorbing materials are rated either by their Sabin absorption coefficients (α_{SAB}) at 125, 500, 1 000, 2 000, and 4 000 Hz or by a single number rating called the noise reduction coefficient (NRC). If a unit area of open window is assumed to transmit all and reflect none of the acoustical energy that reaches it, it is assumed to be 100 percent absorbent. This unit area of totally absorbent surface is called a "sabin" (Sabin, 1942). The absorptive properties of acoustical materials are then compared with this standard. The performance is expressed as a fraction or percentage of the sabin (α_{SAB}). The NRC is the average of the α_{SAB}s at 250, 500, 1 000 and 2 000 Hz rounded to the nearest multiple of 0.05. The NRC has no physical meaning. It is a useful means of comparing similar materials.

Sound-absorbing materials such as acoustical tile, carpets, and drapes placed on ceiling, floor, or wall surfaces can reduce the noise level in most rooms by about 5 to 10 dB for high-frequency sounds, but only by 2 or 3 dB for low-frequency sounds. Unfortunately, such treatment provides no protection to an operator of a noisy machine who is in the midst of the direct noise field. For greatest effectiveness, sound-absorbing materials should be installed as close to the noise source as possible.

If you have a small or limited amount of sound-absorbing material and wish to make the most effective use of it in a noisy room, the best place to put it is in the upper trihedral comers of the room, formed by the ceiling and two walls. Due to the process of reflection, the concentration of sound is greatest in the trihedral comers of a room. Additionally, the upper comer locations also protect the lightweight fragile materials from damage

Because of their light weight and porous nature, acoustical materials are ineffectual in preventing the transmission of either airborne or structure-borne sound from one room to another. In other words, if you can hear people walking or talking in the room or apartment above, installing acoustical tile on your ceiling will not reduce the noise transmission.

3. Acoustical Lining

Noise transmitted through ducts, pipe chases, or electrical channels can be reduced effectively by lining the inside surfaces of such passageways with sound-absorbing materials. In typical duct installations, noise reductions on the order of 10 dB/m for an acoustical lining 2.5 cm thick are well within reason for high frequency noise. A comparable degree of noise reduction for the lower frequency sounds is considerably more difficult to achieve because it usually requires at least a doubling of the thickness and/or length of acoustical treatment.

4. Barriers and Panels

Placing barriers, screens, or deflectors in the noise path can be an effective way of reducing noise transmission, provided that the barriers are large enough in size, and depending upon whether the noise is high frequency or low frequency. High-frequency noise is reduced more effectively than low-frequency noise.

The effectiveness of a barrier depends on its location, its height, and its length. Referring to Figure 5.15, we can see that the noise can follow five different paths. First, the noise follows a direct path to receivers who can see the source well over the top of the barrier. The barrier does not block their line of sight (L/S) and therefore provides no attenuation. No matter how absorptive the barrier is, it cannot pull the sound downward and absorb it. Second, the noise follows a diffracted path to receivers in the shadow zone of the barrier. The noise that passes just over the top edge of the barrier is diffracted (bent) down into the apparent shadow shown in the figure. The larger the angle of diffraction, the more the barrier attenuates the noise in this shadow zone. In other words, less energy is diffracted through large angles than through smaller angles.

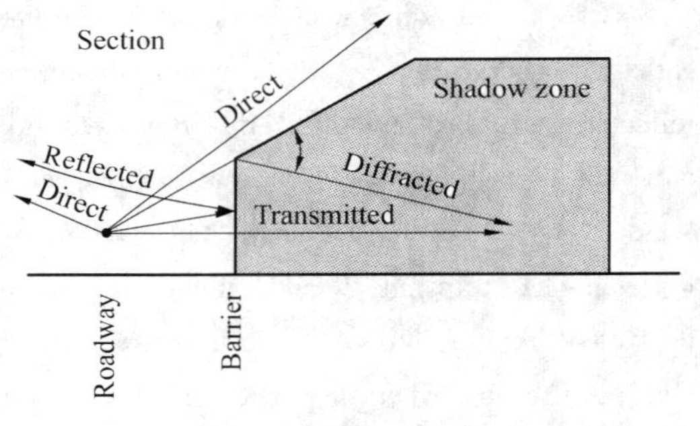

Figure 5.15　Noise paths from a source to a receiver. (Source: Kugler et al., 1976)

Third, in the shadow zone, the noise transmitted directly through the barrier may be significant in some cases. For example, with extremely large angles of diffraction, the diffraction noise may be less than the transmitted noise. In this case, the transmitted noise compromises the performance of the barrier. It can be reduced by constructing a heavier barrier. The allowable amount of transmitted noise depends on the total barrier attenuation desired. More is said about this transmitted noise later.

The fourth path is the reflected path. After reflection, the noise is of concern only to a receiver on the opposite side of the source. For this reason, acoustical absorption on the face of the barrier may sometimes be considered to reduce this reflected noise; however, this treatment will not benefit any receivers in the shadow zone. It should be noted that in most practical cases the reflected noise does not play an important role in barrier design. If the source of noise is represented by a line of noise, another short-circuit path is possible. Part of the source may be unshielded by the barrier. For example, the receiver might see the source beyond the ends of the barrier if the barrier is not long enough. This noise from around the ends may compromise or short-circuit, barrier attenuation. The required barrier length depends on the total net attenuation desired. When 10 to 15 dB attenuation is desired, barriers must, in general, be very long. Therefore, to be effective, barriers must not only break the line of sight to the nearest section of the source, but also to the source far up and down the line.

Of these four paths, the noise diffracted over the barrier into the shadow zone represents most important parameter from the barrier design point of view. Generally, the determination of barrier attenuation or barrier noise reduction involves only calculation of the amount of energy diffracted into the shadow zone. The procedures presented in the barrier nomograph used to predict highway noise are based on this concept.

Another general principle of barrier noise reduction that is worth reviewing at this point is the relation between noise attenuation expressed in (1) decibels, (2) energy terms, and(3)subjective loudness. Table 5.3 gives these relationships for line sources. As indicated in the loudness column, a barrier attenuation of 3 dB will be barely discerned by the receiver.

Table 5.3 Relation between sound level reduction, energy, and loudness for line sources

To reduce A-level by dB	Remove portion of energy (%)	Divide Loudness by
3	50	1.2
6	75	1.5
10	90	2
20	90	4
30	99.9	8
40	99.99	16

(Source: Kugler et al., 1976.)

However to attain this reduction, 50 percent of the acoustical energy must be removed. To cut the loudness of the source in half, a reduction of 10 dB is necessary. That is equivalent to eliminating 90 percent of the energy initially directed toward the receiver. As indicated previously, this drastic reduction in energy requires very long and high barriers. In summary, when designing barrier, you can expect the complexity of the design to be about as follows:

Attention/dB	Complexity
5	Simple
10	Attainable
15	Very difficult
20	Nearly impossible

5. Transmission Loss

When the position of the noise source is very close to the barrier, the diffracted noise is less important than the transmitted noise. If the barrier is in fact a wall panel that is sealed at the edges, the transmitted noise is the only one of concern.

The ratio of the sound energy incident on one surface of a panel to the energy radiated from the opposite surface is called the sound transmission loss (TL). The actual energy loss is partially reflected and partially absorbed. Since TL is frequency-dependent, only a complete octave or one-third octave band curve provides a full description of the performance of the barrier.

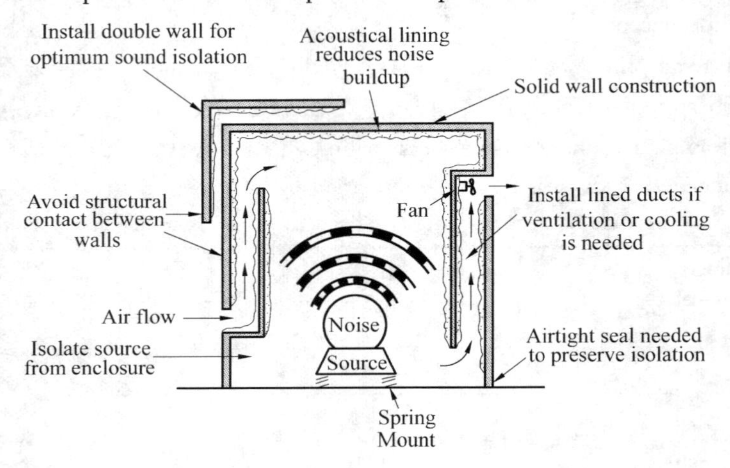

Figure 5.16 Enclosures for controlling noise. (Source: Berendt et al., 1976)

6. Enclosures

Sometimes it is much more practical and economical to enclose a noisy machine in a separate room or box than to quiet it by altering its design, operation, or component parts. The walls of the enclosure should be massive and air tight to contain the sound. Absorbent lining on the interior surfaces of the enclosure will reduce the reverberant buildup of noise within it. Structural contact between the noise source and the enclosure must be avoided, so that the source vibration is not transmitted to the enclosure walls, thus short-circuiting the isolation. For maximum effective noise control, all of the techniques illustrated in Figure5.16 must be employed.

5.5.4 Control of Noise Source by Redress

The best way to solve noise problems is to design them out of the source. However, we are frequently faced with an existing source that, either because of age, abuse, or poor design, is a noise problem. The result is that we must redress, or correct, the problem as it currently exists. The following sections identify some measures that might apply if you are allowed to tinker with the source.

1. Balance Rotating Parts

One of the main sources of machinery noise is structural vibration caused by the rotation of poorly balanced parts, such as fans, fly wheels, pulleys, cams, shafts, and so on. Measures used to correct this condition involve the addition of counterweights to the rotating unit or the removal of some weight from the unit. You are probably familiar with noise caused by imbalance in the high-speed spin cycle of washing machines. The imbalance results from clothes not being distributed evenly in the tube. By redistributing the clothes, balance is achieved and the noise ceases. This same principle of balance can be applied to furnace fans and other common sources of such noise.

2. Reduce Frictional Resistance

A well-designed machine that has been poorly maintained can become a serious source of noise. General cleaning and lubrication of all rotating, sliding, or meshing parts at contact points should go a long way toward fixing the problem.

3. Apply Damping Materials

Since a vibrating body or surface radiates noise, the application of any material that reduces or restrains the vibrational motion of that body will decrease its noise output. Three basic types of redress vibration damping materials are available:

1) Liquid mastics, which are applied with a spray gun and harden into relatively solid materials, the most common being automobile "undercoating"

2) Pads of rubber, felt, plastic foam, leaded vinyls, adhesive tapes, or fibrous blankets which are glued to the vibrating surface.

3) Sheet metal viscoelastic laminates or composites, which are bonded to the vibrating surface.

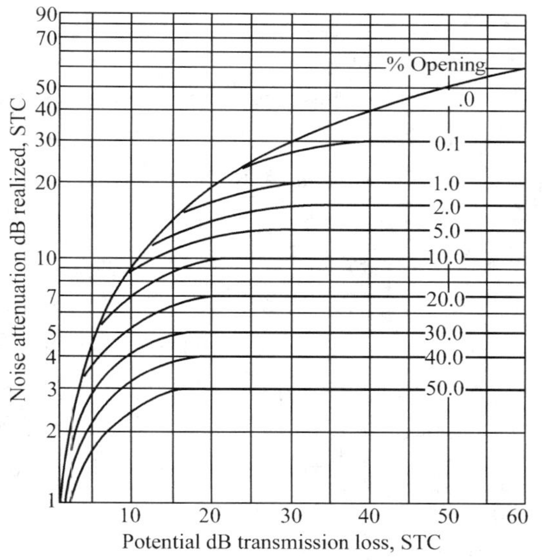

Figure 5.17 Transmission loss potential versus transmission loss realized various opening sizes as a percent of total wall area. STC = sound transmission coefficitent

4. Seal Noise Leaks

Small holes in an otherwise noise-tight structure can reduce the effectiveness of the noise control measures. As seen in Figure 5.17, if the designed transmission loss of an acoustical enclosure is 40 dB, an opening that comprises only 0.1 percent of the surface area will reduce the effectiveness of the enclosure by 10 dB.

5. Perform Routine Maintenance

We all recognize the noise of a worn muffler. Likewise, studies of automobile tire noise in relation to pavement roughness show that maintenance of the pavement surface is essential to keep noise at minimum levels. Normal road wear can yield noise increases on the order of 6 dBA.

5.5.5 Protect the Receiver

When exposure to intense noise fields is required and none of the measures discussed so far is practical, as, for example, for the operator of a chain saw or pavement breaker, then measures must be taken to protect the receiver. The following two techniques are commonly employed.

1. Alter Work Schedule

Limit the amount of continuous exposure to high noise levels. In terms of hearing protection, it is preferable to schedule an intensely noisy operation for a short interval of time each day over a period of several days rather than a continuous eight-hour run for a day or two.

In industrial or construction operations, an intermittent work schedule would benefit not only the operator of the noisy equipment, but also other workers in the vicinity. If an intermittent schedule is not possible, then workers should be given relief time during the day. They should take their relief time at a low-noise-level location, and should be discouraged from trading relief time for dollars, paid vacation, or "early out" at the end of the day.

Inherently noisy operations, such as street repair, municipal trash collection, factory operation, and aircraft traffic, should be curtailed at night and early morning to avoid disturbing the sleep of the community. Remember: operations between 10 P. M. and 7 A. M. ale effectively 10 dBA higher than the measured value.

2. Ear Protection

Molded and pliable earplugs, cup-type protectors, and helmets are commercially available as hearing protectors. Such devices may provide noise reductions ranging from 15 to 35 dB (see Figure 5.18). Earplugs are effective only if they are properly fitted by medical personnel. As shown in Figure5.18, maximum protection can be obtained when both plugs and muffs are employed. Only muffs that have a certification stipulating the attenuation should be used.

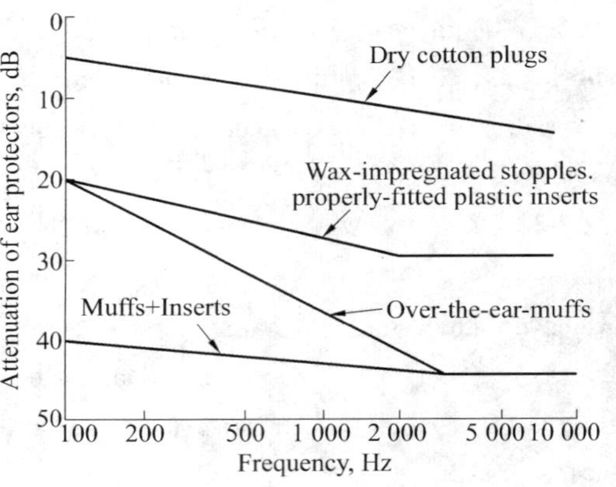

Figure 5.18　Attenuation of ear protectors at various frequencies. (Source: Berendt et al, 1976)

These devices should be used only as a last resort, after all other methods have failed to lower the noise level to acceptable limits. Ear protection devices should be used while operating lawn mowers, mulchers, and chippers, and while firing weapons at target ranges. It should be noted that protective ear devices do interfere with speech communication and can be a hazard in some situations where warning calls may be a routine part of the operation. A modem ear-destructive device is a portable digital music player that uses earphones. In this "reverse" muff, high noise levels are directed at the ear without attenuation. If you can hear someone else's music player, that person is subjecting him or herself to noise levels in excess of 90 ~ 95 dBA.

Review Questions

1. Define sound pressure level.

2. Explain the mechanism by which hearing damage occurs.

3. List five effects of noise other than hearing damage.

4. List the three basic elements that might require alteration or modification to solve a noise problem.

5. Describe two techniques to protect the receiver when design and /or reduces are not practical, that is, when all else fails.

Chapter 6 Radiation Pollution and Protection

6.1 Radiation Fundamentals

Radiation is commonly defined as energy that flows through matter or through a vacuum. Ionizing radiation, in particular, plays invaluable roles in medical diagnosis and therapy, industrial process control, research, and numerous other areas, but it also poses significant public health problems, because the production of ions within tissues can injure people, animals, and plants and cause genetic as well as somatic damage. Ionizing radiation is created by radioactive materials occurring naturally in the environment and by X-ray machines and radioactive substances produced in nuclear reactors and elsewhere. It can be quantified by detecting and measuring the amount of ions arising when it transfers its energy to various radiation-sensitive materials such as photographic film, semiconductor devices, and fluorescent materials. People and equipment can be protected from the harmful effects of exposure to ionizing radiation by confining and isolating the source of radiation, keeping it at the greatest practical distance, minimizing the duration of exposures, and using shields of lead, concrete, or other suitably dense materials.

6.1.1 Radiation and Ionization

Most radiations encountered in the practice of engineering involve the propagation either of mechanical disturbances, as with sound, or of electromagnetic waves or particles. Quantum mechanics informs us that electromagnetic radiation behaves, in some ways, as a beam of particle like quanta, called photons, and that conversely, "particles" like electrons have wavelike characteristics; except where noted otherwise, however, this chapter will adopt the simpler, classical terminology. Electromagnetic and charged or uncharged particulate radiations that are capable of directly or indirectly producing ions by interaction with matter, in particular, are referred to as ionizing radiation. Ionizing radiations are especially important not only because of their widespread use in medicine, industry, and elsewhere but also because they have the potential of causing cancer and other health effects. The forms found in standard application are X-rays and gamma rays, at the highest energy end of the electromagnetic spectrum, and particles such as beta and alpha particles, heavy ions, and neutrons.

For radiation in the form of waves, the wavelength (λ), frequency (f), and velocity (v) are related as

$$\lambda f = v \text{ (velocity of propagation of waves)} \qquad (6.1a)$$

The wavelength is the distance between wave crests. The frequency is the number of waves per second passing per second, measured in hertz (Hz). The velocity depends on the medium through which it travels and, in general, on the frequency and/or energy of the radiation. The velocity of electromagnetic radiation in a vacuum, however, is independent of the frequency, and has been given the special designation c:

$$\lambda f = c = 3 \times 10^8 \text{ m/s (velocity of electromagnetic radiation in vacuum)} \quad (6.1b)$$

A second fundamental expression concerning electromagnetic radiation comes to us from quantum mechanics and builds upon the notion that an electromagnetic wave is in some sense comprised of vast numbers of photons. It was found through experiment that the energy (E) of a (particlelike) photon increases with the frequency of the associated electromagnetic wave:

$$E = hf \quad (6.2)$$

where the constant of proportionality (h) is Planck's constant. Thus the shorter the wavelength, the higher the frequency and energy.

The standard measures of energy of ionizing radiation are the electron volt (eV), a thousand electron volts (keV), and a million electron volts (MeV), as applicable. An electron volt is the energy an electron gains in passing through a difference of potential of 1V, and it is equivalent to 1.6×10^{-12} erg. The intensity of a beam of radiation refers to the quantity of energy passing through a known area per unit of time, usually expressed in ergs per square centimeter per second.

For radiation protection, measures of radiation are based on the deposition of ionizing radiant energy in matter—and in particular, in human tissues. These measures refer to dose in matter, which can be computed from direct measurements, and they are of importance because they can be related to health risks, such as radiation burns and carcinogenesis. Definitions of dose all ultimately involve the concept of ionization of atoms and molecules. Before presenting definitions of dose, a brief overview of atoms and ions is warranted.

6.1.2 Types of Radiation

The common types of ionizing radiation are X-rays, gamma rays, alpha particles, beta particles, and neutrons.

Apart from the fact that they are emitted by radioactive nuclei, alpha particles are the nuclei of ordinary helium atoms, and consist of two protons and two neutrons, with a net positive electric charge. Because they are massive, even highly energetic alpha rays (several MeV) travel steadily but relatively slowly through matter and create many ions per unit of pathlength. Thus, they dissipate their energy rapidly and penetrate only 3 to 5 cm of air. A thin sheet of paper will stop alpha particles, and they rarely pose an external radiation hazard, since they cannot penetrate the outer layers of skin. Alpha-emitting radionuclides are normally of risk only if they enter the body through ingestion, inhalation, injection, or open wounds. Alpha emitters are heavy elements, such as uranium, lead, plutonium, and radium.

Beta particles are generally high- energy electrons emitted by radionuclides of low mass number and carrying a single negative charge. Strontium-90, I-131, and Cs-137 are examples of beta emitters. Beta particles may also consist of positrons—the antiparticle of the electron that has a positive rather than a negative charge. Fluorine-18 is one example of a radionuclide that emits a positron and is currently used extensively in positron emission tomography (PET) imaging. Beta particles have lower specific ionization, dissipate their energies rather quickly, and are moderately penetrative but are stopped by a few millimeters of aluminum. Beta particles cause damage to tissues through either internal or external irradiation. One important property of beta particles is that they may give rise to penetrating X-rays through a process called bremsstrahlung. In particular, when beta particles are stopped by shielding, the rapid deceleration of the beta particles as they are stopped leads to the production of X-rays. Thus, proper precautions must be taken to ensure protection against both the beta particles and these secondary X-rays.

Gamma ray sand X-rays are both photons. X-rays and gamma rays are much more penetrating than alpha and beta particles, so their ionizations tend to be relatively well separated. Unlike alpha and beta particles, which slowdown and come to a halt in a tissue, with a finite range, a beam of X-rays or gamma rays is attenuated nearly exponentially with tissue depth, and a large fraction may pass through the tissue entirely. A typical X-ray beam can be virtually eliminated, however, with lead shielding a fraction of an inch thick.

Neutrons are uncharged and cannot interact with the constituents of an atom through electrical forces and hence are highly penetrating. They are generated primarily by nuclear reactions such as fission or fusion, but some man-made radionuclides, such as Cf-252, emit them. Neutrons are commonly classified by their energy. Thermal neutrons have kinetic energies similar to those of the atoms and molecules surrounding them and can be captured by atomic nuclei to form new nuclides—indeed, activation with thermal neutrons is perhaps the most common mechanism by which substances can be made radioactive. Fast neutrons have energies greater than about 0.1 MeV and typically interact through elastic collisions with atomic nuclei. In this way, they transfer their energy to those nuclei, turning them into highly energetic particles that can create dense trails of secondary ionizations. Neutrons can present major problems in areas around nuclear reactors and particle accelerators but are rarely a source of exposure to the general public.

6.2 Origin of Human Exposure to Ionizing Radiation

6.2.1 External and Internal Radiation Exposure

Ionizing radiation exposure occurs either from a radiation source outside the body (external or direct radiation) or as a result of taking radioactive material into the body (internal radiation). For the general population, natural background radiation is the most significant source of external radiation exposure. X-ray machines are the most commonly encountered man-made sources of external exposure, but in some situations radioactive emitters of gamma rays and beta particles can also be of concern.

Gamma rays as well as alpha and beta radiation are of concern for internally deposited radionuclides. While many gamma rays will escape the body without hitting anything, alpha and beta particles only travel a short distance, so all their ionizing energy is deposited in the body. Where and for how long a radionuclide is retained in the body are of importance. While most medical radionuclides are quickly excreted from the body, I-131 concentrates and stays in the thyroid, presenting a risk of thyroid cancer. Fortunately it has a physical half-life of only 8.14 days. Radium-226, Ra-228, and Sr-90 all tend to go to bone and emit from there for much longer, increasing the risk of bone cancer and leukemia.

Internal radiation exposure results from inhalation, ingestion, dermal contact, and injection of radionuclides. The most common source of inhalation exposure is from natural radon, a gas that results from the decay of natural uranium, thorium, and radium in soil and can seep into basements. Radon-220 and Rn-222 have long chains of daughter radionuclides, and radiation from these daughter products, which attach to dust particles that lodge in the lungs, constitute the primary hazard. For instance, epidemiological studies have shown a high incidence of lung cancer among uranium miners.

Internal exposure due to ingestion may be through the presence of radionuclides in water or food. For instance, some communities have significant natural radium in their drinking water. The U.S. Environmental Protection Agency (EPA) sets standards for the level of radioactivity in drinking water at 15 pCi/l for alpha radioactivity, 5 pCi/l for radium, and a dose limit of 4 mrem/yr for man-made beta/photon radioactivity. Uranium, K-40, and Ra-226 are naturally present in soils and fertilizers and are thus incorporated into foods consumed by people and animals. The Food and Drug Administration (FDA) has developed guidelines for radionuclide levels in food (FDA Derived Intervention Levels) for individuals from three months to adult.

Dermal contact and injection of radionuclides are generally minor contributions to internal exposure. Injected radionuclides, typically for medical diagnosis or therapy, are meant to have short effective half-lives so as to minimize the dose. Technetium-99m, the workhorse of any nuclear medicine department, has a half-life of only 6 hr. As for dermal contact, the percutaneous absorption of radionuclides is typically negligible, especially if the skin is washed immediately after exposure. In addition, except for tritiated water (water in which one of the hydrogen atoms is tritium, the isotope of hydrogen, H-3), any dose is typically limited to the skin.

6.2.2 Sources of Exposure

Important sources of radiation exposure include natural background, radioactive fallout from nuclear testing or use of nuclear devices, radiation from X-ray machines and radionuclides employed in medical diagnosis and treatment, and radiation from industrial and other man-made sources. To help understand and better appreciate the significance of radiation dosages, some radiations to which individuals are or might be exposed to are given in Table 6.1. As a rule of thumb, average natural background radiation doses in the United States are on the order of 3 mSv (300 mrem) per year, of which 2 mSv (200 mrem) comes from indoor radon, but these may vary by a factor of 2 or more depending on the geographical location. On average, medical and consumer products contribute an additional 0.6 mSv (60 mrem) per year.

Table 6.1 Some Sources of Radiation to Which We Are or Might Be Exposed

Source	Dose
Cosmic rays	30 ~ 100 mrem/ year
Naturally occurring radioactive substances in water, air, and soil[a]	36 ~ 110 mrem/ year
Natural environment	
Denver	510 mrem/year including cosmic
San Francisco	145 mrem/year including cosmic
Airport luggage inspection system	2.1 mrem/year[b]
Indoor radiation, radon in air and groundwater	200 ~ 2 400 mrem/ year
Natural gas cooking ranges	5 mrem/year[b]
Natural gas heaters	22 mrem/year[b]
Radioluminous watches and clocks	40 ~ 104 mrem/ year[b]
Radium dial watch (no longer made)	3 mrem/day
Smoke detectors, gas and aerosol	8 mrem/year[b]
Common X-rays, medical and dental[c]	20 ~ 500 mrem/ year
In brick, stone, concrete block homes[d]	8 ~ 13 mrem/ year
Cigarette smoker[b]	8 000 ~ 16 000 mrem/ year
Other man-made sources[e]	10 mrem/year

Note: Exposures are approximate.

a Exposure increases slightly with altitude and varies with latitude, soils, and rock natural radioactivity content. The main sources of background radiation are cosmic rays and potassium 40, thorium, uranium, radium, and their decay products, including radon.

b NCRP Report No. 95, NCRP, Bethesda, MD, Table 5.1, p. 65. Annual dose equivalent.

c May be exceeded with poor practice or equipment or if medically indicated.

d Will vary with radionuclides in the construction materials.

e See Health Effects of Low-Level Radiation, American Council of Science and Health, New York, March 1989, which includes consumer products and radioactivity redistributed by human efforts, including radioactivity present in soot from coal power generation, dust from high-phosphorus fertilizers, etc.

1. Natural Background Radiation

Natural sources of ionizing radiation account for the majority of the radiation dose to the general public. On average, it is estimated that about 80 percent of one's radiation dose is due to natural sources (excluding radiation from smoking tobacco). Radon, potassium, radium, and cosmic radiation are the major sources of natural background radiation exposure.

The most significant source of natural radiation is inhalation of radon and its decay products. As mentioned earlier, it is estimated that more than half of natural radiation exposure, or about 2 mSv (200 mrem) on average, is due to radon. Radon inspection is now a common part of home inspections, and the EPA recommends taking action if radon concentrations are found to be greater than about 0.15 Bq (4 pCi) per liter of air. Because radon emits alpha particles, which are only weakly penetrating, the primary exposure route is through inhalation into the lungs. Once in the lungs, radon and its daughter products can directly irradiate lung tissue. High levels of radon, such as those experienced by uranium miners, are known to cause increased incidence of lung cancer. Studies at high levels of exposure are used to estimate the potential risks from low levels of exposure.

Potassium and radium are two other natural radionuclides that contribute significantly to average radiation exposures. The naturally occurring isotope K-40 is radioactive and emits both beta and gamma rays. Annual dose rates are around 0.15 to 0.19 mSv (15 ~ 19 mrem). Because potassium is an essential nutrient and there is no natural way to separate isotopes, some exposure to K-40 is unavoidable. However, because radiation from potassium is highly penetrating, bulk quantities of potassium may result in additional radiation dose simply due to external exposure. Radium is also naturally occurring and can be taken up in bone tissue if ingested. Average annual doses are about 0.17 mSv (17 mrem) but may be higher in certain geographic areas. For instance, well water in some locations contains relatively high concentrations of radium.

Cosmic radiation is radiation that originates in outer space. At sea level, cosmic radiation dose rate is about 0.3 mSv (30 mrem) per year. But since the atmosphere itself absorbs cosmic radiation, the radiation dose depends significantly on altitude. For instance, on the Colorado Plateau, the dose rate ranges from 0.5 to 1.0 mSv (40 ~ 100 mrem) per year, and flying at 40 000 feet gives a dose rate per year of 15 to 20 mSv (1 500 ~ 20 000 mrem). However, flying round trip across the continental United States (about 10 hr at 40 000 ft) would only give a dose of 0.017 to 0.023 mSv (1.7 ~ 2.3 mrem).

2. X-ray Machines and Equipment

X-rays are used in a wide range of diagnostic and research equipment. They are produced by dental, radiographic (fixed and mobile), mammographic, and fluoroscopic machines, computed tomography (CT) scanners, X-ray therapy units, airport luggage inspection devices, X-ray diffraction devices, industrial X-ray and thickness gauge machines, and other diagnostic equipment. On average, medical X-rays are estimated to result in an annual dose of about 0.4 mSv (40 mrem).

An X-ray machine is essentially a particle accelerator in which electrons "boil" off a hot filament of the X-ray tube and are then accelerated through a vacuum by means of a high voltage until they reach the "target" anode. As they enter the surface layers of the target, they are rapidly slowed down by collisions with the nuclei and electrons in it and are diverted from their original direction of motion. It is well understood that any charged particle that is forced to accelerate or decelerate rapidly will emit electromagnetic radiation; that, in fact, is how electrons driven rapidly back and forth in an antenna give rise to radio waves. The same occurs in an X-ray machine. Each time an electron suffers an abrupt deceleration or change of course, some or all of its kinetic energy is transformed into X-ray energy. If the electron is jerked to rest in a single abrupt collision, then all its energy will reappear as one high-energy, short-wavelength photon. If the electron encounters a less drastic collision, a longer wavelength, lower energy photon will be produced. Thus, these bremsstrahlung X-rays emerging from an X-ray tube span a broad range of energies.

There is a second important process that can result in the emission of X-rays. So-called characteristic X-rays may be generated when some violent event causes a rearrangement of the inner electron orbitals of a heavy atom. Either way, bremsstrahlung and characteristic X-rays arise from what are essentially atomic electronic processes, as opposed to the nuclear deexcitation events that yield gamma rays. Tungsten is usually used as the target material of X-ray tubes and linear accelerators because of its high melting point and high atomic number (for which bremsstrahlung occurs more efficiently). Molybdenum is employed in the anodes of mammography tubes, on the other hand, because of the favorable energies of its characteristic radiation. See Figure 6.1.

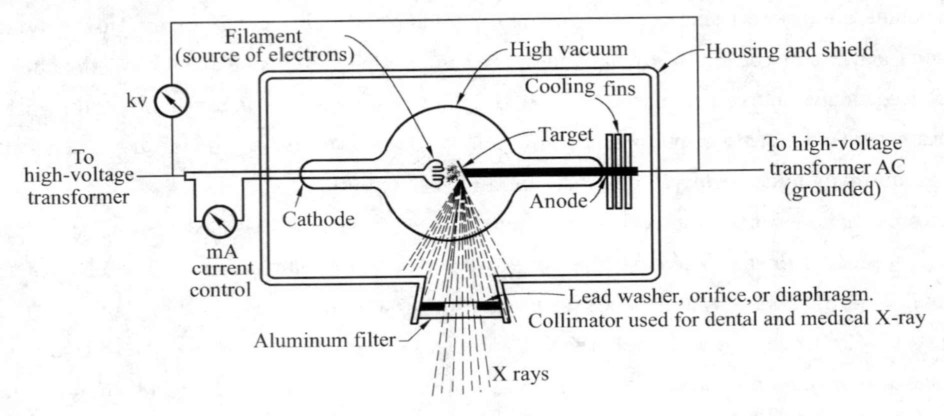

Figure 6.1 An X-ray machine. Diagnostic X-ray machines create images

The strength and quantity of X-rays are usually expressed in terms of electrical current between the filament and target within the X-ray tube. The energy is measured in thousands of volts or kilovolts (kV) in diagnostic machines. The ability of X-rays to penetrate materials usually increases with their energy, which is why the design thickness of concrete or lead shielding for a site is determined, in part, by the highest energy of the radiation to be generated.

The quantity of X-rays produced is directly proportional to the current through the tube and is measured as thousandths of an ampere or milliampere (mA). The dose or energy absorbed by an irradiated object or by shielding is thus a function of both the kilovolt and the milliampere settings of the machine as well as other factors.

As is well known, when X-rays pass through an object, they can produce a shadow picture of it on film or a special screen. Radiographic images (along with those of fluoroscopy and CT) are created by differential attenuation of the beam by the various tissues of the body, which differ in density and effective (i.e., average) atomic number. In dental imaging, the shadowgram of a tooth is recorded directly on film. For a standard radiograph, the film is sandwiched between the two fluorescent screens of a cassette, and it is the light from the screens that exposes the film. In fluoroscopy, the life-sized X-ray pattern emerging from the patient is transformed by an electronic vacuum tube called an image intensifier into a small (2 ~ 3 cm), bright optical image that is either filmed by a still camera or viewed by a television camera.

The CT scanner is a combination of X-ray machine and computer that, together with auxiliary equipment, can produce high-quality, three dimensional anatomic pictures. In many diagnostic situations, CT scanners can show various intracranial or intra-abdominal abnormalities far more clearly than can simple radiographs. Their ability to locate, for example, tumors, blood clots, and anatomical malformations has largely eliminated the need for many kinds of exploratory surgery.

X-rays are also produced as byproducts of machines that operate at high voltages. These include high-voltage television projection systems, color television sets, electron microscopes and power sources, high-power amplifying tubes producing intense microwave fields, radio-transmitting tubes, high-voltage rectifier tubes, and other devices producing penetrating electromagnetic radiation.

3. Radioactive Materials in Medicine

Radioactive materials are used extensively in medicine. About one-third of all patients admitted to U.S. hospitals undergo diagnosis or treatment using radioisotopes, with a sum total of over 100 million procedures annually, and all major hospitals have specific departments dedicated to nuclear medicine. Because of the potential hazards associated with radioactive sources in either diagnosis or therapy, handling and accountability procedures are required at places where they are used and stored.

Radioactive materials used as a diagnostic tool have the virtue of being able to provide information on the physiological status of various organ systems. The radioisotope Tc-99m, a gamma emitter with a 6 hr half-life and 140 keV gamma emission, is used in about 80 percent of nuclear diagnostic procedures. The radioactive material is attached to a tissue-specific carrier or tracer that, if injected, ingested, or inhaled, tends to concentrate in one organ or physiological compartment of the body, from which it emits penetrating gamma rays. The pattern of gamma emission is imaged by a gamma camera, and any region of excess or diminished emission can indicate an improper physiological condition in the region. Some medical therapy applications of radioisotopes are for the treatment of cancer—the radiation emitted (usually gamma rays) by the radionuclide helps to destroy cancerous tumors. Common radioisotopes for such treatment are Cs-137, Ir-192 and I-125.

Positron emission tomography scanning is a technology that results in CT-like images that are created by gamma rays emitted from the body rather than by X-rays transmitted through it. A small amount of a positron-emitting radioisotope, attached to a tissue-specific carrier or tracer is injected into a patient, after which it collects in a specific organ. The positrons emitted travel typically less than 1mm, whereupon they collide with electrons; this gives rise to a pair of 0.511-MeV "annihilation" photons traveling in near-opposite direction, and these can be extracted from any background noise through coincidence detection. The resulting PET image can yield invaluable information on both anatomy (like CT) and physiology (nuclear medicine) that neither a CT nor a gamma camera study alone can provide.

4. Radioactive Materials in Industry

Radioactive materials are also used in many industrial processes. For instance, to determine whether a well drilled deep into the ground has the potential for producing oil, geologists use nuclear well-logging, a technique that employs radiation from a radioisotope inside the well to detect the presence of different materials. Radioisotopes are also used to sterilize instruments, to find flaws in critical steel parts and welds that go into automobiles and modern buildings, to authenticate valuable works of art, and to solve crimes by spotting trace elements of poison. Food irradiation to kill harmful bacteria also typically uses radioactive materials. Radioisotopes can also eliminate dust from film and compact discs as well as static electricity (which may create a fire hazard) from can labels. Radionuclides commonly used in industry include Co-60, Pr-147, and Ir-192.

5. Consumer Products

A number of consumer products contain radioactive materials. Natural radium was used for many years to produce luminous dials and faces for clocks, compasses, and other instruments. Because of the significant exposure hazard, these have since been replaced with less hazardous radioactive materials such as tritium (H-3) and promethium. These new materials result in only a dose of a few microsieverts (0.1 mrem) per year. However, many items containing radium are still in circulation as antiques.

Other consumer products include smoke detectors and certain glazes and coatings. Many smoke detectors contain about a microcurie of americium, an artificially produced radionuclide (some newer models use photoelectric detectors, which do not involve the use of radioactive materials). The resulting dose is less than 0.01 mSv (1 mrem). Some glazing and tinting coatings contain uranium and thorium. Certain uranium glazes in pottery and dishes can produce dose rates of 0.1 to 0.2 mSv (10 ~ 20 mrem) per hour when people dine off of them.

Finally, tobacco typically contains two naturally occurring radionuclides: Pb-210 and Po-210. These nuclides are both decay products of natural radon and tend to stick to the leaves of tobacco plants. When a person smokes a cigarette, these radionuclides enter the lungs and result in a significant dose to lung tissue. The resulting equivalent dose to the respiratory system is estimated to be up to about 160 mSv (16 000 mrem) per year. This corresponds to a whole-body effective dose of about 13 mSv (1 300 mrem), or almost four times the average natural background rate. An issue of interest to radiobiologists is the possible synergistic effects of radiation and chemical carcinogens in the lungs and elsewhere.

6. Nuclear Power

As of 2001, there were over 100 nuclear power reactors operating in the United States and more than 400 such reactors worldwide. Although these reactors differ in design, there are several common processes that may be sources of radiation exposure. These can be divided into three main categories: production of nuclear fuel, power plant operations, and disposition of spent fuel.

The production of nuclear fuel involves the mining and milling of uranium, both of which can lead to exposure. Underground mines, in particular, are associated with elevated levels of radon and its decay products, and working in such mines has been directly correlated with increased incidence of lung cancer. In addition, mine wastes and mill tailings may contain higher than normal concentrations of radioactive materials such as uranium and radium, leading to exposure of nearby populations through airborne dust or leaching and runoff of liquid wastes. In some cases, tailings and mine waste have been used as construction materials for nearby homes, resulting in both direct exposure and inhalation of radon.

Routine nuclear power plant operations have virtually no effect on the public, but they do often lead to exposures to power plant workers. However, because of regulatory requirements, few workers at these plants even approach the annual dose limit of 50 mSv. Some leakage from nuclear power plants in the form of airborne or waterborne releases is inevitable, but this is monitored and limited by Nuclear Regulatory Commission (NRC) regulations. Collective doses from a single reactor have been estimated to be about 3 700 person-Sv (370 000 person-rem) over 45 years, which is less than 0.1 mrem per person per year.

Spent nuclear fuel is high-level radioactive waste and may be processed either for reuse or for disposal. Reprocessing of spent fuel (which, since the 1970s, has not occurred in the United States) involves separating the remaining uranium (and perhaps the plutonium) produced in nuclear power plant operation from the other fission products, which are then recycled into fuel for other nuclear power plants. High-level waste, coming either directly from or as a result of reprocessing of spent fuel, must be disposed of properly in order to prevent radiation exposure to the public.

One major concern of nuclear power is the potential exposure due to nuclear accidents. Nuclear power plants in the United States are designed with many redundant safety features, but two major accidents, Three-Mile Island in Pennsylvania and Chernobyl in Ukraine, have heightened concern about the potential risks of nuclear power. The Three-Mile Island accident led to very little off-site contamination. It has been estimated that the highest doses from the Three-Mile Island accident were about 0.2 to 0.7 mSv and that collective doses were about 33 person-Sv to the neighboring population. This collective dose would imply about one or two excess cancer deaths in the lifetime of the exposed population. This amount of excess risk is far too small to measure in any epidemiological study. On the other hand, the Chernobyl accident leads to measurable global fallout. Although the only observable long-term health effects have been an increase in thyroid cancer among children, especially in Belarus, estimates of global collective doses range from about half to 1×10^6 person-Sv, most of which was received by the populations of the former Soviet Union and Europe. For comparison, it should be noted that natural background radiation leads to collective doses of around 10×10^6 person-Sv per year.

7. Nuclear Weapons Production and Testing

Another source of radiation exposure is from nuclear weapons production and testing. Although older nuclear weapons used uranium, most nuclear weapons today employ plutonium produced in nuclear reactors. The steps leading through the reprocessing of spent reactor fuel are similar to those in producing weapons fuel—and this is the origin of the concern over nuclear proliferation from reprocessing of commercial spent fuel.

The testing of nuclear weapons has historically led to significant releases of radioactive materials into the environment. Radioactive contaminants (with their half-lives) from the detonation of nuclear weapons include I-131 (8 days); Sr-90 (28 years); Sr-89 (53 days); Cs-137 (30 years); and Ce-144 (275 days). From 1945 to 1980, nuclear tests were performed above ground and led to global fallout. The cumulative collective dose from such atmospheric testing as of 1993 was estimated to be about 7×10^6 person-Sv. Although the average dose from such atmospheric nuclear testing is generally relatively small, some local populations were significantly exposed. For instance, several atolls in the Marshall Islands, where the United States conducted many nuclear weapons tests, were evacuated and remain uninhabited. Some of the local residents showed skin lesions characteristic of exposure to acute doses of ionizing radiation. At the Nevada test site, which conducted tests between 1951 and 1962, thyroid doses to children living near the site may have been as high as 1 Gy (100 rad). Since 1963, the majority of nuclear tests have been carried out underground. The environmental effects of these tests have been relatively small, and little exposure to the general population has occurred. Similarly, the dismantling or decommissioning of nuclear weapons generally leads to little environmental release of radioactive materials—but the potential for nuclear proliferation is a significant concern.

6.3 Biological Effects of Radiation

6.3.1 Effects of Ionizing Radiation on Cells

Ionizing radiation causes biological damage in a tissue by bringing about changes in the structure or function of its cells through energy transfer. The transfer of energy to the cell protoplasm can lead to direct change of cell components, such as the direct breaking of one or both strands of a DNA molecule, or to indirect changes, such as through the production of free radicals, which are highly reactive states of atoms or molecules that have just undergone certain chemical or physical reactions. These may cause subsequent chemical changes some distance away from the original ionizing event. High linear energy transfer (LET) radiation, such as alpha particles, in which the trails of the particles are dense with ionizations tend to do more damage to a cell. Low LET (gamma and beta) radiation will dissipate its energy and cause damage over a greater region, since it is more penetrating.

Damage to DNA can lead to reproductive death, point mutation, or even gene deletions, which are related to clinical outcomes such as cancer or birth defects. In many cases, cells can repair DNA damage, especially if only one of the two strands of DNA is damaged. The likelihood of a permanent change increases if the two DNA strands are simultaneously damaged, typically by high LET radiation. If the damage to cell structures is more severe, then cell death may occur. If enough cells in an organ or tissue are killed, as in a radiation burn, the entire organ or tissue may cease to function, and in extreme cases, the organism itself may die.

6.3.2 Biological Effectiveness and Organ Sensitivity

Different types of radiation may be more effective at damaging cells than others. In addition, some organs are more sensitive to radiation damage than others. These have led to the development of units of dose that take into account these different effects (radiation weighting factors) and sensitivities (tissue weighting factors). Note that the approach recommended by the International Commission on Radiation Protection (ICRP, 1977) has been codified in regulations. The ICRP, however, revised their recommended approach in 1991. Both are presented here for completeness. For medical radiation purposes, covering only X-ray and gamma-ray radiation, a dose in sieverts (rems) may be considered numerically equivalent to that in grays (rads). Both radiation and tissue weighting factors are subject to occasional revision. In some cases, laws or regulations specify the weighting factors to use for regulatory purposes. Otherwise, the most recent ICRP recommendations are commonly employed.

6.3.3 Deterministic and Stochastic Effects

The relationship between radiation damage to cells and clinical outcomes is complex, but nearly all ultimate biological effects on an organism may be classified as being either deterministic or stochastic. Stochastic effects involve the increase in the probability of cancer or hereditary disorders, rather than its severity, and may have a long latency period between exposure and effect. These are thought to be caused primarily by damage to DNA. Stochastic effects are generally assumed for regulatory purposes to have no threshold dose, so that even very small amounts of radiation may increase the likelihood of the effect.

Deterministic effects are mainly caused by the death of damaged cells and involve the malfunctioning or loss of function of tissues or organs. These typically involve high levels of radiation exposure, arise not long after exposure, and have threshold doses below which they do not occur. When they do occur, their severity tends to increase with the degree of exposure. Even relatively large doses usually show no immediate apparent injury. Whole-body doses over 5 Gy (500 rad) are typically fatal if untreated, but the time between exposure and death may range up to several months. This is because these high doses of radiation destroy the cells in organs, and death occurs as the organs stop to function properly. Some specific organs are particularly sensitive to radiation—acute exposures of several sieverts (several hundred rems) to testes or ovaries can lead to sterility, and high exposure to the eyes can lead to cataracts.

Whether cells can recover from radiation exposure depends strongly on the dose rate, that is, the dose per unit time. For example, a whole-body exposure of 6 Gy (600 rad) of X-radiation, equivalent to 6 Sv of equivalent dose, administered in 1 day, with no medical treatment, would mean almost certain death within 30 days. On the other hand, 6 Sv in daily increments over a period of 30 years would amount to an annual dose of 0.2 Sv, which is four times the federal occupational exposure limit of 50 mSv/year and may lead to an increased risk of cancer but is most unlikely to produce any deterministic effects. Thus, the difference in dose rate means the difference between certain death in 30 days and possible increase in risk of cancer over a lifetime.

Another major factor in the biological effects of radiation is what part of the body is exposed. Large doses of radiation can be applied to local areas, as in therapy, with little danger. A person could expose a finger to 1 000 rad and experience a localized injury with subsequent healing and scar formation. Cells that divide more rapidly are generally more sensitive to radiation damage. For example, white blood cells and the blood-forming organs, such as the, spleen, lymph nodes, and bone marrow, tend to be radiosensitive.

While the deterministic effects of ionizing radiation are relatively well understood, much uncertainty still surrounds the stochastic effects of radiation, such as the induction of cancer, birth defects, or hereditary effects. For instance, study of the effects on cancer, especially at low dose rates, are complicated by the long latency between exposure and malignancy, the presence of background cancer rates, and confounding exposures that may contribute to increased risk. More is known about the effects and mechanisms of damage from ionizing radiation, however, than about those from chemical carcinogens. For instance, it is known from studies of survivors of the atomic bomb that doses greater than about 0.2 Sv lead to a statistically significant excess in cancer incidence and mortality. However, the effects at lower doses are the subject of some controversy. In particular, a "linear, no-threshold" dose response relationship has been assumed at low doses, and there is some biological and epidemiological evidence for it. As a rule of thumb, the lifetime fatal cancer risk due to radiation exposure under this hypothesis is about 0.05 per Sv (5×10^{-4} per rem). In any case, many believe that assuming a linear, no-threshold relationship is prudent, given the uncertainties. On the other hand, some experts have argued that there is also evidence for homesis—that is, beneficial effects at low doses—and that such effects should rule out the linear, no-threshold hypothesis. Still, a recent National Council on Radiation Protection (NCRP) report has concluded that "there is no conclusive evidence on which to reject the assumption of a linear-non-threshold dose—response relationship." It states, furthermore, that "the probability of effects at very low doses... is so small that it may never be possible to prove or disprove the validity of the linear, non-threshold assumption."

6.4 Methods of Radiation Protection

Three crucial factors involved in effecting external radiation protection are distance from the source, duration of exposure, and shielding. Also important for protecting against internal and external radiation exposure is the control of contamination and the management of waste. These methods are particularly important in implementing the principle of as low as resonably achievable (ALARA) and are implemented by a variety of personnel, engineering, and administrative controls that vary with the type and strength of the sources of radiation. Finally, surveillance and monitoring serve as both warnings of the breakdown of any safeguards as well as a quality check on their effectiveness.

6.4.1 Distance

The further away a person is from a radiation source, the less exposure he or she will receive. This is particularly true for a point source of radiation, such as the point of emission on the target of an X-ray tube; in accord with the familiar inverse square law, the subject exposure decreases inversely with the square of the distance from the source:

$$\frac{I_1}{I_2} = \frac{R_2^2}{R_1^2} \tag{6.3}$$

where R_1, R_2 = any two distances;

I_1, I_2 = values of the intensity at the distances.

For example, if at a distance of 5 ft one is exposed to an exposure rate or dose rate of A units per second, then at a distance of 10 ft the corresponding figure would be $\frac{1}{4}A$. In practice, this applies approximately to other than point sources of radiation, assuming that the radiation source is small relative to its distance from the exposed individual.

6.4.2 Duration of Exposure

When exposure to radiation is necessary or unavoidable, the duration time of such exposure should be kept as short as practicable to accomplish a particular task. In the case of occupational or similar situations, the cumulative exposure shall be kept below an individual's maximum permissible dose. This might mean relocating or reassigning individuals to keep them within the dose limits shown in Table 6.2.

Table 6.2 Summary of Dose Limit Recommendations

	SI Units	Conventional Units
A. Occupational exposures		
1. Effective dose limits		
a. Annual	50 mSv	5 rem
b. Cumulative	10 mSv × age	1 rem × age in years
2. Equivalent dose annual limits for tissues and organs		
a. Lens of eye	150 mSv	15 rem
b. Skin, hands, and feet	500 mSv	50 rem
B. Public exposures (annual)		
1. Effective dose limit, continuous or frequent exposure	1 mSv	0.1 rem
2. Effective dose limit, infrequent exposure	5 mSv	0.5 rem
3. Equivalent dose annual limits for tissues and organs		
a. Lens of eye	15 mSv	1.5 rem
b. Skin, hands, and feet	50 mSv	5 rem
4. Remedial action for natural sources:		
a. Effective dose (excluding radon)	>5 mSv	>0.5 rem

b. Exposure to radon decay products	>0.007 Jh /m^3	>2 WLM
C. Education and training exposures (annual)		
1. Effective dose limit	1 mSv	0.1 rem
2. Equivalent dose annual limits for tissues and organs		
a. Lens of eye	15 mSv	1.5 rem
b. Skin, hands, and feet	50 mSv	5 rem
D. Embryo-fetus exposures (monthly)		
1. Equivalent dose limit	0.5 mSv	0.05 rem
E. Negligible individual dose (annual)	0.01 mSv	0.001 rem

Source: Limitation of Exposure to Ionizing Radiation, Report No. 116, National Council on Radiation Protection, Bethesda, MD, 1993.

Notes: Medical exposures are excluded. Except for B.4, doses are the sum of external and internal exposures excluding natural sources. Working level month (WLM) is the cumulative exposure equivalent to exposure to one working level (WL) for a working month (170 hr). One WLM = 3.5×10^{-3} Joule-hours per cubic meter (Jh/ m^3).

6.4.3 Shielding

Shielding is the interposition of an attenuating material between a source of radiation and the surroundings. A shield is typically some combination of concrete, steel, lead, and lead glass and may not only absorb ionizing radiation but also protect from intense heat. The energy and type of the radiation to be attenuated and the human occupancy or use of the area of exposure are basic factors in the selection of the shielding material, its size, and thickness. The term "half-value layer" (HVL) is used to designate the thickness of a particular material that will reduce the intensity of radiation passing through the material by one-half.

A primary protective barrier is of a material and thickness to attenuate the useful beam to the required degree; a secondary barrier attenuates stray radiation. Radiation, however, will scatter and bounce off the floor and ceiling and from wall to wall. It can therefore be reflected around shields, around corners, over and under doors, and through ventilating transoms and windows. In some cases, the intensity of the scattered radiation may be as large as or greater than the primary radiation source. Shields must be placed to protect anyone who might have access to spaces above or below or on any side of the source. It is generally better to place shields close to the source of radiation because this will reduce the shielded area and the total weight of the shield.

Some of the characteristics of alpha and beta particles, X-rays and gamma rays, and neutrons are given earlier in this chapter. Care also should be taken in designing shielding for beta particles to take into account the secondary bremsstrahlung X-rays that may be produced. Neutron shielding presents special problems around nuclear reactors and particle accelerators.

Additional factors to consider when designing shielding are the workload, the use factor, and the occupancy. The workload, usually expressed in milliampere seconds per week, is the milliamperage of the current passing through the X-ray tube times the sum of the number of seconds the tube is in operation in one week. The use factor is the part of the time a machine is in use that the useful beam may strike the wall, ceiling, or floor to be shielded. The occupancy factor is the time a person on the other side of a shielded wall, floor, or ceiling will be exposed to radiation when a unit is in operation. This factor will vary from 1 when there are people living or working in the adjoining space to 1/16 when people may be exposed in a stairway or elevator.

6.4.4 Contamination Control

Materials that emit alpha and beta particles present a particularly dangerous hazard if ingested or inhaled because their specific ionization is high and they irradiate the body continuously until they are eliminated. The seriousness of the hazard depends on the type of radioactive material, the type of radiation emitted, the energy of the radiation, its physical and biological half-life, and the radiosensitivity of the tissue and body organ where the isotope establishes itself. The primary objective of protection must therefore be to keep the radioactive materials out of the body in the first place. This is accomplished by the use of proper procedures and good practices, such as employing laboratory hoods, air filters, and exhaust systems; eliminating dry sweeping; wearing protective clothing and respirators when indicated; using proper monitoring and survey instruments; and prohibiting eating and smoking where radioactive materials are handled or used.

Entrance to areas that may cause or permit significant exposure to radiation or radioactive materials through occupancy, work, or general access must be controlled. Such areas may be found in hospitals; medical, dental, chiropractic, osteopathic, podiatric, or veterinary institutions, clinics, or offices; educational institutions; commercial, private, or research laboratories performing diagnostic procedures or handling equipment or material for medical uses; or any trucking, storage, messenger, or delivery service establishments or vehicles. High radiation areas are posted "Radiation Area," "Radiation Zone," or "Restricted Area," as required by the NRC and state regulations. These and similar types of installations and areas should be under the supervision of persons qualified by training and experience in radiological health to evaluate the radiation hazards and to establish and administer adequate radiation protection programs.

6.4.5 Waste Management

Radioactive waste from the use of radioactive materials must not endanger individuals, the public health, or the environment. The guiding principles should be prevention (reduction) and control of gaseous, liquid, and solid waste at the source, followed by segregation and the indicated collection, treatment, reuse, and storage or disposal. General principles of radioactive waste management are presented here. The case of waste from nuclear reactors is described in more detail below.

There are two general classifications of radioactive waste, namely highlevel radioactive waste (HLRW or HLW) and low-level radioactive waste (LLRW or LLW). They are largely governed by two different federal laws and covered by two different sets of regulation. Disposal of high-level radioactive waste is the responsibility of the Department of Energy (DOE). The licensing of high-level waste disposal facilities is the responsibility of the Nuclear Regulator Commission (NRC), as specified in 10 CFR Part 60, "Disposal of High-Level Radioactive Waste in Geologic Repositories." The actual disposal of HLW, which is generally taken to include spent nuclear fuel, is the responsibility of the DOE. Disposal of LLW is also subject to licensing by the NRC, the regulations for which are in 10 CFR Part 61, "Licensing Requirements for Land Disposal of Radioactive Waste." The actual disposal facilities are administrated by regional groupings of states called Compacts.

Low-level radioactive wastes consist primarily of contaminated rags, clothing, filters, resins, and activated metals from power plants, laboratories, hospitals, and commercial and industrial facilities. The NRC regulations categorize this LLW into three classes: Class A wastes are materials that degrade to a safe (as determined by the NRC) radiation level within 100 years; class B wastes are materials with higher radioactivity; these degrade to a safe level within 300 years. Class C wastes are materials that do not degrade for up to 500 years. Class A wastes can be buried in solidified or absorbed form in containers. Class B wastes have to be solidified or buried in high-integrity containers. Class C wastes have to be solidified and buried at least 15 m (50 ft) deep.

The waste must also be processed in such a manner as to minimize the risk of exposure to the public. For instance, groundwater contamination should be minimized because it may be used to irrigate crops or fodder used for cows or processed as drinking water. Likewise, radioactive wastes may be taken in by the surface-water biota, which could then be consumed by people. Also, if the individual is in the vicinity of a release, there could be a direct exposure. Such processes have been studied extensively in recent year by means of computer-based environmental pathway/ transport models, in the writing of regulations, the determination of release criteria, and the remediation of specific problems.

Because of the different characteristics of solid, liquid, and gaseous radioactive wastes, each must be processed differently before release or disposal. Liquids are processed to remove the radioactive contaminants. These processes might include filtering, routing through demineralizers, boiling off some of the water (evaporation) and leaving the solids (which are then processed as solid radioactive waste), and/or storing the liquid for a time period to allow the radioactive material to decay. After processing, liquids will be sampled and, if they meet the required standards, they can be placed in storage tanks for reuse in the plant or even be released to the environment. If the samples show the water does not meet the release criteria, it must be processed further. Some materials, such as the evaporator bottoms (solids that remain after the water is evaporated), will be mixed with a substance like concrete to form a solid. This is also sometimes done with spent demineralizer resins. After mixing with a hardener, the material is processed as solid radioactive waste. Gaseous wastes are filtered, often compressed to take up less space, and then allowed to decay for some time period. After the required time has passed, the gases will be sampled. If the required limits are met, the gases will be released to the atmosphere or sometimes they will be reused in specific areas of the plant. Liquid and gaseous radioactive wastes may be released to the environment after processing if they meet the stringent release criteria specified in the site license. Careful monitoring and enforcement of release standards help ensure that members of the public are not put at risk by these releases. Remaining solid wastes are either stored on-site or transported to designated facilities for disposal.

Wastes that are not suitable for release may require further treatment into a solid waste form that can be safely disposed in a geologic repository. For high-level wastes, vitrification is a major solidification process. In this process, highly radioactive liquid and sludge are mixed with glass particles and heated to very high temperatures to produce a molten glass. This molten glass can be formed into glass marbles, blocks, or logs. When the mixture cools, it hardens into a stable glass that traps the radioactive elements and prevents them from moving through the air or water into the environment. Although effective, vitrification is a very expensive process and is performed only if the volume of wastes can be reduced to a reasonable amount.

In addition to proper handling, the proper disposal of radioactive waste will help to minimize the dose received by members of the public. Low-level sites should be isolated, remain stable for 500 years, and include monitoring for 100 years after closure. Although most of the radioactive materials decay rapidly, some radionuclides have long half-lives. These are generally small amounts in large volumes of material. Currently, low-level radioactive waste is accepted at three disposal sites presently operating. Barnwell, South Carolina, can accept all low-level waste except from North Carolina. Hanford, Washington, can accept waste from the northwest and Rocky Mountain compacts. Clive, Utah, is only authorized to accept class A, low-activity, highvolume waste. The Nuclear Waste Policy Act of 1980 gives states the responsibilities for management and disposal of most civilian low-level radioactive waste. Disposal is regulated by a state entering into an agreement with the NRC (agreement state).

There are various ways in which costs can be reduced through efforts to minimize quantities of LLW. But while market forces have had some effect in stimulating the intelligent use of resources, the waste minimization process has received only limited legislative and regulatory attention. There is no direct federal statutory requirement, in particular, for regulations on minimizing the generation of LLW. The Pollution Prevention Act, like ALARA, makes good sense, in general, and should be socially beneficial. Examples include recycling and source reduction. Still, there are a number of procedures one can follow, often quite easily, to reduce the amount of LLW in the first place—the most obvious being not to mix radioactive waste with nonradioactive materials if it is not necessary.

In addition, the Waste Isolation Pilot Plant (also known as the WIPP), located in southeastern New Mexico, is the first geological repository in the United States for the permanent disposal of transuranic wastes. Transuranic waste, which contains man-made elements heavier than uranium (and therefore "beyond uranium" on the periodic chart), is produced during nuclear fuel assembly, nuclear weapons research, productions, and cleanup and as a result of reprocessing spent nuclear fuels. The waste generally consists of protective clothing, tools, glassware, and equipment contaminated with radioactive materials.

6.4.6 Space and Personnel Surveillance and Monitoring

A radiation installation should be responsible for its own monitoring program. A competent health physics/radiation safety unit should be established and given responsibility for safety monitoring and authority to take prompt corrective action whenever indicated. In-house safety rules governing operating procedures are helpful. This should include control of the facility work and storage areas, sources, waste disposal, emergency procedures, and personnel monitoring.

A personnel monitoring system will determine individual exposure received and effectiveness of the control measures being taken. Commonly used dosimeters that allow assessment of beta, gamma, and X-ray exposure are thermoluminescent dosimeters (TLDs), film badges, and finger ring dosimeters. They may be supplemented by "pocket" dosimeters for work in areas where there is a suspicion of high dose rates. Some special instances may call for measurement of the radioactivity in the body or excreta. For instance, urinalysis may be employed for assessing exposure to tritium, C-14, S-35, or P-32, and persons working with radioiodine may require thyroid bioassays. Film badges should be processed and interpreted by a specialized laboratory or government agency. The accuracy of other personnel monitoring devices should be determined by comparison with detectors calibrated at a nationally accredited laboratory approved for the types of radiation monitored. A personnel monitoring system should be in effect for individuals who may be exposed to radiation exceeding one-tenth of the maximum permissible effective dose equivalent values given in Table 6.2.

Instruments for detecting and measuring radiation are necessary for an effective monitoring program, and these must be selected with consideration of their sensitivity and purpose to be served. It is important that all these instruments be properly calibrated. These "survey meter" devices are used to identify areas where contamination may exist. Wipe surveys may then determine the extent of the contamination (and tritium may only be detected through a wipe survey). For instance, appropriate media such as filter paper or cotton swabs are used to wipe a test area, and the presence of contamination may be detected by placing the test wipe in a liquid scintillation counter (with an appropriate "control" wipe to obtain a background count).

6.4.7 Environmental Surveillance and Monitoring

An environmental monitoring program including a preoperational survey in and around a proposed nuclear facility can serve as a check on the various operations that might produce contamination. It also permits evaluation of the exposure to the surrounding area and the need for preventive action. The program should be a cooperative one between health and other regulatory authorities and the facility management.

Industry has responsibility for internal housekeeping and monitoring of the types and quantities of all its radioactive waste discharges. Peripheral monitoring should be shared between the industry and the regulatory authorities. Maximum fallout from a stack will usually occur immediately after emission within a 1-mile (1.6 km) radius. This area should be secured by the plant.

Solid, liquid, and gaseous radioactive materials can be carried considerable distances by air and water and may adversely affect plant and animal life as well as people. It is essential, therefore, that a surveillance and monitoring program be maintained in and around installations from which radiation release is possible. These include nuclear power plants, fuel processing plants, uranium milling industries, university reactors, and certain industries and laboratories. The surveillance monitoring should be carefully planned with the advice and assistance of radiological health specialists and physicists, biologists, meteorologists, environmental engineers, geologists, and others familiar with the problem and the local area. The sampling for resulting contamination should include air, water, milk, food, biota, sediment, soil, and people. Interpretation of the sampling requires mapping of the surrounding area showing salient geographical and topographical features. When coupled with previously prepared air diffusion models and overlays, this information makes possible assessment of the risk associated with any release over a large area under various meteorological conditions. Mapping should show surface and groundwater hydrology, types of soil and vegetation, population centers, transportation systems, sources of water supply, recreational areas, and other land uses. Past meteorological conditions would include prevailing winds and speeds, temperature, and rainfall.

At a nuclear fuel processing plant there is a potential for discharge of I-131, Kr-85, and tritium. Release of I-131 is minimized by storage of fuel elements for at least 100 days before processing. Krypton release is greater than from a nuclear reactor and cannot as yet be effectively removed before release to a stack. Tritium is not yet removable and is discharged as a liquid or gaseous waste. Careful surveillance is necessary to ensure that discharges of these radionuclides are kept to a minimum and within acceptable limits.

Sampling at nuclear facilities may include any combination of the following, with consideration of the facility monitored:

Air: Around the installation and at some distance, as noted above.

Water: At plant outfalls, receiving stream and downstream; also groundwater from nearby wells.

Soil: Immediately around the installation and at some distance, including stream bottom muds.

Biological Specimens: Fish, deer, possibly cows, rodents, plants (especially those used for food); also, where available, shellfish, ducks, plankton, and other plant life.

Milk: From dairy farms in vicinity.

Spaces: Indoor spaces, containers, and conduits.

Sampling stations are usually located on the site being monitored, in the immediate vicinity, and at some distance. Those on-site and in the immediate vicinity should preferably be fixed and continuous monitors. Stations at a distance could be used for periodic grab sampling, selected if possible with reference to other stations maintained by the Weather Bureau, a health department, a university, the U.S. Geological Survey, a radio station, or an airport that could provide supporting data. Stream samples include water, mud, and biota. Tiles, stones, or slides are suitable for the collection of biological attachments.

A monitoring station might contain a continuous water sampler, highvolume air sampler, silver nitrate filter, film badge, adhesive paper, silica gel, rain gauge, ionization chamber, and such other equipment or devices as may be indicated by a safety analysis and the nature of the facility being monitored.

The air monitoring around a boiling-water reactor is described by Thomas. Within a 1-mile radius distributed on a 500 ft grid will be 50 thermoluminescent dosimeters; three air-monitoring stations will sample airborne particulates, radioiodine, heavy particulate fallout, and rainwater. Air filters will be scanned continuously by a beta-gamma-sensitive G-M tube, and readings will be reported at the control center. In addition, four air monitors about 10 miles away in an urban area will provide continuous reports to the control center. Weekly composite samples will also be collected from stations about 40 miles from the plant. Air monitoring is supplemented by the sampling of fish, mussels, algae, and other biota as well as milk, water, vegetation, and soil. State and federal agencies also have responsibilities to protect the public health.

Review Questions

1. Explain the term of ionizing radiation, and list types of ionizing radiation.
2. Distinguish external and internal radiation exposure.
3. Explain the process of radiation exposure caused by X-ray machines.
4. List the principle factors involved in effecting external radiation protection

Chapter 7 Environmental Worldviews and Sustainability

7.1 Environmental Worldviews in Industrial Societies

7.1.1 Environmental Worldview

There are conflicting views about how serious our environmental problems are and what we should do about them. These conflicts arise mostly out of differing environmental worldviews: (1) how people think the world works, (2) what they think their role in the world should be, and (3) what they believe is right and wrong environmental behavior (environmental ethics). People with widely differing environmental worldviews can take the same data, be logically consistent, and arrive at quite different conclusions because they start with different assumptions and values.

The many different types of environmental worldviews are summarized in Figure 7.1. Most can be divided into two groups according to whether they are (1) individual centered (atomistic) or (2) earth centered (holistic). Atomistic environmental worldviews tend to be human centered (anthropocentric) or life centered (biocentric, with the primary focus on individual species or individual organisms). Holistic or ecocentric environmental worldviews focus on sustaining the earth's natural systems (ecosystems), life-forms (biodiversity), and life-support systems (biosphere) for all species.

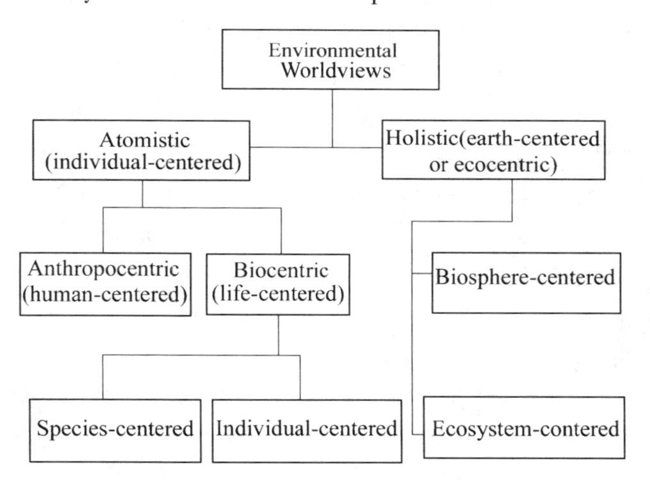

Figure 7.1 General types of environmental worldviews

7.1.2 Instrumental and Intrinsic Values

What we value largely determines how we act. Environmental philosophers normally divide values into two types:

(1) Instrumental, or utilitarian: a value something has because of its usefulness to us or to the biosphere. For example, the concept of preserving natural capital and biodiversity because they

sustain life and support economies is an instrumental value based mostly on the usefulness of these natural goods and services to us.

(2) Intrinsic, or inherent value: the value something has just because it exists, regardless of whether it has any instrumental value to us. There is controversy over whether nonhuman forms of life and nature as a whole have intrinsic value.

The view that a wild species, a biotic community ecosystem, biodiversity or the biosphere has value only because of its usefulness to us is called an anthropocentric (human-centered) instrumental value. According to an anthropocentric worldview, ① humans have intrinsic value, ② the rest of nature has instrumental value, and ③ we are in charge of the earth.

In contrast, the view that these forms of life are valuable simply because they exist, independently of their use to human beings, is called a biocentric (life-centered) intrinsic value. According to a biocentric worldview, ① all species and ecosystems and the biosphere have both intrinsic and instrumental value, ② we are just one of many species, and ③ we have an ethical responsibility not to impair the long-term sustainability and adaptability of the earth's natural systems for all 1ife.

7.1.3 Major Human-Centered Environmental Worldviews

Most people in today's industrial consumer societies have a planetary management worldview, which has become increasingly common during the past 50 years. According to this human-centered environmental worldview, human beings, as the planet's most important and dominant species, can and should manage the planet mostly for their own benefit. Other species and parts of nature are seen as having only instrumental value based on how useful they are to us.

Figure 7.2 (left) summarizes the four major beliefs or assumptions of this one version of this worldview. All or most aspects of this worldview are widely supported because it is said to be the primary driving force behind the major improvements in the human condition since the beginning of the industrial revolution.

There are several variations of this environmental worldview:

(1) The no-problem school. We can solve any environmental, population, or resource problems with more economic growth and development, better management, and better technology.

(2) The free-market school. The best way to manage the planet for human benefit is through a free-market global economy with minimal government interference and regulations. Free-market advocates would convert all public property resources to private property resources and let the global marketplace, governed by pure free-market competition, decide essentially everything.

(3) The responsible planetary management school. We have serious environmental problems, but we can sustain our species with a mixture of market-based competition, better technology, and some government intervention that ① promotes environmentally sustainable forms of economic development, ② protects environmental quality and private property rights, and ③ protects and manages public and common property resources. People holding this view follow the pragmatic

principle of enlightened self-interest: Better earth care is better self-care.

(4) The spaceship-earth school. The earth is seen as a spaceship: a complex machine that we can understand, dominate, change, and manage to prevent environmental overload and provide a good life for everyone. This view developed as a result of photographs taken from space showing the earth as a finite planet or island "floating" in space. This powerful image led many people to see that the earth is our only home and we had better treat it right.

(5) The stewardship school. We have an ethical responsibility to be caring and responsible managers, or stewards, of the earth. According to this view, we can and should make the world a better place for our species and other species through love, care, compassion, knowledge, and technology.

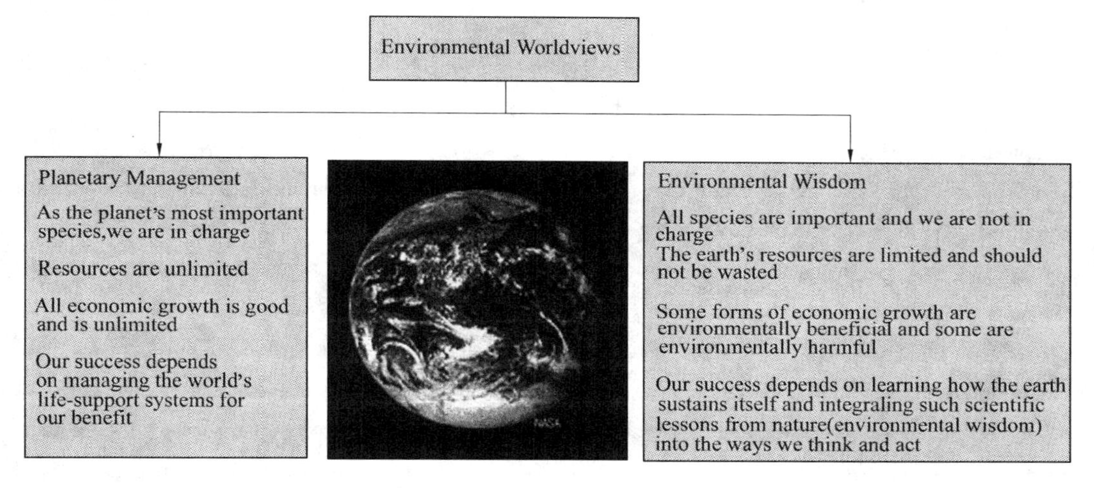

Figure 7.2 Comparison of two opposing environmental worldviews

7.2 Life-Centered Environmental Worldviews

7.2.1 Can We Manage the Planet?

Some people believe any human-centered worldview will eventually fail because it wrongly assumes we now have (or can gain) enough knowledge to become effective managers, or stewards, of the earth.

According to these analysts, the unregulated global free-market approach will not work because it (1) is based on increased losses of natural capital that support all life and economies and (2) focuses on short-term economic benefits regardless of the harmful long-term environmental and social consequences.

The image of the earth as an island or spaceship in space has played an important role in raising global environmental awareness. However, critics argue that thinking of the earth as a spaceship that we should and can manage is an oversimplified and misleading way to view an incredibly complex and ever-changing planet. For example, these critics point out that we do not even know how many species live on the earth, much less what their roles are and how they interact with one another and their nonliving environment. We have only an inkling of what goes on in a handful of soil, a meadow, a patch of forest, a pond or any other part of the earth.

As biologist David Ehrenfeld puts it, "In no important instance have we been able to demonstrate comprehensive successful management of the world, nor do we understand it well enough to manage it even in theory." Environmental educator David Orr says we are losing rather than gaining the knowledge and wisdom needed to adapt creatively to continually changing environmental conditions: "On balance, I think, we are becoming more ignorant because we are losing cultural knowledge about how to inhabit our places on the planet sustainably, while impoverishing the genetic knowledge accumulated through millions of years of evolution."

Even if we had enough knowledge and wisdom to manage spaceship earth, some critics see this approach as requiring us to give up individual freedom to survive. Life on spaceship earth under a comprehensive system of planetary management or world government might be much like the regimented life of astronauts in their capsule. The astronauts have almost no individual freedom with essentially all of their actions dictated by a central command.

7.2.2 Major Biocentric and Ecocentric Worldviews

People disagree over how far we should extend our ethical concerns for various forms or levels of life (see Figure 7.3). Critics of human-centered environmental worldviews believe such worldviews should be expanded to recognize the inherent or intrinsic value of all forms of life regardless of their potential or actual use to us. Another source of conflict between human-centered and life-centered worldviews results from whether emphasis should be on short-term or long-term values. Because of the nature of major environmental problems, many analysts believe they need to be dealt with in terms of both the short-term and long-term future.

Most people with a life-centered (biocentric) worldview believe we have an ethical responsibility not to cause the premature extinction of a species because of our activities. Most people give protection of a species priority over protection of an individual member of a species because each species is a unique storehouse of generic information that should be respected and protected because it exists (intrinsic value), is a potential economic good for human use (instrumental value), and is capable through evolution and speciation of adapting to changing environmental conditions. In this sense, individuals are temporary representatives of a species, and the premature extinction of a species can be regarded as killing future generations of a species and eliminating its possibility for future evolutionary adaptation or speciation.

Trying to decide whether all or only some species should be protected from premature extinction resulting from human activities is a difficult and controversial ethical problem. It is hard to know where to draw the line and be ethically consistent. For example,

(1) Should all species be protected from premature extinction because of their intrinsic value,

or should only certain ones be preserved because of their know or potential instrumental value to us or to their ecosystems.

(2) Should all insect and bacterial species be protected, or should we attempt to exterminate those that eat our crops, harm us, or transmit disease organisms?

(3) Should we emphasize protecting keystone species in ecosystems over other species that play lesser ecological roles?

Others believe we must go beyond this biocentric worldview, which focuses on species and individual organisms. They believe we have an ethical responsibility not to degrade the earth's systems (ecosystems), life-forms (biodiversity) and life-support systems (biosphere) for this and future generations of humans (Connections, below) and other forms of life based on their intrinsic and instrumental values. In other words, they have an earth-centered, or ecocentric, environmental worldview, devoted to preserving the earth's biodiversity and the functioning of its life-support systems (see Figure 7.3) for all forms of life.

Why should we care about the earth's biodiversity? According to the late environmentalist and systems expert Donella Meadows,

Biodiversity contains the accumulated wisdom of nature and the key to its future. If you wanted to destroy a society, you would burn its libraries and kill its intellectuals. You would destroy its knowledge. Nature's knowledge is contained in the DNA within living cells. The variety of genetic information is the driving engine of evolution and the source of adaptability.

According to the ecocentric worldview, we are part of, not apart from, the community of life and the ecological processes that sustain all life. Aldo Leopold summed up this idea in 1948: "All ethics rest upon a single premise: that the individual is a member of a community of interdependent parts."

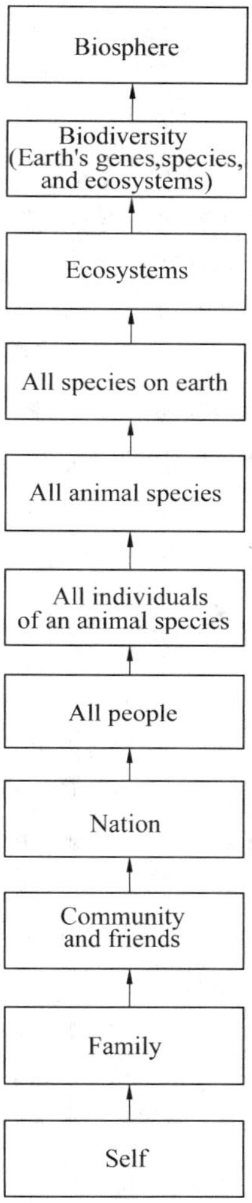

Biosphere
↑
Biodiversity (Earth's genes, species, and ecosystems)
↑
Ecosystems
↑
All species on earth
↑
All animal species
↑
All individuals of an animal species
↑
All people
↑
Nation
↑
Community and friends
↑
Family
↑
Self

Figure 7.3 Levels of ethical concern. People disagree over how far we should extend our ethical concerns on this scale

There are many life-centered and earth-centered environmental worldviews, and several of them overlap in some of their beliefs. Figure 7.2 (right) summarizes the four major beliefs or assumptions of the environmental wisdom worldview, which are the opposite of those making up the planetary management worldview (see Figure 7.2, left). A related ecocentric environmental worldview is the deep ecology worldview.

Others say we do not need to be biocentrists or ecocentrists to value life or the earth. They point out that human-centered stewardship and planetary management environmental worldviews also call for us to value individuals, species, and the earth's life-support systems as part of our responsibility as the earth's caretakers.

7.2.3 Social Ecology Worldview

According to activist and philosopher Murray Bookchin, the ecological crisis we currently face results from the power of our hierarchical and authoritarian social, economic, and political structures and the various technologies used to dominate people and nature. In other words, our current environmental situation has been created by industrialized societies driven by the conventional planetary management environmental worldview (see Figure 7.2, left).

To deal with the ecological problems we face, Bookchin believes we must adopt a social ecology environmental worldview. It is built around creating better versions of democratic communities, new forms of environmentally sustainable production, and types of appropriate technology that are smaller in scale, consume fewer resources and less earth capital, and do not cause environmental degradation of local ecological regions .

7.2.4 Physical and Biological Limits to Human Economic Growth

The planetary management and environmental wisdom worldviews (and related versions of these two types of worldviews) differ over whether there are physical and biological limits to economic growth, beyond which both ecological and economic collapse is likely to occur. This argument over limits has been going on since Thomas Malthus published his book "The Principles of Political Economy" in 1836, with each side insisting it is right at least in the long term.

In 2000, conservation biologist Carlos Davidson (with a background in economics) proposed a way to bridge the gap between these two opposing viewpoints and help motivate the political changes needed to halt or slow the spread of environmental degradation.

He disagrees with the view of some economists that technology will allow continuing economic growth without causing serious environmental damage. However, he also disagrees with the concept that continuing economic growth based on consuming and degrading natural capital will lead to ecological and economic crashes.

Instead of crashes, he suggests we use the metaphor of the gradual unraveling of some of the threads in a woven tapestry to describe the effects of environmental degradation. He views nature as an incredibly diverse and interwoven tapestry of threads consisting of a variety of patterns (biomes, aquatic systems, and ecosystems). As environmental degradation removes threads from different parts of the biosphere, nature's tapestry in these areas becomes worn and tattered.

Clearly, if too many of the threads are removed, the tapestry can be destroyed. However, Davidson contends that as threads are pulled from various parts of the tapestry, local losses of ecological function and occasional local and regional tears occur rather than overall collapse. Global problems such as climate change, ozone depletion, and biodiversity loss may be exceptions to this analogy. However, even with these problems, some areas of the tapestry are damaged more than other areas.

Davidson agrees with the views of ecologists and environmentalists that degradation in parts of the earth's ecological tapestry is occurring, spreading, and must be prevented. He supports doing this by using the pollution prevention and precautionary principles to protect areas of the biosphere from further damage.

However, he believes using catastrophe metaphors such as "ecological collapse" and "going over a cliff" can hinder achievement of these important goals. He argues that repeated predictions of catastrophe, like the boy who cries wolf, initially motivate people's concern. However, when the threats turn out to be less severe than predicted, people tend to ignore future warnings. As a result, they are not politically motivated to prevent more tears in nature's tapestry and look more deeply at the economic, political, and social forces responsible for environmental degradation.

Biologist Kevin S. McCann at McGill University in Montreal, Canada, believes the tapestry metaphor is misleading because it assumes nature is in a static equilibrium state when in fact it undergoes constant dynamic change. He points out that tearing nature's fabric sends "waves of dynamic change through an ecosystem, and the waves get bigger as biological diversity declines."

7.3 Living More Sustainably

7.3.1 Ethical Guidelines for Working with the Earth

Ethicists and philosophers have developed ethical guidelines for living more sustainably on the earth. Anyone can use such guidelines, regardless of their environmental worldview.

1. Biosphere and Ecosystems

We should try to understand and work with the rest of nature to help sustain the natural capital, biodiversity, and adaptability of the earth's life-support systems. When we alter nature to meet our needs or wants, we should choose methods that do the least possible short-and long-term environmental harm. This ethical concept of ahimsa, or avoiding unnecessary harm, is a key element of Hinduism, Buddhism, and Jainism.

2. Species and Cultures

No wild species should become prematurely extinct because of our actions. The best ways to protect species and individuals of species are to (1) protect the places where they live, (2) help restore biological habitats we have degraded, and (3) control the accidental or deliberate introduction of nonnative species into aquatic and terrestrial systems. No human culture should become extinct because of our actions.

3. Individual Responsibility

We should not inflict unnecessary suffering or pain on any animal we raise or hunt for food or use for scientific or other purposes. We should leave the earth in as good as or better condition than we found it. We should use no more of the earth's resources than we need. We should work with the earth to help heal ecological wounds we have inflicted.

In March 2000, the Earth Charter was finalized. More than 100 000 people in 51 countries and

25 global leaders in environment, business, politics, religion, and education took part in creating this charter. It is a document creating an ethical and moral framework to guide the conduct of people and nations to each other and to the earth. Here are its four guiding principles:

(1) Respect earth and life in all its diversity.

(2) Care for life with understanding, love, and compassion.

(3) Build societies that are free, just, participatory, sustainable, and peaceful.

(4) Secure earth's bounty and beauty for present and future generations.

The moral rules of most of the world's religions and ethical systems boil down to a few simple ones. Do not lie. Do not cheat. Do not steal. Do not hurt people. Do not hurt the earth that sustains us. Help each other out. Do not do to someone else what you would not want done to yourself.

7.3.2 Implement of Earth Education

Most environmentalists believe that learning how to live more sustainably takes a foundation of environmental or earth education that relies heavily on an interdisciplinary and holistic approach to learning. According to its proponents, the most important goals of such an education are these:

(1) Develop respect or reverence for all life.

(2) Understand as much as we can about how the earth works and sustains itself, and use such knowledge to guide our lives, communities, and societies.

(3) Use critical thinking skills to become seekers of environmental wisdom instead of vessels of environmental information.

(4) Understand and evaluate our environmental worldview and see this as a lifelong process.

(5) Learn how to evaluate the beneficial and harmful consequences of our choices of lifestyle and profession on the earth, today and in the future.

(6) Use critical thinking skills to evaluate advertising that encourages us to buy more and more things. As humorist Will Rogers said, "Too many people spend money they haven't earned to buy things they don't want, to impress people they don't like." David Orr points out, "Our children, consumers-in-training, can identify more than a thousand corporate logos but only a dozen or so plants and animals native to their region." Each year U.S. corporations spend more than $ 150 billion on advertising, far more than is spent on all secondary education in the country.

(7) Foster a desire to make the world a better place and act on this desire. As David Orr puts it, education should help students "make me leap from 'I know ' to 'I care' to 'I ll do something.'"

According to environmental educator Mitchell Thomashow, four basic questions should be at the heart of environmental education:

(1) Where do the things I consume come from?

(2) What do I know about the place where I live?

(3) How am I connected to the earth and other living things?

(4) What is my purpose and responsibility as a human being?

How we answer these questions determines our ecological identity. In addition to formal education, some analysts believe we need to experience nature directly to help us learn to walk more lightly on the earth. Noel Perrin summarizes some actions various colleges and universities have taken to promote environmental sustainability.

7.3.3 Major Components of the Environmental Revolution

The environmental revolution that many environmentalists call for us to bring about during this century would have several components:

(1) An efficiency revolution that involves not wasting matter and energy resources.

(2) A solar-hydrogen revolution based on decreasing our dependence on carbon-based nonrenewable fossil fuels and increasing our dependence on forms of renewable solar energy that can be used to produce hydrogen fuel from water.

(3) A pollution prevention revolution that reduces pollution and environmental degradation from harmful chemicals by keeping them from being released into the environment by recycling or reusing them within industrial processes, trying to find less polluting substitutes, or not producing them at all.

(4) A biodiversity protection revolution devoted to sustaining the genes, species, ecosystems, and ecological processes that make up the earth's biodiversity.

(5) A sufficiency revolution. This involves trying to meet the basic needs of all people on the planet and asking how many material things we really need to have a decent and meaningful life.

(6) A demographic revolution based on reducing fertility to bring the size and growth rate of the human population into balance with the earth's ability to support humans and other species without serious environmental degradation.

(7) An economic and political revolution in which we use economic systems to reward environmentally beneficial behavior and discourage environmentally harmful behavior.

Opponents of such a cultural change like to paint environmentalists as messengers of gloom, doom, and hopelessness. However, the real message of environmentalism is not gloom and doom, fear, and catastrophe but hope and a positive vision of the future. This is an empowering message of challenge and adventure as we struggle to find better and more responsible ways to live on this planet. We should rejoice in our environmental accomplishments, but the real question is where do we go from here? How do we transfer the environmental advances of the past 35 years to developing countries? How, during this century, do we make a new cultural transition to more environmentally sustainable societies based on learning from and working with nature?

The challenge is to implement such solutions by converting our environmental wisdom and beliefs into political action. This requires us to become involved in making the world a better place. This means recognizing that a major lesson of our brief history on this marvelous planet is that individuals matter. Virtually all of the environmental progress we have made during the last few decades occurred because individuals banded together to insist that we can do better.

It is an incredibly exciting time to be alive as we struggle to enter into a new relationship with the earth that keeps us all alive and supports our economies. Jump in and get involved in guiding this new wave of cultural change.

Envision the earth's life-sustaining processes as a beautiful and diverse web of interrelationships—a kaleidoscope of patterns, rhythms, and connections whose very complexity and multitude of possibilities remind us that cooperation, sharing, honesty, humility, compassion, and love should be the guidelines for our behavior toward one another and the earth.

Review Questions

1. List pieces of good or encouraging news about progress in developing international cooperation and policy on environmental issues.

2. Distinguish between (a) environmental worldviews and environmental ethics, (b) instrumental values and intrinsic values.

3. List seven guidelines of environmental or earth education and four questions that should be answered to determine one's ecological identity.

Common Vocabularies of Environmental Engineering

A.

Abatement 减低；降低；减少
Abiotic 无生命的；非生物的
Absorber Vessel 吸收器；吸收装置
Absorbing Ability 吸收能力
Absorption 吸收作用
Acceptable Daily Intake 容许日摄入量
Accidental Spill 事故性溢漏
Acetate Moiety 乙酸根；乙酸基
Acetylene Dichloride 对称二氯乙烯
Acid(ic) Precipitation 酸性降水
Acid-Affected 受酸影响的；受酸危害的
Acid-Causing Substance 致酸物质
Acid-Forming Substance 成酸物质
Acidification 酸化
Acidifying Substance 酸化物质
Acidogenic Bacteria 产酸细菌
Acids 酸
Acoustic Insulation 隔音
Activated Carbon 活性炭
Activated Sludge 活性污泥
Adiabatic Process 绝热过程
Adsorbent 吸附剂
Adsorption 吸附作用
Adverse Climate Change 不利的气候变化
Adverse Effect 不利效应；不利影响
Aerobic Processes 需氧过程
Aerosol Load 气溶胶浓度；气溶胶含量
Aerosol 气溶胶；烟雾体
Air Injection System 空气喷射系统
Air Monitoring 大气监测
Air Pollutant 空气污染物
Air Pollution Load 空气污染物浓度
Albedo Radiation 反照辐射
Albedo 反照率；反射率
Algal Bloom 水华；藻华
Alkali Lands 碱地
Alkali Metal 碱金属
Alkaline-Earth Metal 碱土金属
Alkalinization 碱化；碱化作用
Alkyd Resins 醇酸树脂
Alkyl Aluminium Halide 卤化烷基铝
Alkyne 炔；链炔烃
Allowable Daily Intake 容许日摄入量
Alternative Fuel 代用燃料
Altitude Distribution 高度分布；垂直方向分布
Alumina 氧化铝
Ammonia (NH_3) 氨
Ammonia Solution (NH_4OH) 氨水
Ammonium Hydrogen Sulphite 亚硫酸氢铵
Ammonium Hydroxide 氢氧化铵；氨溶液；氨水
Ammonium Sulphate 硫酸铵
Ammonolysis 氨解作用
Amplification Factor 放大因子
Anaerobic Conditions 缺氧情况
Anaerobic Processes 厌氧过程
Anion Exchanger 阴离子交换剂
Anomaly 反常；异常
Antarctic Ozone Hole 南极臭氧层空洞
Antarctic Springtime Ozone Depletion 南极春季臭氧消耗
Antarctic Stratospheric Circumpolar Vortex 南极平流层环极涡旋
Aquatic Acidification 水的酸化
Aromatic 芳族的；芳香的；芳香剂
Asbestos 石棉
Atmospheric Absorption 大气吸收
Atmospheric Acidity 大气酸度
Atmospheric Circulation 大气环流
Atmospheric Constituent 大气组分；大气组元

Atmospheric Fluidized Bed Combustion 常压流化床燃烧
Atmospheric Layer 大气层
Attributes 属性
Automatic Detection 自动检测
Autotroph 自养生物
Autotrophic 自养的
Available (Chlorine, Oxygen, Etc.) 有效氯,氧等
Average Total Ozone 臭氧平均总量

B.

Background Concentration 本底污染浓度
Background Pollution 本底污染
Backscattering 后向散射
Back-Up 支持；支持物；备份
Bacteria 细菌
Balance of Nature 自然生态平衡
Base Cations 碱阳离子
Base Metal 碱金属
Basic Reaction 碱性反应
Basic Solution 碱性溶液
Basin 流域；盆地
Bedrock 基岩
Benthic 底栖的；海底的
Benthos 底栖生物；水底生物
Bio(-)Indicator 对环境有指示作用的生物指示品种
Bioaccumulation 生物累积
Biocenose 生物群落
Biochemical Action 生化作用；生物化学作用
Biochemical Oxygen Demand (BOD) 生化需氧量
Biochemical Processes 生物化学过程
Biodegradation 生物降解
Bioethics 生物伦理学
Biofilter 生物过滤器；生物滤池
Biological Cycle 生物循环；生物周期
Biological Diversity 生物多样性
Biological Nitrogen Fixation 生物固氮
Biological Treatment 生物处理
Biomass 生物量
Biome 生物群落
Biomonitoring 生物监测
Biotechnology 生物技术；生物工艺学
Bitrate Radical 硝酸根
Blast Furnaces 高炉
Blowing Agent 发泡剂
Body of Water 水体
Body Tissue 体组织
Bottom-Living Fish 底栖鱼类
Boundary Layer 边界层
Bowl 盆地；区域
Breakdown 分解
Bromide 溴化；溴化物
Brominated Species 含溴产品
Bromination 溴化作用；溴化处理
Bromofluorocarbons 溴氟碳化合物
Bromoform 溴仿；三溴甲烷
Bromomethane 溴甲烷；甲基溴
Bromotrifluoromethane 三氟溴甲烷
Bubbling Fluidized Bed Combustion 沸腾式流化床燃烧
Buffer Solution 缓冲溶液
Buffer(ing) Capacity 缓冲能力；缓冲容量

C.

Cadmium Contamination 镉污染
Carbon Cycle 碳循环
Carbon Dioxide 二氧化碳
Carbon Monoxide 一氧化碳
Carbon Tetrachloride 四氯化碳
Carbon Tetrafluoride 四氟化碳
Carbonic Acid 碳酸
Carcinogens 致癌物
Carrying Capacity 容纳量；装载量；负荷量
Catalysis 催化作用
Catalyst 催化剂
Catalytic Converter 催化转化器
Catalytic Cycle 催化循环
Catalytic Exhaust System 排气催化系统
Cation Exchange Capacity 阳离子交换能力
Cation Exchanger 阳离子交换剂；阳离子交换器
Caves 洞穴
Chemical Conversion 化学转化
Chemical Oxygen Demand (COD) 化学需氧量
Chemical Reactivity 化学反应性
Chemical Residence Time 化学品停留时间；化学品存在时间
Chemical Trace Constituent 化学痕量成分
Chemical Transformation 化学变化

Chemical Transmitter 化学传递介质
Chemically Reactive Substance 化学活性物质
Chemically-Fixed Energy 化学能；化学固定能
Chlorhydrogenation 氯氢化作用
Chloride Prescrubber 氯化物预洗净气剂
Chlorine Dioxide 二氧化氯
Chlorine Monoxide 氧化氯
Chlorine Nitrate 硝酸氯
Chlorine 氯
Chlorofluorocarbon（CFC）氯氟碳化合物；含氯氟烃
Chlorofluorocarbon Refrigerant 氯氟碳化合物制冷剂
Chromatographic Analysis 色谱分析
Chromatophore Cell 色素细胞
Circle of Latitude 纬度圈
Circulating Fluidized Bed Combustor 循环流化床燃烧室
Circulation Pattern of The Atmosphere 大气环流模式
Circumpolar Vortex 环极涡旋；绕极涡旋
Civil Engineering 土木工程
Clean Air Act 空气清洁法
Climate Change 气候变化
Climate Modelling 建立气候模型
Climatic Factors 气候因子
Climatic Forecast 气候预报
Climatic Zone 气候带
Climatography 气候志
Cloud System 云系
Cloud-Radiation Feedback 云-辐射反馈作用
Cloudy 多云
CO_2 Equivalent 二氧化碳当量
Coal 煤
Coastal Areas 沿海地区
Coastal Climate 沿海气候
Coastal Ecosystems 沿海生态系统
Coastal Erosion 海岸侵蚀
Combustion Emissions 燃烧废气
Combustion Engines 内燃机
Combustion Plant 燃烧设施；燃烧车间
Combustion Source 火源；火箱；燃烧室
Component 成分；组分
Composite Pollution 混合污染
Composts 堆肥
Compound Organic Matter 复合有机物
Compression Ratio 压缩比
Compressive Resistance 抗压强度；压应力
Concentration Factor 浓缩系数；富集系数
Concentration Level 浓缩水平；富集水平
Condensate 冷凝；冷凝液
Conservation Area 保护区
Contaminant 污染物
Contaminated 被污染的
Content 含量
Cooling Waters 冷却水
Coral Reefs 珊瑚礁
Cost Impact 成本影响；成本冲击
Cost-Benefit Analysis 成本效益分析
Cost-Effective 成本效率高的
Cost-Effectiveness 成本效率
Cumulative Dose 累积剂量
Cyanide 氰化；氰化物
Cyclone Separator 旋风式分离器

D.

Daily Dose 日剂量
Daily Variation 日变化
Dams 水坝
Danger Label 危险标记；危险标志
Danger Level 危险水平；危险程度
Danger Threshold 危险阈；安全限值
Decay（of A Disturbance）扰动的逐渐减弱
De-Duster 除尘器
Deep Injection（of Wastes）废物的深灌注
Deeper Ocean 深海
Deep-Sea Circulation 深海环流
Deep-Sea Disposal（of Wastes）废物的深海倾置
Deep-Well Injection（of Hazardous Wastes）危险废物的深井灌注
Defoliation 脱叶；去叶
Deforestation 森林砍伐
Dehydrogenation 脱氢作用
Delayed Effect 延迟效应
Denitrfying Agent 脱硝剂；反硝化剂；脱氮剂
Denitrification 反硝化作用；脱氮作用
Denitrogenation 脱氮作用
Denoxing 消除氮的氧化物

Density 密度
Depletion in Numbers 数量减少
Deposition 沉积作用；沉积物
Desalination 脱盐
Desertification 沙漠化
Desulphurisation 脱硫作用
Desulphurization of Fuels 燃料脱硫
Detention Time 滞留时间
Detergency 去垢性；去垢力
Dewatering 脱水
Diatom Ooze 硅藻泥
Dielectric (Substance) 电介质；绝缘物质
Dielectric Oil 绝缘油
Diene 二烯烃
Diesel(-Fuelled) Vehicle 柴油车辆
Digestion Tank 消化罐
Dimensional Stability 尺度稳定性
Dioxin 二噁英
Disaster Preparedness 防灾准备
Disaster Prevention 防灾
Discharge Permit 排放许可
Discharge Pipe 排水管
Discharge Point 排放点
Discharge Rate 排放速度；排放速率
Discharge Record 排放记录
Discharge Standard 排放标准
Discharge 排出物；排放物；排出；排放
Disordered Environment 环境秩序混乱
Disposal Capabilities 废物处理能力；处理容量
Disposal of Wastes 废物处理
Disposal Site 废物处理场；垃圾倾置场
Dissolved Air Flotation 溶气浮选法
Dissolved Oxygen (DO) 溶解氧
Disturbed Ecological Balance 生态平衡失调；生态平衡被扰乱
Divide 分水岭；分水界
Dobson Instrument 多布森分光计
Dobson Unit 多布森单位
Domestic Legislation 国内立法；国内法规
Domestic Sewage 生活污水
Domestic Solid Wastes 生活固态废物
Dose Equivalent 剂量当量
Dose-Effective Curve 剂量效应曲线
Drainage Basin 流域；流域盆地
Drinking Water Standard 饮用水标准
Drinking Water Supply 饮用水供应
Drinking Water Treatment 饮用水处理
Drinking Water 饮用水
Drought Early Warning 干旱预警
Drying 干燥
Dust Collector 集尘器；除尘器
Dust Deposit 粉尘沉积；尘土堆积
Dust Discharge 排尘；粉尘排放
Dust Load 含尘量；尘埃浓度
Dust Separator 除尘器
Duststorm 尘暴
Dynamic Model 动力模型

E.

Early Warning System 预警系统
Earthquakes 地震
Ecolabelling 生态标志
Ecological Balance 生态平衡
Ecological Damage 生态破坏
Ecological Disruption 生态失调；生态破坏
Ecological Factor 生态因素
Ecological Security 生态安全
Ecological Suitability 生态适应性
Ecology 生态学
Eco-Mark 生态标记
Ecosystem 生态系统；生态系
Effluent Discharge 排放废物；排出废液
Effluent Standard 排放标准
Effluent 流出物；废液；污水
El Nino 厄尔尼诺效应
Electric Power Plants 发电厂
Electrodialysis 电渗析
Electrostatic Precipitator (ESP) 静电除尘器
Emission Certification 排放许可证
Emission Concentration 排放浓度
Emission Load 排放量
Emission Performance 排放特性
Emission Point 排放点
Endangered Species 濒危物种
Endocrinology 内分泌学

Energy Conservation 节用能源；节能
Energy Efficiency 能效
Energy Mix 混合能源
Energy-Efficient 节能的；高能效的
Energy-Intensive 耗能的；能源密集的
Energy-Saving 节省能源的；节能的
Enhancement 加强；扩大；增强
Enriched Uranium 浓缩铀
Environment Protection 环境保护；环保
Environmental Assessment 环境评价
Environmental Capacity 环境容量
Environmental Concentration 环境浓度
Environmental Criteria 环境标准
Environmental Degradation 环境退化
Environmental Disaster 环境灾难
Environmental Disorder 环境失调
Environmental Economic Issues 环境经济问题
Environmental Engineering 环境工程学
Environmental Ethics 环境伦理学；环境道德
Environmental Exposure 环境风险
Environmental Hazard 环境公害
Environmental Health Hazards 环境健康危害
Environmental Impact Assessment（EIA）环境影响评价
Environmental Impact Statement（EIS）环境影响报告
Environmental Law 环境法
Environmental Limit Concentration 环境浓度极限
Environmental Management 环境管理
Environmental Policy 环境政策
Environmental Pollution 环境污染
Environmental Protection Agency（EPA）环境保护局
Environmental Quality Indicators 环境质量指标
Environmental Quality Standards 环境质量标准
Environmental Quality 环境质量
Environmental Risk 环境风险
Environmental Science 环境科学
Enzymes 酶
Epidemics 流行病
Epidemiology 流行病学
Eutrophic Waters 富养化水
Eutrophication 富营养化
Evaporation 蒸发作用
Excess Air 过量空气
Exhaust After-Treatment Device 废气后处理净化装置
Exhaust Emission Limit 废气污染极限；废气污染限值
Exhaust Emission(s) 废气污染物；废气污染
Exposed To Air 暴露在空气中
Exposed 暴露的；没有保护的
Exposure Duration 暴露时间
Exposure Limit 暴露极限；暴露限值
Exposure Time 暴露时间
Extinct Species 灭绝品种；灭绝物种
Extinguishing Agent 灭火剂

F.

Fabric Filter 织物过滤器；袋滤器；网状滤器
Feed 给料；进料
Feedback 反馈；回授
Filter（Layer）Separator 过滤除尘器
Filter 过滤器；滤池
Filtration Rate 过滤速度；滤水率
Fixed Ammonia 固定氨
Fixed Combustion Source 固定床燃烧室
Flammable Liquids 可燃液体；易燃液体
Floc 絮凝物；絮状沉淀
Flocculation 絮凝作用；结絮作用
Flood Control System 防洪系统
Flood Forecasting 洪水预报
Flow Area 过水面积
Flow Control 流量控制
Flow Rate 流量
Fluidized Bed Combustion（FBC）流化床燃烧法
Fluoride 氟化；氟化物
Fluorination 氟化作用
Fly Ash 飞灰
Foamed Plastic 泡沫塑料
Foamed Polyurethane 聚氨酯泡沫塑料
Foaming Agent 发泡剂
Foaming 发泡
Fog 雾
Fogging 成雾；喷雾
Food Chain 食物链
Forest Decay 森林退化；森林逐渐失去活力
Forest Deterioration 森林破坏；森林逐渐衰败

Fossil Fuel 矿物燃料
Free Acyl Radical 酰游离基；酰自由基
Freeze-Dry 冷冻干燥法
Freon 氟利昂
Freshwater 淡水；淡水的
Fuel Desulphurisation 燃料脱硫作用
Fuel Gas 气体燃料
Fuelwood 薪材
Functional Group 官能原子团
Fungi 真菌

G.

Gamma Observations 伽马射线观测
Garbage Collection 垃圾收集
Garbage Grinding 垃圾磨碎
Gas Adsorption Chromatography 气相吸附色谱法
Gas Analyser 气体分析仪
Gas Chromatography 气相色谱法
Gas Cleaning 气体净化
Gas Scrubber 气体洗涤器
Gas-Borne Particles 悬浮气体中的微粒
Gaseous Hydrocarbon 气态碳氢化合物
Gaseous Pollutants 气体污染物
Gasification 气化作用
Gas-Liquid (Partition) Chromatography 气液分配色谱法
Gasoline(-Fuelled) Vehicle 汽油车辆
Global Change 全球性变化；全世界性变化
Global Climate Change 全球气候变化
Global Warming 全球变暖
Grain Size Distribution 颗粒大小分布
Grains 谷物
Granulate 颗粒化；颗粒
Granulation 颗粒化
Grassland 草原；草地；牧地
Green Belt 绿化地带；绿带
Greenhouse Effect 温室效应
Greenhouse Gases 温室气体
Grit 砂砾；粗砂
Ground Air 土壤空气；土壤中所含的空气
Ground Level Ozone 地面臭氧
Ground Station 地面站
Ground Water 地下水；潜水
Groundwater Flow 地下水径流
Groundwater Level 地下水位；地下水面；潜水面
Groundwater Runoff 地下水径流
Groundwater Seepage 地下水渗漏；地下水渗透
Growing Period 生长期

H.

Habitat Conservation 生境保护
Habitat Dislocation 生境破坏
Half-Life 半衰期
Haloalkane 卤烷
Halogenide 卤化物
Hard Water 硬水
Harmful Algal Bloom 水华
Hazard Assessment 危害评价
Hazard Classification System 危险性分类制度
Hazard Concentration Limit 危险浓度极限
Hazard Label 危险标记
Hazard Rating 危害等级；危害分级
Hazard 危险；危害；公害
Hazardous Characteristics 危险特性
Hazardous Substance 危险物质；有害物质
Hazardous Wastes 危险废物
Health Risk Assessment 健康风险评价
Heat Content 焓；热函；热含量
Heat Engineering 热力工程；热工学
Heat Flow 热流
Heat Insulation 热绝缘；绝热
Heat Resistance 耐热性；抗热性
Heat Stabilizing 热稳定的；耐热的
Heavy Diesel Vehicle 重型柴油车辆
Heavy Goods Vehicle 重型货车；重型运输车辆
Heavy Metals 重金属
Herbicides 除草剂
Hexachlorobenzene 六氯苯
Hexane 己烷
Hexene 己烯
High Latitudes 高纬度地区
High-Rise Buildings 高层建筑
Homosphere 均质层
Hopper 料斗
Horizon 层,层位；地平,地平线；水平
Horizontal Diffusion 水平扩散

Horizontal Transfer 水平转移
Hospital Wastes 医院废物
Household Waste 家庭垃圾；生活垃圾
Humectant 润湿剂
Humic Acid 腐殖酸
Humid Tropics 湿热带
Humidity 湿度
Hydration 水合作用
Hydraulic Conductivity 导水率；渗透系数
Hydraulic Gradient 水力坡降；水力坡度
Hydraulic Head 压力水头
Hydrocarbons 烃
Hydrochloric Acid 盐酸
Hydrochlorofluorocarbon（HCFC）氢氯氟碳化合物；氟氯烃化合物
Hydrocracking（HC）氢化裂解；加氢裂解
Hydrology 水文学
Hydroperoxide 过氧化氢
Hydroperoxyl Radical 氢过氧游离基
Hydrosphere 水圈
Hydrosulphuric Acid 氢硫酸；硫化氢
Hydrotreating 加氢处理
Hyperchlorous Anhydride 氧化氯
Hypochlorous Acid 次氯酸

I.

Ice Age 冰期
Ice Shelf 冰架
Identification Mark 识别标记
Identification of Pollutants 污染物鉴别
Ignitable chemical 易燃药品
Impairment 恶化；损伤；损害
Impervious Layer 不透水层
Inactive 不活泼的；钝性的
Incineration Of Waste 废物焚烧
Incineration Plant 焚烧工厂
Incinerator 焚化炉；焚烧炉
Incipient Inhibition 初期抑制
Incoming Radiation 入射辐射
Indicator 指示剂
Inducer 诱导物
Industrial Effluents 工业废水
Industrial Fumes 工业烟尘
Industrial Noise 工业噪声
Industrial Wastewaters 工业废水
Infiltration Rate 渗透速率；渗滤速率
Infiltration 渗入；渗透；渗滤
Inflow Pipe 进气管；进水管；流入管
Infrared Radiation（IR Radiation）红外辐射
Infrared Radiometer 红外辐射仪
Injection Well 注入井
Injury 损害；伤害
Inorganic Chemistry 无机化学
Inorganic Fluorine Compounds 无机氟化合物
Inorganic Pollutants 无机污染物
Inorganic Substances 无机物质
Integrated Environmental Controls 综合环境控制
Intermediate Chemical 中间化学品
Intermediate Product 中间产品；半成品；半制成品
Intermediate Radical 中间游离基；瞬间游离基
Intermediate Species 中间物质；瞬间物质
Intermediate Water 中层水；中间水域
Intermediate 中间体；中间物
Internal Combustion Engine 内燃机
Inter-Relations 相互关系
Invertebrates 无脊椎动物
Invisible Radiation 不可见的辐射
Ion Balance 离子平衡
Ion Exchanger 离子交换剂
Ion Source 离子源
Ion-Exchange Chromatography 离子交换色谱法
Ion-Exchange Resin 离子交换树脂
Ionization Constant 电离常数
Ionization Detector 电离检测器
Ionizing Radiation 电离辐射
Ionosphere 电离层
Ion-Selective Electrodes 离子选择电极
Irradiance 辐照度
Irrigation 灌溉
Irritant 刺激剂；刺激物；刺激性的
Isolation of A Pollutant 污染物分离
Isopsophic Index 噪音指数

L.

Label 标记；标志
Lag Effect 滞后效应

Lagoon 氧化塘；环礁；泻湖
Land Disposal Site 地面处理场
Land Mammals 陆地哺乳动物
Land Pollution 土壤污染
Land Reclamation 土地改良；垦荒
Land Restoration 土地恢复；土壤改良
Land Station 地面站
Land Surface Temperature 地面温度
Land Treatment (of Wastes) 废物的地面处理
Landfill (Site) 垃圾填埋地；填埋
Landfill Dumping 倾弃垃圾填地
Landfill Treatment 垃圾填埋处理
Landscaped Area 风景区
Lapse Rate 直减率；递减率
Latitude 纬度
Latitudinal Distribution 纬度分布
Latitudinal Zone 纬度地带
Leachate 沥滤液
Leaching 沥滤
Leaded 含铅的
Lead-Free 无铅的
Lethal Concentration (LC) 致死浓度
Lethal Concentration Low (LC LO) 最低致死浓度
Lethal Dose (LD) 致死剂量
Lethal Time 致死时间
Lethal 致死的；致命的；死亡的
Life Cycle 生活周期；生活史
Life Support(Ing) System 生命维持系统
Light Diesel Vehicle 轻型柴油车辆
Light Radiation 可见光辐射
Lime Treatment 用石灰处理
Limnetic Zone 湖沼带；湖泊透光层
Liquefied Gas 液化气
Liquefied Petroleum Gas 液化石油气
Litter Collection 垃圾收集
Load Per Unit of Mass 单位质量的含量
Load Per Unit of Volume 单位体积的含量
Load 负荷；负担；负载；载荷；浓度；含量
Loading 负荷；浓度；含量
Local (Emission) Source 本地排放来源
Local Environment 局部环境；本地环境
Longitude 经度
Lower Limit of Detectability 可检出的最低限值；最低检出量
Low-Sulphur Fuel 低硫燃料
Low-Temperature Chemistry 低温化学
Low-Waste Technology 低废技术

M.

Malaria 疟疾
Mammals 哺乳动物
Marine Incineration 海上焚化；海上焚烧
Marine Pollution 海洋污染
Marine Resources Conservation 海洋资源保护
Mass Concentration 质量浓度
Mass Mixing Ratio 质量混合比
Maximum Emission Concentration 最大排放浓度
Maximum Tolerated Dose 最大耐受剂量
Metabolism 新陈代谢
Metal Finishing 金属加工
Metal Plating 金属电镀
Metal Smelting 金属冶炼
Methane 甲烷
Methyl 甲基
Microbiology 微生物学
Microclimate Effects 微气候影响
Microorganisms 微生物
Micropollutants 微污染物
Molecular Scattering 分子散射
Molecule Breaking 分子离解
Motor Vehicle Emissions 机动车辆排放物
Municipal Refuse Disposal Area 城市垃圾处理场；城市垃圾场
Municipal Solid Waste (MSW) 城市固态废物
Municipal Waste 城市废物
Municipal Water Distribution Systems 城市配水系统
Mutagen 诱变剂；致变物；诱变因素
Mutants 突变体
Mutated Microorganisms Release 突变微生物释放

N.

National Ambient Air Quality Objective 国家环境空气质量目标
National Ambient Air Quality Standard 国家环境空气质量标准
National Conservation Programmes 国家保护计划

Natural Degradation 自然降解；自然分解；自然退化
Natural Environment Rehabilitation 自然环境复原；恢复自然环境
Natural Fertilizers 天然肥料
Natural Fibres 天然纤维
Natural Gas Extraction 天然气开采
Natural Gas 天然气
Natural Purification 自然净化作用；自净
Natural-Occurring Substance 天然物质
Nature Conservation 自然保护
Nature Laboratory 自然实验室
Nature Reserve 自然保护区；自然保留地
Needle-Leaved Forest 针叶林
Net Radiation 净辐射
Net Resource Depletion 资源净损耗
Neutralizer 中和剂
Nitrate Nitrogen 硝态氮
Nitrate 硝酸盐
Nitric Acid 硝酸
Nitric Oxide Catalysis 氧化氮催化作用
Nitrites 亚硝酸盐
Nitrobacteria 硝化细菌
Nitrocompound 硝基化合物
Nitrogen Cycle 氮循环
Nitrogen Fertilizer 氮肥肥料
Nitrogen Fixing Plants 固氮植物
Nitrogen Monoxide 氧化氮；一氧化一氮
Nitrogen Oxide Control 氮氧化物控制
Nitrogen Oxide Radical 亚硝酰基；亚硝基
Nitrogen Oxide 氧化氮；氮的氧化物
Nitrogenouz Wastes 含氮废物
Nitrosamines 亚硝胺
Noise Abatement 噪音治理
Noise Limits 噪声限度
Noise Pollution Level (NPL) 噪声污染级
Noise Pollution 噪声污染
Noise Screen 噪声屏蔽
Non-Waste Technology (NWT) 无废技术
North Latitude 北纬
Northern Polar Zone 北极区
Nuclear Energy 核能
Nuclear Fuels 核燃料
Nuclear Power Plants 核电站
Nuisance 公害；损害
Nutrient (Material) 营养物质；养料；养分
Nutrient Depletion 养分枯竭
Nutrient Leaching 浸出养分；养分沥滤
Nutrient Requirement 营养需要量
Nutrient Salts 盐类营养物
Nutrition 营养

O.

Objectionable Odour 令人不愉快的气味；恶臭
Obscuration (of Smoke) 烟雾的不透光性
Ocean Circulation 大洋环流；海洋环流
Ocean Currents 洋流
Ocean Disposal 海洋废物处理
Ocean Dumping 海洋倾倒
Ocean Warming 海洋水温升高
Oceanic Absorption 海洋吸收
Odorant 香料；香味剂；气味剂
Odour Nuisance 令人讨厌的气味
Offensive Odour 恶臭；令人不愉快的气味
Off-Gases 废气；尾气
Oil Collection Vessel 浮油回收船
Oil Extraction 原油开采
Oil Slick 浮油；油膜
Oil Spills 石油泄漏
Oils 油类
Olefiant Gas 成油气；乙烯气
Oligotrophic Lakes 贫养湖泊
On-Site Processing 就地处理
On-Site 现场；就地
Opacity 不透明度；阻光度
Open Air 户外；露天
Open Burning 露天焚烧
Open Dump 露天垃圾场
Open Dumping 露天倾弃
Open Sea 公海
Optimization Model 最优模型
Optimum 气候适宜期
Organic Chemistry 有机化学
Organic Content of Soils 土壤的有机物含量
Organic Farming 有机农业
Organic Halogen Compound 有机卤化合物

Organic Phosphorus Compound 有机磷化合物
Organic Pollutants 有机物污染
Organic Solvents 有机溶剂
Organic Substances 有机物质
Organic 有机的；有机
Organizing Committee 组织委员会
Outdoor Air 户外空气；室外空气
Out-Emission 向外排放
Outersphere 外逸层
Outfall 污水排出口
Outflow 外流；流出；流出物
Outlet 出口；出水口；排水口
Outside Air 户外空气；室外空气
Overburden 覆盖层
Overcropping 耕种过度
Overcrowding 过度拥挤
Overcultivation 耕种过度
Overcutting 砍伐过度
Overequipment 装备过度
Overfertilization 过度施肥
Overfire Air 从上部引入帮助燃烧的空气
Overfishing 过度捕捞
Overflow Filter 溢流滤池
Overflow 溢流
Overgrazing 过度放牧
Overintensive Agriculture 过度密集农业
Overland Flow 地表径流
Overloading 超负荷；超载
Overwintering Plant 越冬植物；多年生植物
Oxidant 氧化剂
Oxidation Catalyst 氧化催化剂
Oxidation Pond 氧化塘
Oxidation Rate 氧化率；氧化速率
Oxidation Tank 氧化槽
Oxidation 氧化作用
Oxidation-Reduction Potential（RH）氧化还原电势
Oxidizer 助燃剂；氧化剂
Oxidizing Agent 氧化剂
Oxidizing Air 助燃空气
Oxidizing Solids 固体助燃剂
Oxygen Depletion 缺氧；氧气耗竭
Oxygen 氧气
Oxygen-Consuming Capacity 耗氧能力
Ozone（O_3）臭氧
Ozone Depletion 臭氧消耗；臭氧枯竭
Ozone Hole 臭氧层空洞
Ozone Layer 臭氧层
Ozone Loss 臭氧损耗
Ozone Unit 臭氧单位
Ozone Value 臭氧值（某地的臭氧浓度，一般以多布森单位表示）
Ozone-Forming Species 产生臭氧物质
Ozone-Modifying Substance 臭氧改性物质
Ozonizer 臭氧发生器
Ozonosphere 臭氧层

P.

Package Plant 移动式污水处理装置
Packaging Waste 打包废物
Packed Tower 填充塔；填料塔
Packer Truck 装有液压压紧设备的垃圾收集汽车
Paint Solvent 涂料溶剂
Paints 涂料
Parasites 寄生生物
Particle-Size Distribution 粒度分布
Particulate Emission 微粒排放
Particulate Pollutant 微粒污染物；粒状污染物
Particulate Precipitator 微粒沉淀器；除尘器
Pathogenic Organisms 病原生物体
Pathway of A Pollutant 污染物的路径
Pattern of Pollution 污染的模式
Peak 峰；峰值；极大值
Pedosphere 土壤圈；土壤层
Pelagic Pollution 远洋污染
Penetration Time 渗透时间；穿透时间
Percolation 渗滤；渗漏；渗透
Percutaneous 经皮的；通过皮肤的
Perfluorinated Hydrocarbon 全氟烃；全氟碳化物
Perfluorinated 全氟化的
Performance of The Adsorbent 吸附剂效率；吸附剂性能
Permafrost 永冻层；永久冻土
Permanent Meadow 永久性草地；永久性草原
Permanent Waste Storage 废物的永久性储藏
Permissible Exposure Limit 容许曝露限度

Permit to Discharge 排放许可证；倾卸许可证
Peroxides 过氧化物
Peroxy Radical 过氧游离基；过氧自由基
Persistence 持久性；持续性；长期留存
Persistent Organic Pollutants 难降解有机污染物
Personal Safety Equipment 个人防护设备
Pest Management 病虫害控制
Pest Outbreak 暴发病虫害
Pesticide 农药；杀虫剂
Petrol 汽油
Pharmaceutical Wastes 医药废物
Phenol Compounds 酚类化合物
Phenol-Formaldehyde Resin 酚醛树脂
Phenolphthalein 酚酞
Phenols 酚类化合物
Phosphoric Acid 磷酸
Phosphorus 磷
Phosphorylation 磷酸化作用
Photoautotroph 光合自养生物；光能自养生物
Photocatalysis 光催化作用
Photochemical Agents 光化学试剂
Photochemical Effects 光化学效应
Photochemical Pollution 光化学污染
Photochemical Smog 光化学烟雾
Photodegradation 光致降解
Photodissociation 光致解离作用
Photosynthesis 光合作用
Physical Environment 自然环境；物理环境
Physical Magnification Factor 物理放大因数
Physico-Chemical Processes 物理化学过程
Phytocide 除莠剂；除草剂
Phytocoenosis 植物群落
Phytomass 植物生物量
Phytoplankton Bloom 浮游植物大量繁殖
Phytoplankton 浮游植物
Phytotoxic 植物毒性的；对植物有毒的
Pipelines 管道
Plague 瘟疫
Plankton 浮游生物
Plasma 等离子体
Plasmapause 等离子体层顶
Plasmasphere 等离子体层
Plastic Fabric 塑料织物
Plastic Foam 泡沫塑料
Plate Filter 压滤器；压滤机
Plume 烟缕；烟羽
Point Pollution 点污染
Point Source 点污染源
Poison 毒；有毒物质
Polar Anticyclone 极地反气旋
Polar Aurora 极光
Polar Cyclone 极地涡旋；极涡
Polar Stratospheric Cloud 极地平流层云
Polar Vortex 极地涡旋；极涡
Pollutant Levels 污染物浓度
Pollutant Load 污染物负荷；污染物浓度
Pollutant 污染物
Polluter Pays Principle（PPP）污染者付清理费原则
Polyvinyl Chloride（PVC）聚氯乙烯
Polyaromatic Hydrocarbon 多环芳烃
Polybrominated Biphenyls（PBB）多溴联苯
Polychlorinated Biphenyls（PCBs）多氯联苯
Polycyclic Aromatic Hydrocarbon（PAH）多环芳烃
Polyelectrolytes 聚合电解质
Polyethylene（Pe）聚乙烯
Polypropylene（PP）聚丙烯
Polystyrene（PS）聚苯乙烯
Power Plant 发电厂
ppm（Parts Per Million）百万分之一
ppt（Parts Per Trillion）万亿分之一
Precipitator 除尘器；吸尘器
Precursor 化学前体
Pricing Policies of Resources 资源的定价政策
Primary Consumers 初级消费者
Primary Forest 原始林；原生林
Primary Formation 原始地形；一级结构
Primary Hazard 主要危险
Primary Pollutant 原生污染物；一次污染物
Primary Standard 一级标准
Primary Treatment 废水的初级处理；一级处理
Priority Chemical 首要化学品；优先化学品
Probe 探针；探头；探测器
Procedure 程序；过程；步骤
Process Flows 工艺流程

Profile 轮廓；外形；纵剖面；分布图；示意曲线
Prokaryotic Micro-Organisms 原核微生物
Propylene 丙烯
Protected Areas 保护区
Protected Landscape Area 自然保护区
Protected Species 受保护的物种
Protozoa 原生生物
Public Health 公共卫生
Pulverization 粉碎；研磨
Pumping Station 泵站
Pungent Odor 刺激性臭味
Purified Water 净化水
Purity 纯度；纯净

Q.

Quality Assurance 质量保证
Quality Control 质量控制
Quality Objectives 质量目标
Quality Standards 质量标准
Quarry 采石场；石矿

R.

Race Relations 种族关系
Radiance 辐射率
Radiation Accident 放射性事故
Radiation Damage 辐射线损伤
Radiation Protection 辐射防护
Radiative Effect 辐射效应
Radical 基；根；原子团；游离基；自由基
Radioactive Substances 放射性物质
Radioactive Tracer Techniques 放射性示踪技术
Radioactive Waste Management 放射性废物管理
Radioactivity 放射性
Rainfall Amount 雨量
Rainfall Season 雨季
Rainfall 降雨
Rare Earth Metals 稀土金属
Rare Taxa 稀有分类群
Rate 率；速率；速度
Rating 校准；定标
Raw Radiance 原始辐射率
Raw Refuse 未经任何处理的垃圾
Raw Sewage 原污水；未经任何处理的污水
Reaction Time 反应时间
Reactive Gas 活性气体
Reactive Hydrocarbon 活性烃
Reactivity 反应性；反应作用；反应速度
Receiving Waters 承受水体；承受水域
Receptor Area 受体面积；接受面积
Reclaimable Waste 可再利用废物；可回收废物
Reclamation 回收；开垦；改良；驯化
Recommended Exposure Limit 建议的曝露限值
Recovery Phase 复原阶段
Recovery Plant 资源回收厂
Red Tide 红潮；赤潮
Redox Potential 氧化还原电势
Reducing Agent 还原剂
Refuse 废物,垃圾
Refining 净化；精炼；使成熟
Refractories 耐火材料
Refrigerant Fluid 制冷液
Refrigerant 制冷剂
Regenerative Capacity 再生能力
Regenerative Heat Exchanger 再生式热交换器
Regular Gas 普通汽油
Regulated Flow 调节流量；调节径流
Regulatory Control 法规控制
Remote Sensing 遥感
Remote Sensor 遥感传感器
Removal Efficiency 去除效率
Removal of Pollutants 消除污染物；去除污染物
Renewable Resources 可再生资源
Reservoir 水库；吸收库；储层；蓄水池
Residence Time 停留时间；阻滞时间
Residential Areas 居民区
Residual 残留物；残积物
Resource Recovery 资源回收
Respirable Suspended Particulates (RSPs) 可因呼吸进入人体的悬浮微粒
Response Time (of A Measuring Instrument) 测量仪器的响应时间
Retention Time 停留时间；阻滞时间
Retention 滞留水量
Reuse of Materials 材料再利用
Reuse 重复使用；再使用
Revegetation 植被恢复

Reverse Osmosis 反渗透；逆向渗透
Risk Analysis 风险分析
Risk Label 危险标记；危险标志
Risk Management 风险管理
Riverine Input (of Pollutants) 沿河排入的污染物
Rotating Biological Contactor (RBC) 生物转盘
Rotating Fluidized Bed Combustion 旋转式流化床燃烧
Runoff Water 径流水；流水；活水
Rural Areas 农村地区
Rural Water Supply 农村供水

S.

Safe Storage 安全储藏；安全储存
Safe Water 安全给水
Safeguards 防护设施；保障措施
Saline Fields 盐碱地；盐渍地
Salinification 盐化作用；盐渍化
Salinity 含盐量；盐浓度；咸度
Sampling 取样；采样
Sandblasting 喷砂清洁处理
Sandstorm 沙暴；沙漠风暴
Sandy Soil 沙质土
Sanitary Landfill 卫生填地；垃圾填坑；卫生掩埋
Sanitary Sewers 生活污水管道；下水道
Sanitation Facility 卫生设施
Saturated Air 饱和空气
Saturated Zone 饱和层；饱和带；饱和区
Scrap 废料；残渣；碎屑
Screening 屏蔽；掩蔽；隔离
Screenings 筛选过的物质
Scrubber 洗涤器；涤气器
Scrubbing 洗涤；涤气
Sea Level 海平面
Sea Water Conversion 海水转化淡水；海水淡化
Sea Water Desalinization 海水脱盐；海水淡化
Seasonal Deviation 季节性偏差
Seasonal Fluctuation 季节性起伏
Secondary Pollutant 次生污染物；二次污染物
Secondary Raw Material 次生原料
Secondary Standard 二级标准
Secondary Treatment 废水的二级处理
Sediment 沉降物；沉积物；沉淀物
Sedimentation Rate 沉降速度
Sedimentation Tank 沉降槽；沉降池
Sedimentation 沉降作用；沉积作用；沉淀作用
Seepage 渗漏；渗出；渗流
Selective Catalytic Reduction 选择性催化还原
Self-Purification 自净作用
Self-Restoring Capacity of Nature 自然的自我恢复能力
Sensitive Species 敏感物种
Sensitivity 敏感性；灵敏度
Sensor 传感器；探测器；敏感元件
Separate Sewer System 分流排污系统
Separating Efficiency 分离效率
Separators 分离器
Septic Tanks 化粪池
Settleable Solids 可沉降的固体
Settler 沉降器；澄清器
Settling Tank 沉降槽
Settling Time 沉降时间
Sewage Disposal 污水处置
Sewage Lagoon 污水氧化塘
Sewage Sludge 污水污泥
Sewage Treatment Plants 污水处理厂
Sewage Treatment Systems 污水处理系统
Sewage 污水
Sewer 下水道；污水管
Sewerage (System) 下水道系统；污水排水工程
Shellfish 水生贝壳类动物
Short-Wave Radiation 短波辐射
Side Effect 副作用
Side Reaction 副反应；支反应
Silt 淤泥；粉沙；河道沉积泥沙
Sinking Time 沉降时间
Skimmer 撇沫器；撇油器
Skimming 从水面上撇取浮油；
Slag 炉渣；矿渣
Sludge Digestion 污泥消化法
Sludge 污泥；淤泥
Slurry 泥浆
Smog 烟雾
Social Surveys 社会调查
Socio-Economic Factors 社会-经济因素
Soda (Ash) 苏打灰；钠碱灰；纯碱；碳酸钠

Sodium Carbonate 碳酸钠
Sodium Chloride 氯化钠；食盐
Sodium Hydroxide 氢氧化钠
Soil Contamination 土壤污染
Soil Degradation 土壤退化
Soil Erosion 土壤侵蚀
Soil Pores 土壤孔隙
Solar Energy 太阳能
Solid Waste Disposal 固体废物处置
Solid Waste Management 固体废物管理
Solid Waste 固体废物
Soluble Solids 可溶性固体
Solution 溶液；溶体；
Solvent Extraction 溶剂萃取；
Sorption 吸着作用
Sorting of Household Refuse 家庭垃圾的分类
Species 物种；种；类；核素；原子团
Specific Weight 比重
Spill Hazard 溢漏风险
Spill Incident 溢漏事故
Stack Effluents 烟囱排放物
Stack Gas Cleaning 烟道气净化
Standard Conditions 标准状态
Standard Deviation 标准差
Standard Dobson (Spectrophotometer) 标准多布森分光光度计
Standard Solution 标准溶液；规定液
Stationary Source 固定污染源；静止污染源
Steady State 稳态；稳定状态；常定状态
Sterilant 消毒剂；杀菌剂
Storage Site 储藏地；储藏场所
Stratopause 平流层顶
Stratosphere 平流层
Stratospheric Ozone 平流层臭氧
Streamflow Regulation 流量调节；径流调节
Substantiation 具体化；实质化
Sulphates 硫酸盐
Sulphur Compound 硫化合物
Surface Soil 表土
Surface Waters 地表水
Surfactant 表面活性剂
Survival Time 生存时间；存活期
Suspended Matter 悬浮物；
Suspended Particulate Matter (SPM) 悬浮微粒物质
Suspended Solids (SS) 悬浮固体
Suspension Agent 悬浮剂
Sustainable Development 可持续发展

T.

Tag 作标记；标志
Tagged 标记的；示踪的
Technological Requirements 技术要求；技术规格
Temperate Ecosystems 温带生态系统
Teratogency 产生畸形
Terrestrial Ecosystem 陆地生态系统
Terrestrial Radiation 地球辐射
Tertiary Treatment 废水的三级处理
Thermal Capacity 热容量
Thermal Infrared Spectrum (TIR) 红外热光谱
Thermal Load 热负荷
Thermosphere 热层
Three-Way Catalytic Converter 三元催化排气净化器
Threshold Dose 阈剂量；最低有效剂量
Threshold Limit Value (TLV) 阈限值；最低限值
Threshold Value 阈值
Tide 潮，潮汐
Time Belt 时区
Time Zone 时区
Time-Scale 时标；时间尺度
Time-Series 时间序列；时间先后次序
Tipping Site 垃圾场；倾置场
Tipping 倾弃
Total Head 总水头；总能头
Total Nitrogen 氮的总量
Total Organic Carbon (TOC) 有机碳总量
Total Oxygen Demand (TOD) 总需氧量
Total Suspended Particulates (TSP) 悬浮微粒总量
Total Suspended Solids (TSS) 悬浮固体总量
Toxic Waste 有毒废物
Toxicity 毒性
Trace Element 痕量元素
Traffic Noise 交通噪音
Traffic Pollution 交通污染
Transboundary Air Pollution 越界空气污染
Transfer Station 转移站
Transition Phase 转变阶段；过渡阶段
Transpiration 蒸腾作用
Transport Systems 运输系统
Trickling Filter 滴滤池；散水滤床

Trihalomethane(THM)三卤甲烷
Trophic Factors 营养因子
Tropical Belt 热带
Troposphere 对流层
Turbidimeter 浊度计
Turbidity 浊度
Typhus 斑疹;伤寒

U.

Ultraviolet (Solar) Radiation (UVR) 紫外太阳辐射
Ultraviolet Range 紫外线波长范围
Underflow 潜流；底流
Underground Flow 地下水流；渗流
Underprediction 预测偏低
Uniform Flow 均匀流；等速流
Uniform Hazard 均匀危险因素；均匀风险
Unsaturated Derivative 不饱和衍生物
Unsaturated Zone 不饱和带；非饱水带；包气带
Unsteady Flow 不稳定流；非恒定流
Uptake 摄取；吸收
Urban (Ambient) Air Quality 城市环境空气质量
Urban Stress 城区压力
Urban Traffic 城市交通
Urban Water Supply 城市供水
Urbane Solid Waste 城市固体废物
Utilities 公用事业

V.

Vacuum Insulation 真空绝缘
Vapour Pressure 水汽压力；蒸气压力
Varied Flow 变速流；非均匀流
Vector 矢量
Vegetation 植被
Venting 通风；透气；排气；风干
Verdant Zone 无霜带
Vertical Gradient of Pollutant 污染物的铅直梯度
Vibration 振动
Violator 违犯者
Viruses 病毒
Visible Radiation 可见光辐射
Visible 可看见的；能见的；显而易见的；宏观的
Volatile Organic Compounds (VOCs) 挥发性有机化合物
Volatile 挥发性的；易挥发的
Volcanoes 火山
Vortex 涡旋
Vulnerability Analysis 弱点分析
Vulnerable to Pollution 易受污染损害的

W.

Waste Conversion Techniques 废物转化技术
Waste Disposal 废物处置
Waste Material 废料
Waste Minimization 废物最少化
Waste Plant 废物处理厂
Waste Reclamation 废物回收利用
Waste Recovery 废物回收
Waste Reduction 减少废物
Wastewater 废水
Water Body 水体；水域
Water Devide 分水岭；分水线；分水界
Water Erosion 水蚀
Water Pumps 水泵
Water Resource Development 水资源开发
Water Resources 水资源；水利资源
Water Reticulation Network 水系；水网系统
Water Supply Engineering 给水工程；供水工程
Water Supply 供水；给水；给水工程
Water Surface 水面
Water Table 地下水位;潜水位
Water Treatment 水处理
Water Vapour 水蒸气；水汽
Waterborne Disease 水传播疾病
Watercourse 水道；河道；航道
Water-Related Disease 与水有关的疾病
Watershed Management 流域管理
Watershed 分水岭；分水线；分水界
Water-Table Gradient 地下水面比降
Wet Screening 湿法筛滤过
Wet Scrubbing 湿法洗涤
Wind Energy 风能
Wind Erosion 风蚀
Windmill 风车；风力发动机
Windstorm 风暴
Wood Alcohol 木醇；甲醇
Work Procedure 操作程序

Y.

Yard Waste 庭园废物
Yeasts 酵母

Z.

Zenith Column 大气气柱
Zone Time (Zt) 区时
Zooplankton 浮游动物

References

[1] DAVIS M L, CORNWELL D A. Introduction to Environmental Engineering[M]. 4th ed. New York: McGraw-Hill, 2007.

[2] NATHANSON J A. Basic Environmental Technology-Water Supply, Waste Management, and Pollution Control [M]. 4th ed. New York: Prentice-Hall, 2004.

[3] ENGER E D, SMITH B F. Environmental Science-A Study of Interrelationships[M]. 9th ed. New York: McGraw-Hill, 2004.

[4] TYLER M J G. Living in the Environment[M] .3rd ed. Belmont: Thomson Learning, 2004.

[5] METCALF, eddy. Wastewater Engineering Treatment and Reuse [M]. 4th ed. New York: McGraw-Hill, 2003.

[6] Theisen H, Vigil S. Integrated Solid Waste Management-Engineering Principles and Management Issues[M]. New York: McGraw-Hill, 2000.

[7] NEVERS N D. Air Pollution Control Engineering [M]. 2nd ed. New York: McGraw-Hill, 2000.

[8] RUBIN E S, DAVIDSON C. Introduction to Engineering & the Environment[M]. New York: McGraw-Hill, 2001.

[9] NIGEL H. Suspended growth processes, Handbook of Water and Wastewater Microbiology[M]. UK: Academic Press, 2003.

[10] WEINER R F, MATTHEWS R A. Environmental Engineering [M]. 4th ed. USA: Butterworth-Heinemann, 2003.

[11] SALVATO J A, NEMEROW N L, AGARDY F J. Environmental Engineering[M]. 5th ed. New Jersey: John Wiley & Sons, Inc., Hoboken, 2003.